Mechanisms and Machine Science

Volume 150

This book series establishes a well-defined forum for monographs, edited Books, and proceedings on mechanical engineering with particular emphasis on MMS (Mechanism and Machine Science). The final goal is the publication of research that shows the development of mechanical engineering and particularly MMS in all technical aspects, even in very recent assessments. Published works share an approach by which technical details and formulation are discussed, and discuss modern formalisms with the aim to circulate research and technical achievements for use in professional, research, academic, and teaching activities.

This technical approach is an essential characteristic of the series. By discussing technical details and formulations in terms of modern formalisms, the possibility is created not only to show technical developments but also to explain achievements for technical teaching and research activity today and for the future.

The book series is intended to collect technical views on developments of the broad field of MMS in a unique frame that can be seen in its totality as an Encyclopaedia of MMS but with the additional purpose of archiving and teaching MMS achievements. Therefore, the book series will be of use not only for researchers and teachers in Mechanical Engineering but also for professionals and students for their formation and future work.

The series is promoted under the auspices of International Federation for the Promotion of Mechanism and Machine Science (IFToMM).

Prospective authors and editors can contact Mr. Pierpaolo Riva (publishing editor, Springer) at: pierpaolo.riva@springer.com

Indexed by SCOPUS and Google Scholar.

Marco Ceccarelli · Juan Carlos Jauregui-Correa
Editors

State-of-the-Art and Innovations in Mechanism and Machine Science

A Tribute to Carlos López-Cajún

Springer

Editors
Marco Ceccarelli
Department of Industrial Engineering
University of Rome Tor Vergata
Roma, Italy

Juan Carlos Jauregui-Correa
Centro Universitario
Autonomous University of Queretaro
Santiago de Querétaro, Mexico

ISSN 2211-0984 ISSN 2211-0992 (electronic)
Mechanisms and Machine Science
ISBN 978-3-031-47039-4 ISBN 978-3-031-47040-0 (eBook)
https://doi.org/10.1007/978-3-031-47040-0

This Springer imprint is published by the registered company Springer Nature Switzerland AG
The registered company address is: Gewerbestrasse 11, 6330 Cham, Switzerland

Paper in this product is recyclable.

Preface

This volume contains the contributions of those IFToMM members who wanted to give special recognition to Professor Carlos Santiago López-Cajún by summarizing the most interesting aspects of the latest Advances in MMS areas as a tribute to the memory of his areas of interest. Professor Carlos Santiago López-Cajún died unexpectedly in December 2020. The volume was programmed as documentation to support the celebratory event held on 11 May 2023 at the Autonomous University of Queretaro, Campus San Juan del Rio, where Professor Carlos Santiago López-Cajún has taught since 1991. The event was very well participated with colleagues and former students who have remembered the figure of Professor Carlos Santiago López-Cajún with his distinguished characters from scientific, teaching, and human viewpoints. A plate was posted on a wall of a hall of the Campus San Juan del Rio of the Autonomous University of Queretaro.

a)

b)

(a) plate in memory of Professor Carlos Santiago López-Cajún; (b) colleagues and family of Professor Carlos Santiago López-Cajún posting the plate in his memory at the Campus San Juan del Rio of the Autonomous University of Queretaro

The addressed topics in the volume have been programmed in the areas of interest to Prof. Carlos Santiago López-Cajún with significant activity and contributions. Contributions were received from authors from around the world who personally knew Prof. Carlos Santiago López-Cajún at different levels of collaboration and familiarity with the typical IFToMM spirit within the IFToMM community. In fact, Prof. Carlos Santiago López-Cajún represented an emblematic personality of IFToM-Mist is characterized by activities and attitudes based on cordial personal collaborations centered on the development and technology transfer of results in MMS for the growth of the scientific and technical community for the benefit of the welfare of mankind and peace in the world.

In particular, the contributions in the volume as a tribute to the memory of Prof. Carlos Santiago López-Cajún span from education and theoretical works up to design procedures and prototypes even with experience in practical applications of several areas of MMS. In particular, after the first two chapters outlining the figure of Prof. Carlos Santiago López-Cajún and reporting on the event on his memory on 11 May 2023, the chapters of the volume refer to the history of machines and mechanisms even referring to robotics; education of MMS in general and with special focus on cam transmissions and on robotics; design procedures and solutions as with innovation features in mechanism design and on specific areas such gear systems, actuation of robots, and cable systems; robotics as in motion planning of humanoids and testing with drones; control of vibrations; biomechanics with solutions for healthcare and rehabilitation; and power source design in the form of battery packages.

It should be noted that this volume also intends to be an editorial initiative as a series of books to remember and recognize IFToMMist personalities who, due to their contributions and humanity, can be a reference and inspiration for the entire technical-scientific community, not only in IFToMM but especially for the new generations of MMS engineers and scientists. And, this volume on Prof. Carlos Santiago López-Cajún is an excellent beginning to consolidating the first attempt,[1] which we hope will be followed by many other volumes on other distinguished figures from the recent past.

Roma, Italy
Santiago de Querétaro, Mexico
July 2023

Marco Ceccarelli
Juan Carlos Jauregui-Correa

[1] Volume on Prof. Veniamin Goldfarb: Trubachev, E.S., Barmina, N.A., Malina, O.V. (2021). Professor V. I. Goldfarb: Life Activity and Contribution to Gearing Science. In: Barmina, N., Trubachev, E. (eds) Gears in Design, Production and Education—A Tribute to Prof. Veniamin Goldfarb, Mechanisms and Machine Science, vol 101. Springer, Cham. https://doi.org/10.1007/978-3-030-73022-2_1.

Contents

Chapter 1
Carlos López-Cajún: IFToMMist MMS Scientist

Marco Ceccarelli and Juan Carlos Jauregui Correa

Abstract Prof. Carlos López-Cajún has been an IFToMMist MMS scientist, who is recognized worldwide for his great scientific reputation for his valuable achievements in several fields of MMS and for his friendly attitude with the true spirit of IFToMM in collaboration, sharing, and improving MMS. Prof. López-Cajún activity was centered on mechanical design of systems in several areas such as robotics, medical systems, and industry devices with significant achievements. As IFToMMist he contributed significantly to the activities of the Permanent Commission of History of MMS in organization of events and publications. He served very actively IFToMM as Member of the PC on History of MMS, member of the Executive Council and as Secretary General. Moreover, he participated in the organization of many initiatives and conferences such as the 13th IFToMM World Congress in 2011, and the HMM2016 Symposium in his city Queretaro in 2016.

Keywords History of MMS · History of IFToMM · History of mechanism design · Carlos López-Cajún

1.1 Biographical Notes

Carlos Santiago López-Cajún, Fig. 1.1, (in short Carlos) was born on January 18, 1948, in Campeche, Mexico and passed away suddenly on Sunday 20 December 2020 in hospital in Queretaro, Mexico, [1–3].

Prof. López-Cajún spent his childhood in Campeche where the family is still based, Fig. 1.2. There he formed his warm character with a calm behavior as always

M. Ceccarelli (✉)
Tor Vergata Rome University, 10133 Rome, Italy
e-mail: marco.ceccarelli@uniroma2.it

J. C. J. Correa
Autonomous University of Querétaro, Querétaro, México
e-mail: jc.jauregui@uaq.mx

M. Ceccarelli and J. C. Jauregui-Correa (eds.), *State-of-the-Art and Innovations in Mechanism and Machine Science*, Mechanisms and Machine Science 150,
https://doi.org/10.1007/978-3-031-47040-0_1

(a) (b)

Fig. 1.1 Prof Carlos S. López-Cajún (1948–2020): **a** in his maturity as professor at University of Queretaro; **b** when graduated at UNAM in Mexico City

available to interact and help people, while starting to have interest in technology and particularly to machinery.

Prof. López Cajún graduated Bachelor and Master degrees on Mechanical Engineering in 1969 and 1977 from Universidad Nacional Autónoma de México (UNAM) in Mexico City, Fig. 1.1b). Much later, he got Ph.D. degree at Case Western Reserve University, USA in 1982, after having worked in Industry. The experience in USA for his Ph.D. degree was important not only for his formation in other fields than mechanisms, but also to have experiences abroad with a first international vision. He was employed from 1969 to 1972 at the company Panamerican Valves and from 1972 to 1975 at the Mexican Bus Company in Mexico City. While working in Industry he got the position also part-time Professor at the Faculty of Engineering of UNAM in Mexico City. During the student period he was also at the Department of Mechanical Engineering of UNAM in Mexico City where he met and collaborated with Professors Ricardo and Enrique Chicurel who were later his mentors and supporters in his activity for promoting internationalization of the Mexican community, within IFToMM. In this period at UNAM, he also started a long-life friendship with Jorge Angeles, who was at that time senior student, and later, while Jorge Angeles was professor at McGill University in Montreal, Canada, he worked out a significant scientific collaboration mainly in the topics of cam transmissions.

Once back in Mexico after the Ph.D. degree, Prof. López-Cajún worked again in Industry from 1988 to 1995 being employed at Mexican Institute of Transportation in Queretaro where he acquired more professional experience, even with leader positions.

(a)

(b)

Fig. 1.2 Prof López-Cajún and family: **a** with his father and mother; **b** (on left behind) with his full family in Campeche, Mexico

Then upon invitation by Prof. Jorge Angeles he spent sabbatical leave in 1986–87 as Visiting Researcher at McGill University, Quebec, CA, and a period returning to University of Notre Dame, USA.

In 1996 he got the position of Full Professor in Mechanical Engineering at Universidad Autónoma de Querétaro, serving at the campus San Juan del Río, by starting regular and continuous activity in teaching and research attaching several topics and problems in MMS also with collaboration and supervising students. Prof. López-Cajún was very active until his last days, with significant activity not only on local and national frames but also at international levels and mainly with collaboration from

IFToMM community by joint projects, attending conference events, and spending period of visits.

1.2 IFToMMist Figure

An IFToMMist is identified as an MMS scientist involved in IFToMM with the spirit of IFToMM summarized in friendly attitude for collaboration in promoting community (and colleague) and MMS developments, teaching, research, and expertise transfer. Carlos Santiago López-Cajún well behaved with such a IFToMM attitude, even in leadership roles.

In teaching Prof. López-Cajún addressed main attention to Mechanics of Transmissions and Mechanism Design giving courses both on basic formation and advanced topics for bachelor, master, and Ph.D. students at the most in his university in addition to invited lectures and seminars abroad all round the world. He was also interested in other subjects with special focus on Mechanism Science, such as the Robotics, Automation, Diagnostics of machinery, and History of mechanical engineering. This last topic was his passion outside of regular teaching frames although he always commented history hints in his lectures to make the students aware not only of the past efforts and inventions but to explain that advances in technology and particularly in MMS are due to efforts and dedication of persons, inventors, academics, and professionals, during several years of activity.

His dedication and clarity in teaching together with his kind and humanity attitude made him as a very appreciated teacher by students who still remember him as a clever teacher. He spent teaching activity not only in lecturing but also in assisting the students with tutorials and meetings for better understanding of subjects and concepts by the students even with numerical examples if not possible experimental practice. Because of his friendly attitude he was very often considered as a father for those students, who were in strict relationship with him during of thesis projects and even more as Ph.D. students. He was also used to share his teaching with lectures by other colleagues with the aim to show team experience to students with different viewpoints and cultural backgrounds. While he was never aggressive with students, and he was available to help for their understanding, at the exams he was rigorous and objective in the evaluations showing the serious of the subject consequences.

As Professor at Universidad Autónoma de Querétaro he lived in Queretaro where he took care of his nephew who lost both the parents quite early. Only in 2008 he got married with Maria Agueda Amador after many years of relationship with her. At their wedding Prof. López-Cajún invited his friends from IFToMM as weeding wisdom with the vision spirit that the IFToMM community is like a family. In Fig. 1.3 the photo shows the just married Carlos and Maria together with Ineke and Teun Koetsier (from Netherlands), and Brunella and Marco Ceccarelli (from Italy).

Prof. López-Cajún lived and participated in the IFToMM community by combining technical-scientific activities with friendly social relationships that improve and make of long duration technical-scientific collaborations.

Fig. 1.3 Wedding memory photo of Carlos López-Cajún and Maria Agueda Amador with Ineke and Teun Koetsier (at their right) and Marco and Brunella Ceccarelli (at their left)

With his personal involvement, he was a successful promoter of aggregation in national and international contexts. His communicative capacity in community aggregation was successful mainly in cordial meetings around a table during a lunch or a dinner both in congress and non-congress frames. Figure 1.4 shows an example of such friendly meetings during which Prof. López-Cajún helped Mexican colleagues, including his former Ph.D. students now professors, both to discuss and plan greater aggregation of the Mexican community in national academic and research programs for more adequate participation and visibility in international frames, especially in IFToMM. In fact, his strong actions as promoter of IFToMM Mexico, besides being a very active member of ASME, gave him the chance to be a Founder Member of the Mexican Society of Mechanical Engineering (SOMIM) as a national aggregation of a community well recognized within Mexico and abroad.

Thanks to his reputation he was granted of member of the Mexican Academy of Engineering and awarded with several honors by local, national, and international institutions, among which it may be highlighted the Mexican National Researcher award.

Prof. López-Cajún worked out intense international activity not only in congress events but also in promotion and management of international community institutions. He was a very active ASME member by participating to association initiatives including international congresses in the American continent. As a member and co-founder of SOMIM he was strongly dedicated in the growth of the community both academically and professionally with interactions at international levels.

But his most significant international activity was within the IFToMM, International Federation for the Promotion of Machine and Mechanism Science, from simple participation in congresses up to leadership roles. He served IFToMM as Chair of Mexico IFToMM member organization and member of the Permanent Commission on History of MMS for several years and then 2003–2007 member of the Executive

Fig. 1.4 A meeting organized by Prof. Carlos López-Cajún with Mexican colleagues in Mexico City in 2018 (from the left to right: Carlos López-Cajún, Marco Ceccarelli -invited guest-, Eusebio Hernandez and Ivan Valdez from IPN-Ticoman, Mario Acevedo from Panamericana University, and Christopher Torres from IPN-Zapeteco

Council and Secretary General in 2008–2011. Moreover, he proposed and contributed to the organization of the 13th IFToMM World Congress held at Guanajuato, México in 2011 that was initially planned in Queretaro. When he recognized the difficulties of local convenient Convention Center he facilitated the possibility to have yet in Mexico but in a different attractive location. He was Chair of the HMM2016 Symposium on History of Mechanism and Machines held at Querétaro, México in 2016 with a very successfully international participation.

Prof. López-Cajún joined the Commission for History of the MMS during the 1999 IFToMM World Congress in Oulu, Finland, thanks to Teun Koetsier and Marco Ceccarelli, past and current Chair of the Commission at that time, [4, 5], because of his interests on history of mechanical engineering as also a means for training engineer and researcher in MMS fields. Prof. López-Cajún was very active in the commission by participating both in meetings and interactions, also through email and telemeetings, and in specific research and interpretation of inventions, events, and personalities in the history of mechanical engineering. In fact, it has contributed since the first event in 2000 to the HMM symposium with significant contributions that have attracted great attention from participants and the from experts in finding documentation and published past works.

Figure 1.5 shows examples of his participation in the commission meetings with contributions in the discussion of the commission issues and plans to meet the IFToMM constitutional requirements but also to promote a wider international participation of colleagues (fundamental characteristic of the commission) not only from the member organizations of IFToMM but also from other communities interested in the history of the machines. In this active participation, Prof. López-Cajún has also established relationships and collaborations with IFToMM spirit at international levels, as for example in promoting the interest in the history of mechanical machinery by also young researchers in MMS areas. His contributions were presented and published in the HMM symposium ranging from historical and interpretative studies of machines also of the recent past up to theoretical and design analysis for reconstruction of machines not only Mexican technology.

One other major contribution of Prof. López-Cajún in the Permanent Commission, after having actively participate to all the HHM symposia since the beginning 2000, was the organization of the 2016 HMM symposium in Queretaro, as by his capacity of member of the scientific committee for HMM. The 2016 Symposium was very well participated with 23 papers presented by authors from all around the world and published in the Springer book series on History of MMS, [6]. He organized also a social cultural program, as in the tradition of the HMM symposia, that permitted the participants to strengthen or start their friendship with more plans of collaborations.

Prof. López-Cajún worked a significant role in IFToMM in different positions, staring as simple IFToMMist participating to IFToMM thematic conferences and

(a)

(b)

Fig. 1.5 Participation of Prof. Carlos López-Cajún at the meetings of the IFToMM PC for History of MMS: **a** in 2000 during HMM2000 symposium in Cassino, Italy; **b** in 2011 during IFToMM World Congress in Guanajuato, Mexico in the world congresses to get a well understanding of the IFToMM community. After being an active member of the IFToMM Member Organization Mexico, he was elected Chair of it in 1999, reinforcing participation of Mexican colleagues in the IFToMM technical committees with a more international impact of the Mexican community including the Mexican association of mechanical engineering SOMIM. In this promotional activity of the Mexican community, it is to note his commitment with a proposal to host the IFToMM World Congress in 2011 in Mexico which was well accepted by the IFToMM General Assembly in 2007. Initially, the proposal agreed within the Mexican community was to host the Congress at the University of Querétaro, the seat of Prof. López-Cajún who, after having quickly checked the efficiency and the possibility of suitable infrastructures (a large convention center promised in Queretaro was postponed by the Mexican local government), was able to involve his Mexican colleagues to find an optimal solution in the city of Guanajuato, with characteristics of social and tourist attractiveness combined with a network of surrounding technical universities to which the University of Querétaro also referred. As member of the organizing committee, Prof. López-Cajún supported Professor Riccardo Chicurel, president of the Mexico IFToMM Member Organization, to chair the organization of the 2011 IFToMM World Congress with a continuous dedication and presence as shown in the example in Fig. 1.6.

Following the participation at different IFToMM levels and leadership as Chair of an IFToMM Member Organization, Prof. López-Cajún was then elected a member of the IFToMM Executive Council for the 2003–2007 term, during which he contributed to the improvement and growth of the IFToMM activities, with regular participation in the meetings of the IFToMM Executive Council chaired by Professor Kenneth Waldron, Fig. 1.7. During this he coordinated with the members of the IFToMM Executive Council, with appreciation and recognition for his dedication and results in promoting IFToMM especially in the South American world, by also involving the Iberoamerican Federation of Mechanical Engineering FebIM which in those years

Fig. 1.6 A meeting for planning the 2011 IFToMM World Congress on 23 September 2010 (from the left): Adolfo Pamanes, Marco Ceccarelli Ricardo Chicurel, Juan Carlos Jauregui, and Carlos Lopez-Cajun

Fig. 1.7 Meeting participation of Prof. Carlos López-Cajún as EC member at the meeting of the IFToMM Executive Council in 2005 in Besancon, France

established a bilateral agreement with IFToMM for concrete collaborations. A first result of this agreement was the MUSME Congress in 2002 in Mexico City that was sponsored by both federations on issues of Multibody Dynamics and Mechatronics. In this frame, the two federations also shared the activities of their technical committees on Robotics and Mechatronics, being Prof. López-Cajún the liaison person for many years.

At the General Assembly of the IFToMM in 2007 he was unanimously elected General Secretary of IFToMM that worked out very successfully in coordination with the president Marco Ceccarelli and the treasurer Joseph Rooney who, together with the past president Kenneth Waldron, formed the Presidential. This leadership dedication and its results are reported in documents that are archived at the IFToMM archive at the CISM in Udine, Italy, and published partially in [7–10] as survey summary. Prof. López-Cajún as General Secretary, was in agreement with constitutional tasks, a point of reference for the management of IFToMM affairs, editor of the official documents of the IFToMM, manager of relations and communications with the bodies (PC, TC, Chair of GA) of the IFToMM and with the IFToMM community in general, and collaborator of the president in coordinating the activities and co-chairing the meetings of the Executive Council in presence and in telematic form. Prof. López-Cajún has carried out his duties as secretary general with discretion and never without exaggerating in a leading role but leaving plenty of space, in agreement with the president, for discussion by all the present IFToMM leaders and representatives. Figure 1.8 shows examples of the people participating at the Executive Council meetings.

In 2009 in occasion of the fortieth anniversary of the IFToMM Prof. López-Cajún organized the ceremony at the historic site of the University of Guanajuato with an event limited to a single day and a few interventions by the past presidents and the few founding fathers still alive.

Figure 1.9 shows examples of the management of Executive Council meetings by Prof. López-Cajún carried with efficiency also in fulfilling timed agenda although several topics to be discussed. He can be noted in Fig. 1.9a while recording the interventions near the treasurer and the president. In Fig. 1.9b Prof. López-Cajún at the center of the coordination table between the president and the treasurer coordinates the interventions assisted with multimedia systems and projector allowing interventions also online.

(a)

(b)

(c)

(d)

Fig. 1.8 Examples of participation of Prof. Carlos López-Cajún at the meetings of the IFToMM Executive Council: **a** as EC member in 2005 in Besancon, France; **b** in 2008 in Tokyo; **c** in 2009 in Guanajuato, Mexico, checking the 2011 World Congress site; **d** in 2010 in Sousse, Tunisia; **e** in 2011 during IFToMM World Congress in Guanajuato, Mexico

Fig. 1.9 Prof. Carlos López-Cajún as IFToMM Secretary General, recording the discussion at the meeting of EC meeting: **a** in 2008 in Tokyo (first in the left); **b** in 2011 during IFToMM World Congress in Guanajuato, Mexico (at the center of chair desk)

Figure 1.10a shows the presidential desk meeting, also held in a convivial form, during which to adequately prepare an Executive Council meeting was prepared with preliminary discussions to facilitate Executive Council meeting outcomes. In this task Prof. López-Cajún has played an important role, always being punctual in the preparation of the agenda points but also in the cordiality of the discussion of each point even in critical terms. Figure 1.11b, on the other hand, shows moments of work with the president to prepare reports and official documents following not only the meetings of the Executive Council but also to produce suitable and complete documents for the IFToMM archive as prescribed by constitutional obligations.

(a)

(b)

Fig. 1.10 Prof. Carlos López-Cajún at work as IFToMM Secretary General in 2011 during IFToMM World Congress in Guanajuato, Mexico: **a** meeting of the Presidential Desk; **b** revising the minutes of the 2011 EC meeting with the IFToMM President

The IFToMMist character of Prof. López-Cajún can be summarized during his full career in participating to IFToMM activities in any aspects such as participation, organization, and leadership of initiatives, events, community promotion, and MMS developments. Throughout his life, Prof. López-Cajún has combined professional scientific activity, characterized by dedication and intelligence, with an aptitude for relating with people, students or researcher colleagues, with an empathy that has made his figure attractive both at a scientific and didactic level, leaving an imperishable memory of his spirit and his results both in scientific and social relations.

1.3 MMS Scientist

The prolific scientific activity of Prof. López-Cajún covered many aspects of mechanical engineering and especially in the areas of MMS with particular interests and significant results especially in analysis and synthesis of mechanisms, cam transmissions, design of service robots, machine diagnostics, vehicle mechanics, vibration mechanics, and history of mechanisms and machines, [2]. The most significant contributions are documented in more than 500 papers (that are still available not only in Google) that are published in international journals and international conferences with a significant number of citations by other authors documenting a significant impact in respective thematic areas. Furthermore, significant are co-authored books among which it is to note:

- Optimization of Cam Mechanisms in 1991, [11], co-authored with prof Jorge Angeles Fig. 1.11a)
- Mechanisms (in Spanish) in 2008, [12], coauthored with prof Marco Ceccarelli Fig. 1.11b)
- Proceedings of 2016 HMM symposium held in Queretaro in 2016, Mexico, [6].

As results of his reputation and expertise Prof. López-Cajún was appointed member of several scientific committees for international conferences and journals. He contributed to many projects in different topics, such as Mechanism Design, like in Robotics, Biomedical devices, Space systems, and History of Engineering and Machines with achievements in highly disseminated publications.

The above-mentioned books are representative of the level of expertise of Prof. López-Cajún both from the perspective of research results and as a dissemination with the aim of transferring and teaching the results achieved and their usefulness. In particular, the first book, [11], refers to the topic on transmission cams with significant results already at the beginning of his scientific curriculum, in the wake of which Prof. López-Cajún continued during his life to investigate problems for design, functionality and efficiency, deepening solutions for the design of cam profiles, of the transmission cam system, of the dynamics, of the regulation and control of the operation, also for possible and innovative applications.

The second book, [12], can be considered a synthesis not only of knowledge on aspects of transmission mechanics but also and above all of experiences and

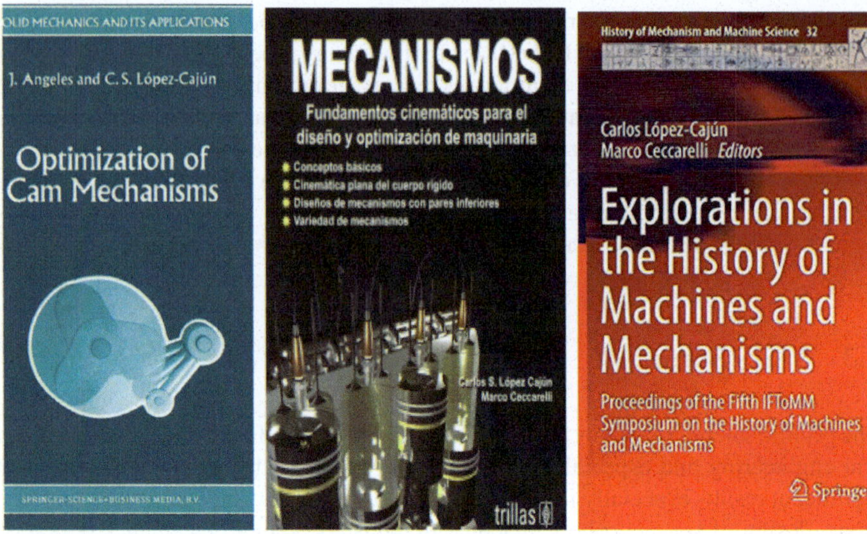

(a) (b)

Fig. 1.11 Books by Prof. Carlos López-Cajún co-authored: **a** with Jorge Angeles on Cam Mechanisms in 1991, [11]; with Marco Ceccarelli in 2008 on Mechanism teaching, [12]; **c** with Marco Ceccarelli in 2016 on Proceedings of the HMM symposium

maturity in teaching activities with subjects and methodologies leaning towards the future, especially with the use of computer procedures. The book is however organized still considering fundamental the basic concepts as well as the conceptual graphic modeling which allow a student to have a concrete vision of the mechanics of mechanical systems. The book, in fact, is organized in such a way as to proceed from a structural description with a conceptual modeling up to outlining calculation procedures for the design and analysis of the mechanical elements of transmissions for modern machines.

The third book, [6], wants to be representative of Prof. López-Cajún's editorial ability to aggregate contributions, in this case in the context of a conference event related to the history of machines in order to share problems and solutions for better collaboration and activities for development of a proper knowledge in the historical developments of machines. The book is centered, being the proceedings of the HMM symposium on the history of machines and mechanisms, on research and interpretation results of historical developments that have allowed inventions of machines, development of theories, and definition of technical-scientific personalities, who have contributed significantly to the development of machine technology both in the past and in recent times.

During his long activity Prof. López-Cajún authored and/or coauthored more than 500 papers in international journals and conferences, most of them are still of reference and background for the last advances in specific topics. The main contributions by Prof. López-Cajún can be recognized mainly in mechanism design and cam transmissions, machine diagnostics and vibration mechanics, analysis and design of robots, and history of mechanisms and machines. The literature of his works is still available and cited as significant background for the typical procedure of considering the problems for identification of the open issue and significant parameters that can help in solving or proposing solutions that can be evaluated numerically and partially with prototypes. The MMS scientist character of Prof. López-Cajún can be indeed summarized not only by his multidisciplinary interests on MMS topics but also by the approaches by which he attached the challenges in proposing solutions and testing them for a future transfer in practical implementations.

1.4 Conclusions

The distinguished figure of Prof. Carlos López-Cajún can be summarized mainly in the following aspects:

- achievements in several MMS fields with impact for future developments
- methods of teaching and interacting with students
- social attitude with kind character open to promote scientific-technical collaborations and technical-scientific community
- activity with IFToMM sprit in international frames as per the University concept with a significant dedication to IFToMM.

Unforgettable are the events, facts, collaborations and interactions that Prof. López-Cajún experienced within the IFToMM community, making him a IFToMMist reference figure also for future generations, having fully embodied the original spirit of IFToMM in the mission of promoting the science of the machines of the mechanisms not only for the improvement of technology but mainly for the well-being and peace in human society. Prof. López-Cajún well embodied the spirit of the IFToMM which is based on no barriers of any kind in facing technical-scientific problems to start and sustain collaborations that can be useful for mutual cultural and technical growth as well as for the promotion of areas and structures that are less capable or less structured, also due to lack of funds.

The legacy left by Prof. López-Cajún in the field of teaching and academic training is demonstrated by the good memory for a long time by students and those who had have him as mentor or supervisor.

Prof. Carlos López-Cajún will be remembered as a distinguished figure with multi-faceted skills and significant results of activity not only in the history of IFToMM and of the disciplinary areas of Mechanisms and Machines Science.

References

1. López-Cajún, C.S.: Curriculum and list of publications. University to Queretaro, Queretaro, Personal archive (2018)
2. Ceccarelli, M., Jauregui Correa, J.C., Carlos López-Cajún: (1948–2020) In: Ceccarelli M., Gasparetto A. (eds.) Distinguished figures in mechanism and machine science – Part 5: legacy and contribution of the IFToMM community, pp. 111–135. Springer, Cham (2023). https://doi.org/10.1007/978-3-031-18288-4_4
3. Acevedo M., Ceccarelli M.: In: Ceccarelli M., López-García, R. (eds.), Memory of Past PC Members – Prof Carlos López-Cajún (1948–2020), Explorations in the History and Heritage of Machines and Mechanisms – 7th International Symposium on History of Machines and Mechanisms (HMM), pp. 4–5. Springer Nature Switzerland AG, 2022. https://doi.org/10.1007/978-3-030-98499-1_1
4. Ceccarelli M., Koetsier T.: On the IFToMM Permanent Commission for History of MMS. In: Proceedings of International Symposium on History of Machines and Mechanisms HMM2004, pp.10–25. Kluwer, Dordrecht (2004)
5. Gasparetto, A., Ceccarelli, M.: Recent advances and challenges in the IFToMM PC for history of MMS. In: Ceccarelli, M., López-García, R. (eds.) Explorations in the history and heritage of machines and mechanisms. HMM 2022. History of Mechanism and Machine Science, vol 40. Springer, Cham (2022). https://doi.org/10.1007/978-3-030-98499-1_2
6. López-Cajún C.S., Ceccarelli M. (eds.) Explorations in the history of machines and mechanisms, history of mechanism and machine science 32. Springer International Publishing Switzerland, Cham (2016). DOI https://doi.org/10.1007/978-3-319-31184-5_1
7. López-Cajún C.S.: Minutes with appendices of IFToMM EC meetings and telemeetings 2008–2011, items: Ab 02–2008, Ab 01–2009, Ab 01–2010, Ab 02–2010, 2010 Ab 03–2010, Ab 01–2011, Ab 02–2011, Ab 03–2011, IFToMM Archives, CISM Udine
8. Ceccarelli, M., López-Cajún, C.S., Rooney, J.: IFToMM Executive Council Meeting. Mech. Mach. Theory **48**(2012), 140–143 (2011). https://doi.org/10.1016/j.mechmachtheory.2011.10.002
9. Ceccarelli M.: Twenty-five year of activity in IFToMM, journal Theory of Mechanism and Machine (http//tmm.spbstu.ru). St Petersburg State University, Vol.11. No.2, pp. 3–14 (2013)
10. Ceccarelli, M., Experiences in Leadership IFToMM: Achievements and Challenges, Advances in Mechanism Design III – Proceedings of TMM 2020, Springer Nature Switzerland AG, 2022, pp. 3–16. https://doi.org/10.1007/978-3-030-83594-1_1
11. Angeles, J., López-Cajún, C.S.: Optimization of Cam Mechanisms. Kluwer Academic Publishers, Dordrecht (1991)
12. López-Cajún, C.S., Ceccarelli, M.: Mechanisms (Ed.) Trillas, Mexico, 2008 – 2nd edition 2013 (in Spanish)

Chapter 2
First in Memoriam Seminar for Carlos Lopez Cajun (Summary of Presentations)

Juan Primo Benitez-Rangel, Aurelio Dominguez-Gonzalez, Alejandro Lozano-Guzman, and Wenceslao Ortiz-Vargas

Abstract This chapter includes a summary of the presentations and activities conducted during the First In Memoriam Seminar for Carlos Lopez-Cajun. The chapter consists of an overview of the event's activities, pictures, and the written presentations submitted by three of the participants. The Seminar was held at the Autonomous University of Queretaro, in the San Juan del Rio's campus on May 11th, 2023. The Seminar included oral presentations and the unveiling of a recognition plaque. It was attended by graduate and undergraduate students, Lopez-Cajun's relatives, and his widow.

Keywords Memorial · Distinguished figures · History of MMS · Teaching MMS

2.1 Introduction

This chapter summarizes the "First in Memoriam Seminar for Carlos Lopez Cajun." The Seminar was held at the Autonomous University of Queretaro in San Juan del Rio's campus on May 11th, 2023, and it was sponsored by the Permanent Commission for the History of Mechanism and Machine Science and the School of Engineering and was organized by Profs. Marco Ceccarelli and Juan Carlos Jauregui.

J. P. Benitez-Rangel (✉) · A. Dominguez-Gonzalez · A. Lozano-Guzman · W. Ortiz-Vargas
Universidad Autonoma de Queretaro, Santiago de Querétaro, Mexico
e-mail: benitez@uaq.mx

A. Dominguez-Gonzalez
e-mail: auredgz@uaq.mx

A. Lozano-Guzman
e-mail: alejandro.lozanog@uaq.mx

W. Ortiz-Vargas
e-mail: wov44@hotmail.com

M. Ceccarelli and J. C. Jauregui-Correa (eds.), *State-of-the-Art and Innovations in Mechanism and Machine Science*, Mechanisms and Machine Science 150,
https://doi.org/10.1007/978-3-031-47040-0_2

The Seminar's objective was to:

Recognize the contribution and legacy of Distinguish Figures, in this case, Carlos Lopez.
Present and Highlight technical contributions of research activities realized in the fields where Carlos Lopez made a contribution
Share the legacy of Distinguish Figures with young researchers and students
Generate a network of active researchers to promote IFToMM's goals
Foster dialogue between young and consolidated researchers.

The program included oral presentations of colleges, students, university staff, and graduate students. The Seminar described different topics, such as his lecturing skills, research abilities, and academic outputs, but mainly his human sense of teaching young generations and encouraging them to be excellent engineers. Everybody remembers him as a great Professor and an excellent college. The public was basically undergraduate and graduate students who could learn about the lives of distinguished persons who worked to set the basis for good engineering schools.

The Seminar ended by unveiling of the recognition plaque. His widow, relatives, and the school authorities formalized the unveiling.

This chapter includes only the written presentations of Dr. Juan Primo Benitez-Rangel and M.C. Wenseslao Oritz-Vargas. They were Carlos Lopez-Cajun's student and college. Dr. Aurelio Dominguez-Gonzalez was also his student and college, and Dr. Alejandro Lozano-Guzman was his college and personal friend.

The following pictures illustrate the event; they show the number of participants and the different aspects of the Seminar (Figs. 2.1, 2.2, 2.3, 2.4, and 2.5).

Following are the written presentations submitted for this chapter.

2.2 Juan Primo Benitez-Rangel and Wenceslao Ortiz-Vargas

2.2.1 Summary

Dr. Carlos Santiago López Cajún, to whom consequently we will call only for his last name. During his working life, he participated in several activities in the Autonomous University of Queretaro (UAQ) in San Juan del Rio's Campus, where he was attached since his beginning in 1996 until his retirement in 2018.

His activities went from teaching to the management of Ph.D. programs. The following brief information will show his national and international production and contribution to UAQ, as well as his pedagogical practice.

Fig. 2.1 Opening ceremony by Juan Carlos Jauregui

Fig. 2.2 Panoramic view of the auditorium

Fig. 2.3 Group photo. From left to right on the first row: Dr. Matin Valtierra, head of the Mechatronic graduate studies. Dr. Christopher Torres, from Iftomm. Dr. Eduardo Hernandez, former Dr. Lopez-Cajun's Ph.D. student. Dr. Juan Carlos Jauregui, organizer. Dr. Marco Ceccarelli, former Iftomm's president. Dr. Luis Morales, head of the Electro-mechanic department. Dr. Ageda Lopez, widow. Dr. Aurelio Dominguez, professor and college. M.C. Wenceslao Ortiz, college. Dr. Angel Perez, head of the Automotive department. M.C. Manual Quijada, college. In the back, students and attendees

Fig. 2.4 Dr. Luis Morales and Dr. Ageda Lopez (widow) unveiling of the recognition plaque

Fig. 2.5 Authorities during the unveiling of the recognition plaque

2.2.2 *Introduction*

Dr. Carlos Santiago López Cajún participated in several activities in the university, this paper gives a quick summary of all his labor in San Juan del Rio's Campus.

2.2.3 *His Beginning*

Dr Lopez Cajún came to the Engineering Department in 1991, with some other professors, among them: Dr Alejandro Lozano Guzman, Dr Ricardo Tapia Armas, and later M. I. José Antonio Romero Navarrete currently with a Ph.D. degree. All of them researchers of Mexican Institute of Transportation (IMT).

Another professor who contributed in the development of the Engineering Department was M.C. Jaime Hernández Rivera.

In February 1996, Dr López Cajún participated to get his full-time contract in UAQ and got it to be attached in San Juan del Rio's Campus. During this time, he taught in both Bachelor Program and Ph.D. program. At the same time, he collaborated in Engineering Ph.D.s and coordinated the Master of Instrumentation and Control from 1997 to 1999 in Queretaro City.

In 2002, the Master Program of Instrumentation and Control started and in 2005 the Engineering Ph.D. in San Juan del Rio's Campus . In these both programs he collaborated in teaching and management of thesis.

The following chart shows the subjects he used to teach in the San Juan del Rio's Campus (Table 2.1).

Table 2.1 Taught subjects by Dr. Lopez Cajún in UAQ

Subject	Degree
Dynamics	Bachelor
Statics	Bachelor
Mechanical of Materials	Bachelor
Mechanical Vibrations	Bachelor
Machinery Design	Bachelor
Kinematics of Machinery	Bachelor
Design-Aided by Computer	Bachelor
Project and Design	Bachelor
Robotics	Master Degree

Also, he participated as synod in several theses in Bachelor Degree and Ph.D. (Fig. 2.6).

All the time, his professorship was characterized for being strict and demanding based on Physics and Mathematics to improve the development of the subject. In spite of his strong character and personality, he always had a great attitude in the classroom that contributed to educate excellent professionals.

As an anecdote, in 1995 in the subject called design-aided by computer, there was a student named Hugo Mendoza, who always got paralyzed when Dr. Cajún stood behind him to observe the process of the practices and told him: "Draw, Hugo"! And the student replied: "Yes, but move, because you make me feel nervous and I cannot work". ¡So, Dr. Cajún just laughed!

Full of excitement, Dr. Cajún started with the participation in congresses with some students for their first time. These congresses were organized by him most of

Fig. 2.6 Engineering Week Celebration, 1993. From right to left, Alvaro student and president of Student Union, then Engineer Wenceslao Ortiz Vargas, Coordinator of Electromechanical Engineering Department and Dr. Lopez Cajún

the times. The first one was in Mexican Society of Mechanical Engineering (SOMIM) in Guanajuato City in 1996, one of the most important international congress in the mechanical area in Mexico. Later, from this previous experience in congress some more others came over. Then, "The International Engineering Congress came to UAQ, and now in 2023 there was the XIX Edition.

2.2.4 *Professional Environment*

Dr. López Cajún did not develop research assignments in San Juan del Rio's Campus because the career was emerging with few students, few professors, few researchers, and no Ph.D. existing.

With no doubt, he always participated in his academic commitments. In 1999, there was the CIEES evaluation in the Engineering Department and he was the one who represented and defended the project in the Engineering Department, achieving the approval of the certification. This certification is maintained up to now (Figs. 2.7 and 2.8).

Fig. 2.7 In the Engineering Week in 1996, there was the opening of the laboratories in Engineering Department. From right to left, Dr. Carlos Santiago López Cajún, LAE Jorge León Hernández, General Coordinator San Juan del Rio's Campus and Engineer Wenceslao Ortiz Vargas, Coordinator of de Engineering Department SJR

Fig. 2.8 Metting with co-workers and assessors of CIEES in 1999

Also, he participated in Queretaro City, in Central Campus as Coordinator of the Master Degree in Sciences (Instrumentation and Automatic Control) and Coordinator of Engineering Ph.D. in 1998.

2.2.5 Sabbatical Year

In 2007, Dr. López Cajún, asked for permission to take his sabbatical year to the Engineering Department. He attended to a residency in the University of Cassino in Italy to work in a collaboration with Dr. Marco Ceccarelli and had as team members to Dr. Juan Carlos Jauregui Correa from Research Center and Technical Assistance from Queretaro State (CIATEQ-AGS) and Eduardo Castillo Castañeda from Engineering Department in UAQ.

In 2008, from the collaboration with Dr. Ceccarelli, he published a book called "Kinematic Fundaments for designing and optimizing of machinery" (Trillas) that shows al his work during this year (Fig. 2.9).

Mechanisms is one of the books that became a reference for all the subjects related to mechanical design in Bachelor Degree and Ph.D. For Robotics subject, Dr. Jorge Angeles was the one who taught it.

Dr. Cajun helped to correct the book called: "*Fundamentals of Robotic Mechanical*" where he was also mentioned in the appreciations.

Fig. 2.9 Book published by Dr. Lopez-Cajún and Dr. Ceccarelli

Later, from the collaboration between Dr. López Cajún and Dr. Jorge Angeles emerged a book called: "*Optimization of Cam Mechanisms*" (Fig. 2.10).

2.2.6 *Didactic Changes*

As it was mentioned before, Dr. López Cajún was very strict in his classes. However, from 2005, there was a notorious change in his teaching process and the integration of new and pedagogical strategies and methodologies that were transforming his professorship to give the students a constructive and nice lesson.

Following with the same ideology of teching, Dr. López Cajún changed also the way he interacted with the others being really charismatic, worried for the learning process. Before this change his office was quiet and deserted, instead his office was

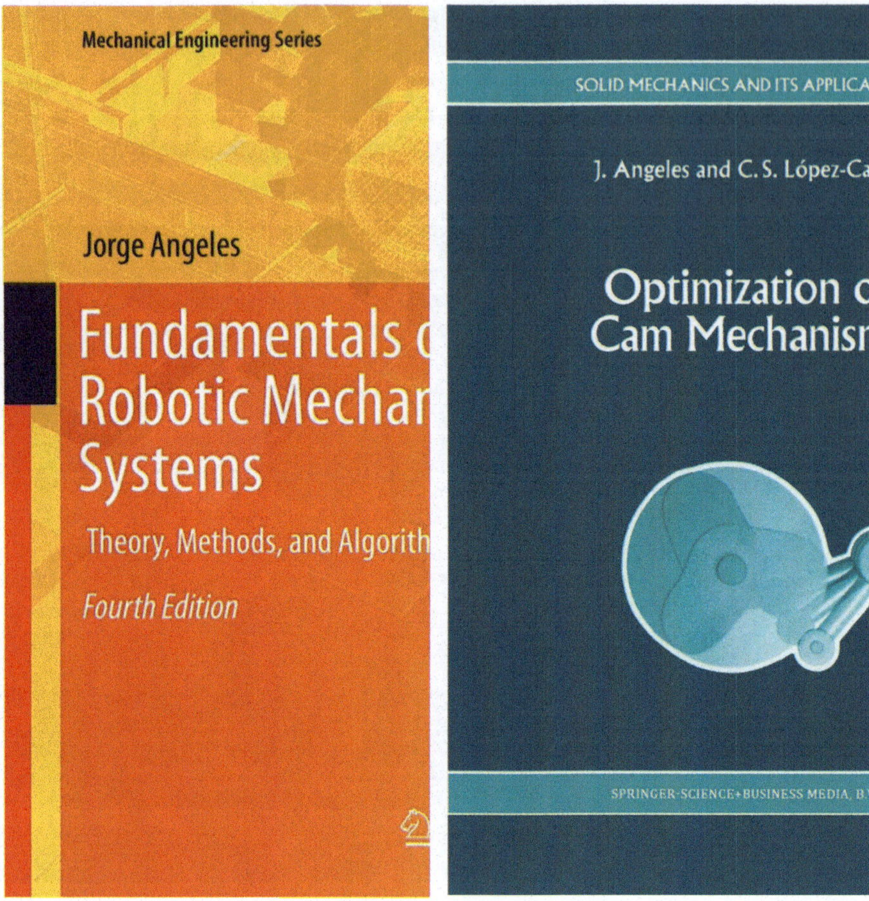

Fig. 2.10 Books where Dr. López Cajún participated

full of students asking for answers to their questions about the class, comments about their projects, and suggestions to assignments (Fig. 2.11).

Finally, we want to think that this outstanding change was influenced by his wife, professor Ma. Agueda Amador Mena, for two reasons. First, after his wedding time. Second, his wife was a Pedagogy specialist.

Fig. 2.11 Dr. Carlos Santiago López Cajún's wedding in 2005. (Marco Ceccarelli & Juan Carlos Jauregui Correa, 2022, Distinguished Figures in Mechanism and Machine Science)

2.3 Aurelio Dominguez-Gonzalez

2.3.1 A Little History with Dr. López Cajún

Dr. Carlos de Santiago López Cajun was an internationally recognized professor-researcher who worked for the Mexican Institute of Transportation (IMT) before joining the Faculty of Engineering. Here, I will mention the facts that made Dr. López Cajúm begin to collaborate with the Autonomous University of Querétaro (UAQ).

In 1987, the Faculty of Engineering extended to San Juan del Río, opening the degree program in Electromechanical Engineering. For its inauguration, there were only 4 classrooms, offices, and bathrooms, and together with the Faculty of Accounting and Administration, these were the only two located on the San Juan del Rio's campus. The decision of why electromechanics is easy to deduce. San Juan del Río was the second pole of development in the state, and this municipality was already at that time of industrial vocation.

For those years, the number of engineering professionals was minimal; many came from other states given the job offer. In this context, hiring professors who could teach the courses established for the Electromechanical Engineering program was really complicated for the program coordinator, Engineer Arturo Bautista Ángeles. Given the first generation's progress and the classes were increasingly specialized, Eng. Bautista requested support from the IMT together with the Faculty Management. At that time, the Faculty was headed by Eng. José Luis Mendoza Cedillo. Dr. Alejandro Lozano Guzmán, director of a Department at the IMT, invited engineers and researchers to support this cause.

The proposal was not very attractive to researchers, given the salaries paid at the UAQ. Additionally, the vast majority had their residence in the city of Querétaro. For this reason, it is necessary to recognize the support received from three researchers at the beginning, who were Eng. Ricardo Tapia Armas, M. C. José Antonio Romero Navarrete and Dr. Carlos López Cajúm. Later, they also had the support of Dr. Alejandro Lozano Guzmán. Indeed, they saw something that the majority did not about that small, lonely engineering school located on rough terrain and with minimal facilities to teach engineering. Who would have imagined that, over time, this would be the workplace of Dr. Romero Navarrete and Dr. López Cajún.

Dr. López Cajún distinguished himself by being very formal in his class, where he liked to make the students think of him. In fact, he was a teacher respected by everyone, to put it elegantly, or rather, he was feared for the level he imposed on his professorships. Being able to accredit his courses was a triumph, although he always left a door open so that students could pass and assimilate the course. The formality of Dr. López Cajún disappeared when you had the opportunity to spend time with him outside of class. In that sense, the Doctor was very accessible and excellent in sharing his experiences as a student.

I remember an anecdote about Dr. López Cajum. For a while, he taught the Autocad course. During the course, I remember that his students commented that the Doctor was very insistent that they continually keep his documents; he was so insistent that I would say suspiciously insistent. I remember that once he was taking an exam, and we saw him leave his room, and he went to the main switch. The power mysteriously went out, and the students left disconsolately to ask the Doctor for a chance since the exam was about to end. I don't think that after that lesson, young people would ever have a problem endorsing their work. It should be clarified that everyone passed at the end of the course, but the class remained.

Finally, I want to mention that Dr. López Cajun inspired some of his students to continue preparing. He was an example of how a humble boy from Campeche overcame barriers, and thanks to constant work and study, he was able to open doors and do a doctorate abroad at a time when he was extraordinarily exceptional.

Although synchronization can be understood with analytical models and simulations, the results are not easily interpreted because they depend on two-time scales: the oscillation period and the time until the elements move in phase. Additionally, synchronization occurs in systems having many components and complex configurations. When identifying synchronization using experimental data, the analysis is more complicated than only analyzing simulation results. Therefore, there are different approaches to analysis synchronization.

2.4 Alejandro Lozano-Guzmán

Talking about Carlos Santiago López Cajún refers to a great human being who, from his position as an undergraduate and postgraduate professor, knew how to guide many master's and doctoral students. With his fellow teachers, he has also consistently maintained an attitude of constant collaboration for more than 40 years. Reflecting on those years makes me feel quite a veteran, a privilege and a bad thing that I share at the same time with several of those present here, and that makes me refer at the same time, with regret, that Carlos is no longer physically here, but with great joy to remember the friend and professor López Cajún. Carlos graduated with a Bachelor's and a Master's degree from the Faculty of Engineering of the UNAM and a doctorate from Case Western Reserve University, which Carlos referred to as the Reserve University of the Western Case. The name Case comes from the last name of Leonard Case Jr., promoter of the Western Reserve College, which was effectively a territorial reservation southwest of Cleveland, Ohio.

From UNAM and from Case, an outstanding professor comes out who, according to Scopus, appears with an index of 10 with 47 published documents. This data illustrates the quality of Professor López's work since he reached a very acceptable h-index for Mexican institutions with a relatively low number of publications. It also reflects the quality of his writings. That is, Dr. Lopez's publications are read and cited. Thus, in Carlos' curriculum published by the Faculty of Engineering of the UAQ, the works that he published as author or co-author with researchers of recognized prestige stand out, and to mention just a few, Mihir Sen from the University of Notre Dame, Juan Carlos Jáuregui from the UAQ, Marco Ceccarelli from the University of Rome, Víctor Manuel Castaño from UNAM and Jorge Angeles from Mc Gill University, as well as numerous professors and students from the Autonomous University of Querétaro and the Mexican Institute of Transportation.

Dr. López published topics on theories of machines and mechanisms, especially his work on the kinematics and dynamics of cams is relevant. Among his contributions, the publication of the book MECHANISMS, co-authored with Dr. Ceccarelli, is noteworthy, in which the background of the theory of machines and mechanisms, the fundamentals of the movement of a rigid body, modeling for analysis and synthesis of mechanisms are presented. And the kinematics of cams and gears under the sign of the Trillas publishing house. What is embodied in this book constitutes an excellent introductory basis for studying Robotics.

During his time at the Mexican Institute of Transportation, Dr. López also contributed to many publications on energy-saving and anti-pollution devices for internal combustion vehicles, among other topics. While reviewing these publications, I was struck by their validity even though they were written in 1995 on quality, productivity, and technology transfer from research centers to potential end users. I quote from this work,

> There is a large body of literature on technology transfer. However, in most of it, it is considered that this transfer does not have a cost, undermining the action represented by the sale of technology. This section presents some considerations to establish this transfer as a transaction that must be subject to the law of supply and demand; that is, technology is a good that must be managed according to commercial considerations. However, the potential buyer must be sufficiently prepared to decide if the purchase is better than developing their own technology. Around this idea, the role that research institutes could play is described, and examples of how the link between institutes and industry is developing are presented.

The above described is a problem that is still in force, which, with great vision, Dr. López clearly raised. From this same publication, he took the following paragraph:

"Using the expression "technology transfer" has created a feeling that such transfer does not cost. However, one cannot lose sight of the fact that technology that is worthwhile, that is, commercially competitive, is owned by companies or individuals. Unlike basic research in the public domain, technology is not transferred; it is sold.

On the other hand, technology transfer models seem not to be applicable in an underdeveloped country since modern technologies were developed in industrialized countries under their specific conditions.

It is also important to mention that transmitting technology also transmits a cultural impact, a way of life.

Therefore, it is necessary to promote the technological culture in local companies so that they can play a role.

Chapter 3
A Note on the History of the Science of Machines

Teun Koetsier

Abstract In the present paper I consider the science of machines as encompassing machine theories covering different aspects of machines. The history of the science of machines is then the history of these machine theories, the relations between them and their relation to the practice of machine building. Such an approach to the science of machines generates interesting questions. The first successful machine theory was the theory of simple machines. It reached maturity with Galilei. The second successful machine theory was kinematics. It reached maturity with Reuleaux and Burmester. Dynamics applied to machines is another example of a machine theory. This paper is dedicated to the memory of my friend Carlos Lopez Cajun. We shared an interest in the history of machines.

Keywords Theory of machines · Simple machines · Kinematics · History of MMS · Heron of Alexandria · Galilei · Gaspard Monge · Robert Willis · Franz Reuleaux

3.1 Introduction

Machines are manmade entities. In the course of time continuously new machines are invented and existing machines are modified. That is why the science of machines, defined as the whole of all theories that deal with different aspects of machines, inevitably differs from natural science.

By definition a machine theory is a coherent theory that covers a major aspect of a considerable class of machines. In the course of time other aspects of machines gain importance, which require new theories. This way of looking at the science of machines and its history has its own merits. The obvious questions are: When and why did a particular theory appear?, How did it develop?, What was its relation with the practice of machine building?, How did it relate to other machine theories?

T. Koetsier (✉)
Department of Mathematics, Vrije Universiteit Amsterdam, Amsterdam, Netherlands
e-mail: t.koetsier@vu.nl

M. Ceccarelli and J. C. Jauregui-Correa (eds.), *State-of-the-Art and Innovations in Mechanism and Machine Science*, Mechanisms and Machine Science 150,
https://doi.org/10.1007/978-3-031-47040-0_3

The first successful machine theory was the theory of simple machines and their combinations. In my view the second successful theory was kinematics. In this paper I will restrict myself to a sketch of the history of the theory of simple machines and the history of kinematics.

The period around 1900 was crucial in the sense that the role of science in the practice of mechanical engineering gained momentum. Other important theories that came to play a role in the science of machines were, for example, machine dynamics, control theory, strength of materials, machine thermodynamics, tribology and robotics.

3.2 Simple Machines 1: *Mechanical Problems*

In the Iliad, Homer allegedly uses the Greek word mechane (μηχανῆ), which means machine, to describe political manipulation [4]. This meaning of the word has survived up to the present day in the use of the word machination for a plot or an intrigue.

The word was already used with its more common meaning by Aeschylus. Aeschylus (c. 525/524—c. 456/455 BC) was a Greek tragedian. He used the word mechane in the theatre to refer to "a piece of stage machinery in the form of a large crane which enabled characters, and objects such as chariots, to appear to fly through the air. It was also used to lower a character, usually a god, from the top of the stage building so that he could resolve the complexities of the plot—the deus ex machina" [10].

What did this piece of stage machinery look like? Chondros et al. have proposed a machine consisting of a 14 m long beam with a pivot point for vertical and horizontal rotation, 10 m from the load and 4 m from the counterweight. The beam is rotated upwards, then the load is lifted by means of two tread wheels and finally the beam is rotated horizontally until the load can be lowered onto the stage. Using the machine safely definitely required skilled operators [4] (Fig. 3.1).

Fig. 3.1 The reconstruction of a Deus ex Machina by Chondros et al. Courtesy of Thomas Chondros

Although initially the word may have only referred to this piece of stage machinery this did not last long. Soon also other hoisting devices and pieces of artillery were called machines. Machines also became the object of scientific research. At a certain moment the Greeks started to distinguish the art or science of machines, mechanike techne (μηχανικὴ τέχνη) or mechanics, which encompassed not only, on the one hand, the practice of the building machines, but, on the other hand, also the theoretical analysis of machines.

The oldest extant book about mechanics is *Mechanical Problems*, usually included in the Aristotelian corpus. It is often assumed that it was written by a pupil of Aristotle in the time of Strato, who was a contemporary of Euclid.

Mechanical Problems is about devices, machines, that enable man to produce things "contrary to nature [...] by skill, for the benefit of mankind". The author wrote: "Of this kind are those in which the less master the greater, and things possessing little weight move heavy weights, and all similar devices which we term mechanical problems" [1, pp. 330–331].

The author of *Mechanical Problems* explains the functioning of the devices that he discusses by means of the functioning of the lever. He discusses wheels, ropes, pulleys and the wedge, but not the screw. In one way or another the devices all enable the operator to exert a big force by means of a smaller force. The author of the book does not discuss more complex machines. Yet it is obvious that, for example, the machine reconstructed by Chondros et al. is a combination of the devices treated in *Mechanical Problems*: the beam is a lever, the load is lifted by means of the treadmill and a wheel and axle. Possibly also pulleys were used.

Very probably only after *Mechanical Problems* was written the screw was invented or became only then well known.

3.3 Simple Machines 2: Heron of Alexandria

A very important book is Heron of Alexandria's *Mechanics*. The engineer and mathematician Heron, who lived in the first century CE, restricts himself in the book to hoisting machines. Heron does not use the word 'machine', but he uses the word 'power'. There are in Heron's view five simple powers and combinations of these five simple machines give more complex machines that can be used to lift heavy things. The theory of the five simple powers is the core of Heron's theory. The theory is introduced in this way: "Since the powers by which a given burden is moved by a given power are five, we must of necessity present their theory and their names, because these powers are all related to the same principle, though they are very different in form; and as for their names they are as follows: the axle going through a wheel (the windlass), the lever, the pulley, the wedge, the screw" [7, p. 50].

In his book Heron first describes the five powers.

As for the wedge, Heron writes: "The fourth power, which comes after this, is one that is called the wedge. And this is used for some machines for perfume and for the joining of great works of carpentry, and it has many uses; and its greatest use is when we want to split the bottom of the stone which we want to cut and we have its sides free from the mountain from which we cut it. And in this sort of work the rest of the powers cannot do anything, even if we combine them all. But as for the wedge, it will do it by itself alone. [...] And the smaller the angle of the wedge, the easier is it working, as we shall explain" [7, p. 55].

The fifth power is the screw, which is defined as a wedge wrapped around a cylinder.

After having described the five powers Heron announces that he will give the reason why each of the engines is able to move a great weight by a small power. His explanations, although with a reference to Archimedes' proof of the law of the balance, are only slightly more extended than those offered by the Aristotelian *Mechanical Problems*.

In Heron's *Mechanics* the description and explanation of the five powers is accompanied by a discussion of their combinations. A simple combination is the endless screw: a screw is combined with a cogwheel on an axle. Given a screw Heron carefully wrote how to determine the size of the teeth of the wheel.

Heron's theory is correct. The only thing missing is a precise determination of the mechanical advantage of the wedge and the screw.

Heron is very much aware of the fact that he is discussing machines in their idealized form and that he does not consider friction, which of course plays an important role in reality.

We cannot discuss here the relation of Heron's work with the great flourishing of technological practice in Alexandria at the time. See, for example, [15].

3.4 Simple Machines 3: Galilei

In the theory of simple machines things started to move again when in the middle of the sixteenth century Latin translations of the works of Archimedes and Pappus became available. Archimedes was really rediscovered after the publication in 1558 of the first critical Latin translation of several of his works by Federico Commandino (1509–1575). Commandino also translated the *Collection* of Pappus of Alexandria, of which Book IIX contains parts on simple machines based upon Heron's Mechanics.

The marquis Guido Ubaldo del Monte (1545–1607), as Commandino's pupil, did the final editing of Pappus' *Collection*, and realized that a revision of Heron's theory of simple machines using Archimedean rigor was necessary. The result was his *Mechanicorum Liber* published in 1577 [5]. The Italian translation by Filippo Pigafetta appeared in 1581 in Venice under the title *Le Mechaniche*.

Del Monte's work represented a different approach. The book starts with a very long discussion of the balance and after that the lever, the pulleys, the wheel and axle, the wedge, and the screw are discussed. The approach is deductive starting with definitions, axioms and postulates and is in general an attempt to reach Archimedean clarity.

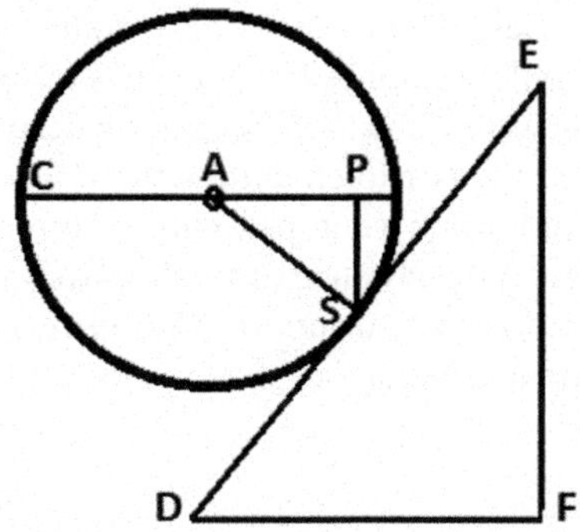

Fig. 3.2 Galilei studying the inclined plane

Del Monte wrote the first serious treatise on the theory of simple machines after Heron and Pappus. Del Monte's *Mechanicorum Liber* did not contain the final word, but it contained a precise definition of the problem: the only thing missing was a theory of the inclined plane based upon the law of the lever. Heron had not been able to determine the mechanical advantage of the wedge and the screw. Del Monte could not either.

When Galilei read Del Monte's book he must have realized its shortcomings. In circa 1690 Galilei in the text called *De Motu* came up with a solution to the problem of the inclined plane (Fig. 3.2).

Consider a balance CAS with fulcrum A. Arm AC is horizontal and arm AS tilts. Weight W_1 is in C and weight W_2 is in S. The law of the balance says that we have equilibrium if and only of the ratio of the weights is inversely proportional to the ratio of the distances of the two weights to the vertical line through the fulcrum. This means that we have equilibrium if

$$\frac{W_1}{W_2} = \frac{AP}{AC} = \frac{AP}{AS} = \frac{EF}{ED}$$

So weight W_1 can be kept in position C on the balance by means of a weight W_2 in S. This means that W_2 in S exerts a force equal to W_1 downward in the direction of the tangent in S to the circle with center A and radius AC. So W_1 directed upward along the inclined plane DE will keep W_2 in position.

When he was a professor in Padua Galilei decided to apply this result in a treatise on all simple machines. He wrote *Le Meccaniche* ca. 1600. He starts with the general law of the balance. Then Galilei proceeds to discuss the lever and the windlass or capstan. In his treatment of the pulleys he follows Del Monte. The discussion of the screw is preceded by the investigation of the inclined plane "which, though it may seem at first somewhat remote from the consideration of this instrument, is nevertheless its basis and foundation". His discussion of the inclined plane is similar to his discussion in *De Motu*. Because a screw is a wedge wrapped around a cylinder he can also derive the mechanical advantage of the screw.

Galilei's *Le Meccaniche* contains the first in every respect satisfactory theory of simple machines [8]. The laws of lever, the capstan, the pulleys, the inclined plane, the wedge and the screw are all derived in a unified way. In Del Monte's work we still find the Aristotelian theory that in order to bring about motion from a situation of equilibrium a non-negligible extra force is needed. Galilei has reached the conviction that in the absence of friction from a situation of equilibrium a minimal force added will bring about motion. Galilei explicitly studies the machines without taking friction into account.

3.5 Simple Machines 4: The Encyclopedists

With Galilei Heron's theory had reached maturity. It is truly remarkable that in the early modern period Heron's ideas about simple machines and their combinations were turned into a definition of what a machine is. As far as I know neither Heron, nor Pappus presented the theory as a theory valid for all machines. Yet this happened in the eighteenth century.

On p. 794 of Volume IX of Diderot and d'Alembert's *Encyclopédie* published in the period 1751–1772 we read about the word Machine: "In a general sense the word means that which serves to increase and regulate moving forces, or some instrument intended to produce motion so as to save either time or force while doing so. [...] *The machines are divided into simple and compound; there are six simple machines to which all other machines can be reduced,* the balance and the lever, of which we make only one species, the winch, the pulley, the inclined plane, the wedge and the screw. [...] M. Varignon adds a seventh which he calls a funicular machine [...]. *A compound machine is one which is in effect composed of several simple machines combined together*. The number of compound machines is now almost infinite, and yet the ancients seem in some way to have far surpassed the moderns in this respect; for their machines of war, architecture, etc. as they are described to us, appear superior to ours" (italic is mine—TK).

About the seventh simple machine that Pierre Varignon added he wrote: "it is made only of cords capable of supporting weights without the help of any other Machine, and in that it is as independent of these as they are of each other" (Varignon, 1687, p. 1). The *Encyclopédie* says about the funiculars:" an assemblage of cords, by means of which two or more powers support one or more weights" (Fig. 3.3).

3.6 Kinematics 1: Leupold

It is remarkable that Heron's theory of the five powers turned into a definition of what a machine is in general. One of the reasons is undoubtedly that, abstracting from friction, the theory is correct if a machine is composed of only simple machines. Moreover there was no other theory. Many machines however, are not composed of

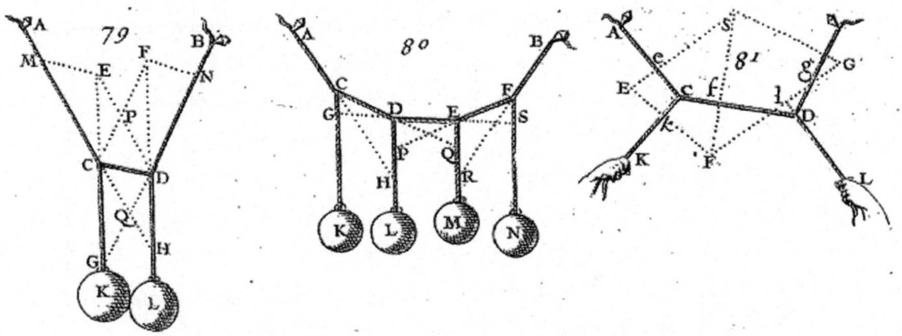

Fig. 3.3 Figures from Varignon's 'Nouvelle mécanique' from the section on funiculars

only simple machines. When Diderot and d'Alembert wrote in the *Encyclopédie* "The number of compound machines is now almost infinite" they seem to have assumed that in principle the theory of the simple machines was sufficient to understand all of these compound machines. Yet, much more can be said about machines. This is, for example, clear when one studies the works of Jacob Leupold (1674–1727). His *Theatrum Machinarum* consists of 10 illustrated volumes. They were published between 1724 and 1739, most of them posthumously. It is a systematic study of mechanical engineering at the time.

In the first volume Leupold literally writes: "Machines are either simple or composed. The simple machines are the so-called five powers, lever, pulleys, the wheel and axle, the wedge and the screw. Composed machines consist of two or more similar or different simple machines" [11, p. 2].

Yet, as Robert Willis would point out more than 100 years later, that though Leupold accepts the idea that all machines are combinations of the simple machines, the insight that a machine is not only a modifier of force but also a modifier of motion is already present in his work. In the first volume he has, for example, a section on how to generate a rectilinear motion from a circular motion. An operator makes a flywheel rotate. On its axis is a wheel with five teeth, which makes a piston rod move up and down in a straight line (Fig. 3.4).

3.7 Kinematics 2: Lazare Carnot

Lazare Carnot (1753–1823), born in Nolay in Burgundy, was trained at the military school in Mézières where Gaspard Monge (1746–1818) was one of his teachers. Monge was born in Beaune, also in the Bourgogne, 20 km from Nolay. So Carnot and Monge were both from Burgundy. That created a bond.

As a lieutenant in the Royal Corps of Engineers, after having graduated on January 1, 1773, Carnot wrote *An essay on machines in general* (Essai sur les machines en general) published in 1783. Carnot's approach to machines is unique in its generality.

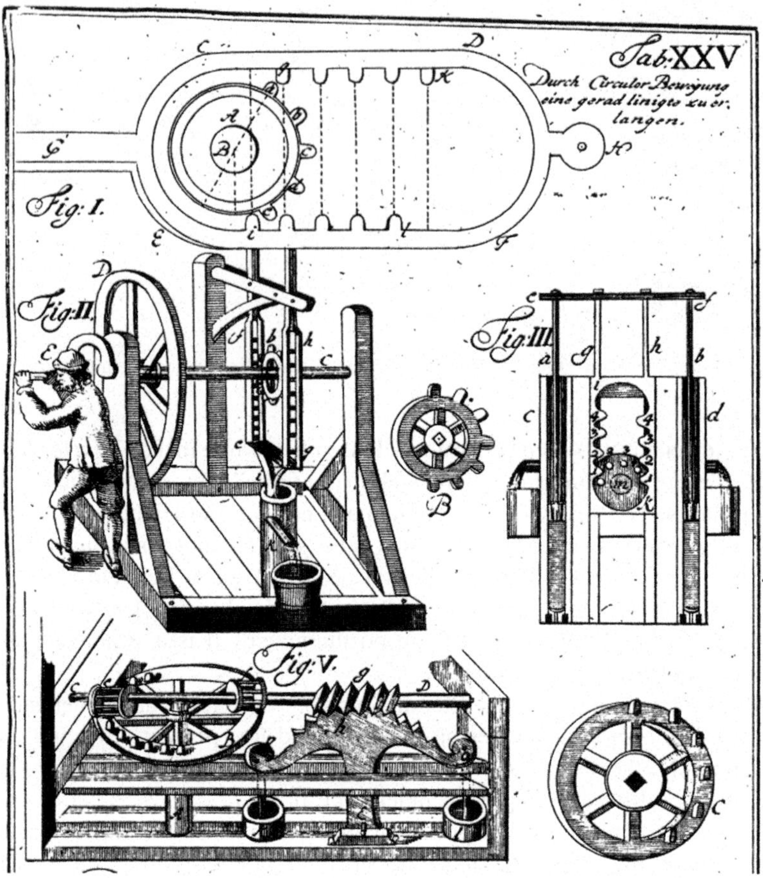

Fig. 3.4 How to get a rectilinear movement by means of a circular movement. *Source* [11], Plate XXV

He argued that a theory of machines should be based upon a theory of the communication of movements. This implied for him that the world needed a theory dealing with the possible motions of machines. He called it a theory of geometrical movements. He left it to others to develop this theory. We see a new way of describing a machine being born here. The machine is no longer a combination of simple machines, but a coherent whole of elements that describe curves in space.

3.8 Kinematics 3: Gaspard Monge

In 1794 the École centrale des travaux publics was created in Paris. It got the name École Polytechnique one year later. Carnot and Monge were the first directors of the school. Monge decided that a course on machine elements had to be included in the curriculum of the school. Jean Nicolas Pierre Hachette (1769–1834) was given the task of preparing a text. Hachette's textbook, the *Elementary Treatise on Machines* (Traité élémentaire des machines) appeared in 1811. Hachette's book is one of the first attempts to combine a systematic description of existing machines with mathematical considerations. Carnot wrote a preface.

Let us have a look at the fourth edition of Hachette's book, (Hachette 1828). On page 5 he writes: "The simple or elementary machines are those which transform into each other one of the four circular and rectilinear motions of a moving point, into another of these same motions. The most complex machines are combinations of them, and the examination of the first leads naturally to the knowledge of all that has been invented up to the present in mechanics" [9, p. 5].

Roughly speaking almost one fourth of the book, in particular Chap. 2, is devoted to the topological-geometrical aspects of the elementary machines, i.e. the kinematical aspects. First there is a system of classification of the elementary machines, which assumes a central position in the book. There are four types of movements of the input and of the output:

Circular continuous (circulaire continu).
Circular alternating (circulaire alternatif).
Rectilinear continuous (rectiligne continu).
Rectilinear alternating (rectiligne alternatif).

This yields ten types of elementary machines: 6 machines where input and output differ plus 4 machines where input and output belong to the same type. This system of classification, invented presumably by Monge, was first published in 1808 under the title *Essay on the composition of the Machines* (Essai sur la composition des Machines), by two Spaniards José María de Lanz y Zaldívar (1764–1839) et Agustín de Bétancourt y Molina (1758–1824) under the supervision of Hachette.

3.9 Kinematics 4: Robert Willis

Robert Willis (1800–1875) was a Cambridge professor who had developed an interest in mechanical engineering. In 1937 he started with lectures on 'mechanism' based on the idea that mechanisms should be studied independently of forces. In 1841 he published *Principles of mechanism*, in fact a book on kinematics of mechanisms.

In the preface he writes that machinery is primarily a modifier of motion. "For every machine will be found to consist of a train of pieces connected together in various ways, so that if one be made to move they all receive a motion, the relation of which to that of the first is governed by the nature of the connexion. The work which the machine has to do will require that the pieces appropriated to this work shall move with respect to each other in some given manner, and the forces applied to the machine to set it in motion must also move the piece which receives them in some other manner. Thus the question of contriving a machine by which a given kind of power may be made to perform given work, is reduced to a problem of mere motion." According to Willis Leupold made a first step in this new direction [17, p. iv].

Willis was critical of Monge's motion of elementary machine. Willis calls it "a merely popular arrangement" and a "mere enumeration and description of the elements, without containing a provision for the calculation of the laws of the motion, or changes of motion produced".

Willis proposes a classification based upon two criteria: the input–output relation and the mode in which the motion is transmitted. The input–output relation concerns the ratio of the input and output velocities and their directions. The transmission mode concerns the nature of the contact between the pieces: immediate contact (rolling, sliding) and intermediate contact (wrapping connectors, links, 'reduplication'). The term reduplication refers to the fact that when we tie a rope to a fixed point, double it over a pulley P and use the rope to pull the pulley towards us in the direction of the fixed point, we move twice as fast as the pulley P (Fig. 3.5).

Willis' book is definitely a big step forward in comparison with the books by Lanz and Bétancourt and Hachette. In the first part of the book of over 300 pages Willis studies single combinations of pieces in considerable detail. In the second part of the book of over 100 pages he discusses aggregate combinations, that is combinations of two or more single combinations. Wherever possible Willis attempts to express the way in which the motions are transmitted in a precise mathematical way.

3.10 Kinematics 5: Franz Reuleaux

Reuleaux took Willis' ideas as his starting point, but made a decisive step forward and after Reuleaux published his results kinematics of mechanisms would never be the same. Willis viewed a machine as a combination of parts so that, when one moves in a certain way, the others are forced to move in other particular ways. Reuleaux took this definition as his starting point: a machine is a connected whole of resistant bodies, which is built in such a way that by means of it mechanical forces of nature can be compelled to work under certain conditions.

The definition is similar to Willis'. However, Reuleaux focussed on the first element in the definition, the connected whole of resistant bodies, and he developed an original theory to classify and compare such 'connected wholes'. Although he had the ideas already in Zürich, he published the results only after his move to Berlin, in a series of articles and then in 1875 in his book *Theoretical Kinematics* (Theoretische Kinematik). The book was almost immediately translated in Italian, French and English.

Reuleaux uses the term kinematics in a very specific way. For him it is 'the science of constrained motion' ('Zwanglauflehre' in German). He starts his analysis

SYNOPTICAL TABLE

OF THE

ELEMENTARY COMBINATIONS OF PURE MECHANISM.

	DIRECTIONAL RELATION CONSTANT.		DIRECTIONAL RELATION CHANGING PERIODICALLY.
	Velocity-Ratio Constant.	*Velocity-Ratio Varying.*	*Velocity-Ratio Constant or Varying.*
	CLASS A.	CLASS B.	CLASS C.
DIVISION A. By Rolling Contact.	Rolling cylinders, cones, and hyperboloids. General arrangement and form of toothed wheels. Pitch.	Rolling curves and rolling curve-wheels. Roëmer's & Huyghens' wheels, &c. Wheels with intermitted teeth. Rolling-curve levers.	Mangle-wheels. Mangle-racks. Escaping geerings.
DIVISION B. By Sliding Contact.	Forms of the individual teeth of wheels. Cams. Screws. Endless screws or worms and their wheels.	Pin and slit lever. Cams. Unequal worm. Geneva stop and other intermittent motions.	Pin and slit lever. Cams in general. Swash plate. Double screw. Spiral and solid cams. Escapements.
DIVISION C. By Wrapping Connectors.	Arrangement and material of bands. Form of their pullies. Guide pullies. Geering chains. Arrangements for limited motions.	Curvilinear pully. Fusees.	Curvilinear pully and lever.
DIVISION D. By Link-work.	Cranks and link-work for equal rotations. Cranks for limited motions. Bell crank-work.	Link-work. Hooke's joints.	Cranks, excentrics, and other link-work. Ratchet wheels and clicks. Intermittent link-work.
DIVISION E. By Reduplication.	Tackle of all kinds, with parallel cords and in trains.	Tackle with unparallel cords.	

Fig. 3.5 Willis' classification of mechanisms

of mechanisms with the consideration of two connected rigid bodies, which he calls a 'kinematical pair'. The two bodies may be joined in different ways, for example by means of a cylindrical or prismatic joint. By adding new rigid elements a kinematical pair becomes a "kinematical chain". The rigid elements of the chain are called "limbs" by Reuleaux ('Glieder' in German). In English they are called links and the connections between the links are called joints.

A kinematical chain can be closed or lead back into itself in different ways. If this is done in such a way that a change of position of a link with respect to an adjacent link leads to a uniquely determined change of position of all the other links Reuleaux calls such chains "compulsory closed kinematical chain" ('zwangläufig geschlossene kinematische Kette' in German). In such a chain each link has only "one relative movement".with respect to each other link. We would here say "one degree of freedom". The word compulsory is left out when no misunderstanding is possible.

When he has the notion of kinematical chain Reuleaux can define what a mechanism is ('Mechanismus' or 'Getriebe' in German). A mechanism is a closed kinematical chain of which one link is fixed. In this way by fixing different links one and the same kinematical chain leads to different mechanisms that are all kinematically equivalent in the sense that the elements move with respect to each other in exactly the same way.

Let us consider an example: a four-link chain in which four bars are connected by means of cylindrical joints. If we fix, for example, the bar at the bottom we get a four-bar mechanism.

Kinematical chains can be connected to each other, which gives 'compound kinematical chains'. Obviously also compound kinematical chains can be turned into 'compound kinematical mechanisms' by fixing one of the links (Fig. 3.6).

A central element in the theory of kinematical pairs or chains is that the links should be able to move with respect to each other with one degree of freedom. When this is the case Reuleaux talks about closed pairs and closed chains. In practice often pairs or chains are used in which this is not the case. Reuleaux calls them incomplete. An incomplete pair is a pair in which an extra force is used to keep two elements

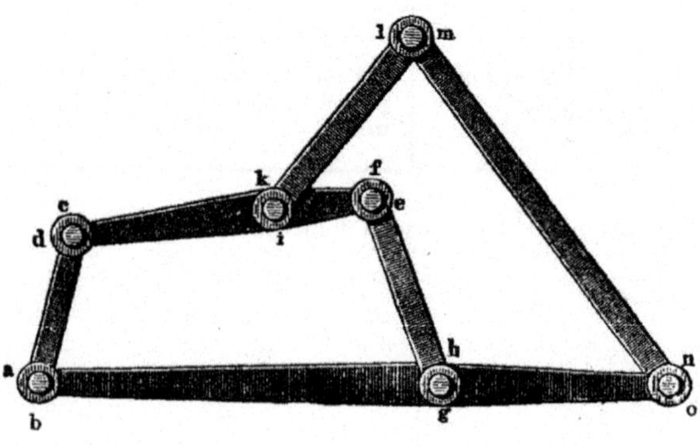

Fig. 3.6 A compound kinematical chain

joined. Reuleaux calls them force-closed pairs. The extra force is often gravity. Reuleaux gives the example of the bearings of water wheels where the weight of the wheel prevents vertical motion of the axle. The extra force may also be friction, like in the case of the driving wheels of a locomotive. Rails and driving wheels form an incomplete kinematic pair. The wheel has three degrees of freedom with respect to the rails: it can move upwards, it can roll and it can slide. The upward movement is prevented by gravity, sliding is prevented by friction, of course indirectly caused by gravity. In this context Reuleaux introduces flexible elements like ropes and belts. He calls them 'tension organs'. The right force closure creates tension and makes them behave like rigid elements. Opposed to tension-organs are pressure organs like a liquid or a gas. Enclosed in a vessel of the right form they can bring about motion. The hydraulic press is a nice example.

There is another form of incompleteness. The occurrence of dead points in a kinematic chain is for Reuleaux also an example of an incomplete mechanism. It can be solved by means of a fly-wheel. Reuleaux extensively discusses kinematical solutions. The idea is to add another mechanism that carries the original mechanism over its dead point.

In Reuleaux' view the history of machines started with force closure. The history of modern machinery is one of the replacement of force closure by pair- and chain-closure. He considers it as an essential characteristic of the development of machines in the future. In Newcomen's steam engine, force closure still pre-dominated. Watt introduced pair- and chain-closure by degrees.

A mechanism becomes a 'machine' if a force is applied to one of its movable links such that mechanical work is performed accompanied by determinate motions. This is Reuleaux' detailed answer to the question What is a machine?

We saw above that the Encylopedists presented the theory of simple machines as the core of the science of machines. It is remarkable that Reuleaux did something similar. For him kinematics of mechanisms as he defined it was the core of the science of machines.

With the completely new classification of mechanisms by means of kinematical chains Reuleaux introduced a new abstract point of view in kinematics of mechanisms. For a mathematician this boiled down to viewing a planar mechanism as a collection of coinciding Euclidean planes moving all (with one degree of freedom) with respect to each other. This was a major and not trivial step forward which helped to turn kinematics of mechanisms into a more coherent discipline. As we have seen it involves two related elements: considering the frame of a mechanism as a link and, moreover, abstraction from the particular shape of the links in a mechanism and concentration on the way in which the links are connected.

A book like *Der Konstrukteur* [14] nicely shows the state of the art at the end of the nineteenth century. On more than 1000 pages Reuleaux discusses the numerous ways in which the elements of a machine can be connected. The book starts with a section on strength of materials and a section on graphostatics. The section on graphostatics can be considered as a remnant of the theory of simple machines.

An important next step was made by the mathematician Ludwig Burmester (1840–1927). Burmester's *Lehrbuch der Kinematik* of 1888 contains the first far-reaching attempt at a synthesis of planar theoretical kinematics and planar kinematics of mechanisms ([2] and [3]). With Burmester and Reuleaux kinematics of machines had reached a state of maturity.

3.11 Dynamics, Concluding Remarks

The next machine theory that reached maturity was dynamics of machines and mechanisms. In 2008 Francis Moon has given a sketch of its development [12]. Although its roots go way back, only in the twentieth century the theory reached maturity with, for example, the work of Jacob Pieter den Hartog (1901–1989) and Steven Timoshenko (1878–1972).

Why did it take so long before dynamics of machinery was developed? That is an interesting question. The difficulty of the subject will have played a role. But also a lack of urgency. The early steam engines were rather slow and the geometrical and kinematical approach was seemed sufficient. When the velocities became higher, for example vibrations became a problem. As for the history of dynamics of machines also [13] is an important book.

Strength of materials is also a machine theory within the scope of the science of machines. For its history see [16]. Another relevant machine theory in this context is tribology. Dowson [6], which I have not seen, might be an interesting source.

References

1. Hett, W.S. (trans.): Aristotle: mechanical problems. In: Aristotle Minor Works, on colours etc., pp. 327–411. William Heinemann Ltd., London. Harvard University Press, Cambridge Massachusetts (1963)
2. Burmester, L.: Lehrbuch der Kinematik, Band 1, Die ebene Bewegung, Leipzig (1888)
3. Burmester, L.: Atlas zum Lehrbuch der Kinematik, Band 1, Die ebene Bewegung, Leipzig (1888)
4. Chondros, T.G., Milidonis, K., Vitzilaios, G., Vaitsis, J.: "Deus-Ex-Machina" reconstruction in the Athens theater of Dionysus. Mech. Mach. Theory **67**, 172–191 (2013)
5. Del Monte, G. U.: Le Mechaniche [...] tradotte in volgare dal Sig. Filippo Pigafetta, Venetia (1581)
6. Dowson, D.: History of tribology. Longman, New York (1979)
7. Drachmann, A.G.: The Mechanical Technology of Greek and Roman Antiquity, A Study of the Literary Sources. Copenhagen—Munksgaard, Madison—the University of Wisconsin Press, London—Hafner Publishing Co. (1963)
8. Galilei, G.: On Motion and On Mechanics, Comprising De Motu (Annotated translation by I. E. Drabkin) and Le Meccaniche (Annotated translation by Stillman Drake). The University of Wisconsin Press, Madison (1960)
9. Hachette, J.N.P.: Traité élémentaire des machines, quatrième édition. Corby, Paris (1828)
10. Phyllis, H., Peter, F. (eds.): The Concise Oxford Companion to the Theatre (2nd ed.). Oxford University Press (2003)
11. Leupold, J.: Theatrum Machinarum Generale, SchauPlatz des Grundes Mechanischer Wissenschaften, Leipzig (1724)

12. Moon, F.: History of dynamics of machines and mechanisms from Leonardo to Timoshenko. In: Yan, H.-S., Ceccarelli, M. (eds.) International Symposium on History of Machines and Mechanisms Proceedings of HMM 2008, pp. 1–20. Springer (2009)
13. Rao, J.S.: History of Rotating Machinery Dynamics. Springer (2011)
14. Reuleaux, F.: Der Konstrukteur, Ein Handbuch zum Gebrauch beim Maschinen-Entwerfen. Friedrich Vieweg und Sohn, Braunschweig, 4e umgearbeitete und erweiterte Auflage (1894)
15. Schiefsky, M.J.: Theory and practice in Heron's mechanics. In: Laird, W.R., Roux, S. (eds.) Mechanics and Natural Philosophy before the Scientific Revolution, Boston Studies in the Philosophy of Science 254. Springer, New York (2007)
16. Timoshenko, S.: History of strength of materials: with a brief account of the history of theory of elasticity and theory of structures, New York (1953)
17. Willis, R.: Principles of Mechanism, Cambridge (1841)

Chapter 4
Motion Synthesis: From the Classical Work of Reuleaux to the More Modern Robot Motion Planning

Bahram Ravani

Abstract This paper presents some of the methods for motion synthesis or geometric motion design based on the previous work of the author and some of his collaborators. The aim of much of the work discussed in this paper has been to either transform the motion design problem to a curve design problem or to generalize techniques from planar to spatial kinematics. The paper does not provide a complete survey of all the relevant methods and literature or discuss details of the methods presented. Such details and much of the references to some of the relevant works can be found in the references provided. The paper is written in memory of Prof. Carlos López-Cajún of Mexico who was active in IFToMM and a contributor to the field of Mechanisms and Machine Science.

Keywords Mechanism synthesis · Geometric motion design · Reuleaux's method · Motion interpolation · Rigid body guidance · Computational kinematics · Computational geometry · Lie groups · Riemannian manifolds

4.1 Introduction

Motion synthesis involves determining the parameters defining a rigid body motion based on a given set of desired poses of a rigid object which is intended to go through the motion. Motion synthesis is important in several fields including mechanical design of mechanisms, robot motion planning, and computer graphics. In mechanical design of linkages, motion synthesis involves determining link dimensions of a mechanism such that one of its links is guided through a set of desired poses representing or approximating the desired motion. In such a context, the problem has usually been referred to as a rigid body guidance problem where a rigid body

B. Ravani (✉)
Department of Mechanical and Aerospace Engineering, University of CA—Davis, One Shields Avenue, Davis, CA 95616, USA
e-mail: bravani@ucdavis.edu

M. Ceccarelli and J. C. Jauregui-Correa (eds.), *State-of-the-Art and Innovations in Mechanism and Machine Science*, Mechanisms and Machine Science 150,
https://doi.org/10.1007/978-3-031-47040-0_4

is guided through several desired poses. In kinematics of mechanisms the work of Burmester [1] is probably one of the more important fundamental and original works in this area. It involves finding points in the moving rigid body (the motion of which is going to be guided by the mechanism) which would remain on a circle through the desired set of poses. When four desired poses are considered the method results in two curves referred to as the "circle point" and the "center point" curves respectively in the moving and the fixed bodies. These two curves are the loci of points that can be used to put the moving and the fixed pivots of a linkage to construct a mechanism which would guide the body through the desired four poses. In this manner the motion of the guided body is synthesized through construction of a mechanical linkage.

When the rigid body is guided through precision poses one can think of the motion of the moving link interpolating the desired poses and the process can be viewed as designing a linkage for motion interpolation. The work of Burmester in planar kinematics has also been extended to five or more desired poses. In the case of five poses, rather than having the two circle and center point curves, the technique results in a number of discrete points which can be considered for the locations of the fixed and the moving pivots of a linkage. Such points are referred to as Burmester points. In the case of more than five desired poses then the resulting motion approximately guides the moving body through the desired poses and the problems can be viewed as motion approximation. It should be pointed out that there has been extensive literature on extending the work of Brumester for planar mechanism design for rigid body guidance. For a more updated treatment of the method the readers are referred to the book by McCarthy [2].

The intention here, however, is not to provide a detailed review of the literature but only to point to some of the most classical and fundamental works in the area. The initial work on some generalizations of the Burmester fundamental work to three dimensions has been published by Roth and his co-workers (see, for example [3–8]). These works have looked at synthesis of 3D dyads (two link mechanisms) with a comprehensive study of synthesis of these dyads provided in [7]. A relatively different approach for design of such dyads can be found in [9]. A rather detailed study of Burmester theory for spherical linkage is given in [10] with a more modern treatment in [2].

Motion synthesis or rigid body guidance for mechanisms can be considered as constrained motion synthesis since the motion of the moving body is not free in the plane or in the 3D space and it is in a constrained manifold due to the linkage constraining the motion being generated.

In the case of free motion of a rigid body in a planar motion, Reuleaux was probably the first in kinematics who developed a simple method (known as Reuleaux's method [11]) of guiding a body between two poses. In planar kinematics since the general motion of a rigid body or link from one pose to a second pose can be achieved by a single rotation about a specific point referred to as the pole of the rotation, Reuleaux's method constructs this point from two distinct corresponding points (called homologous points (see [12] p. 35) in the two desired poses of the rigid body or the moving link. The angle of rotation is also determined from a simple geometric construction and in this manner the motion of the body between the two poses is synthesized in terms of a pure rotation about this geometrically constructed pole point. The simple rotational motion is the minimal form of the motion which can be synthesized between the two poses.

In spatial kinematics, the minimal form of the motion of a rigid body from one pose to a second pose can be achieved by a screw displacement in accordance with Chases theorem [12] which is a generalization of Euler's theorem [13] in spherical kinematics. A screw displacement involves a rotation about and a translation along the axis of a screw. Although Chases only used geometrical reasoning for his theorem, it was Rodrigues [14] who developed the mathematical equations which can be used to reconstruct or synthesize the minimal screw motion described by Chasles theorem between two desired poses of a moving body (see for example, [15] pp. 56–61). Positions of three non-colinear corresponding (homologous) points in the two poses of the body are used and the solution is obtained by first solving a transcendental equation for the orientation of the screw axis and the rotation around it and then an algebraic equation determining the location of the screw axis and the translation along it in terms of the rotation and the axis orientation. If more than three homologous points are used, the problem becomes over-constrained in terms of a redundant set of homologous points and a least-square type solution can be obtained as an approximation of the motion.

Horn [16] was the first to obtain a linear solution to the rotation problem which can be used in spherical motion synthesis. A similar linear solution for the full rigid body displacement was derived in [17] where the authors also considered other homologous features such as corresponding lines or planes in addition to homologous points. The solutions to the 3D problem for both the regular and the over-constrained cases are also obtained in [18] by a computational geometric construction utilizing the 2D Reuleaux's method. Furthermore, a generalization of Reuleaux's method to 3D kinematics is given in [19] using computational line geometry. The parameters of a screw motion consist of a line (which is the screw axis) and a pitch which consists of a rotation angle and a translation along the screw axis. An interesting aspect of this last work [19], which is contrary to all the previous work, is that the solution results first in the location of the screw axis and then the rotation angle and translation along the screw axis are computed.

A totally different computational geometric method for motion synthesis involving interpolating or approximating a desired motion of a moving body is in terms of kinematic mappings. This concept was first developed by Ravani and Roth [20] for synthesizing planar linkages for guiding their coupler links through a set of desired poses. In this work, they used a mapping of planar kinematics developed by Blaschke [21] (see also [22]). In this manner, the motion of the coupler of a planar mechanism and the desired poses are mapped, respectively, into a curve and a set of desired points in the image space of the mapping and the problem of synthesis becomes that of a curve fitting in this image space. This method has been extended to spatial and spherical kinematics [23, 24] as well as synthesis of 3D linkages using the constrained manifolds resulting from the motion of the linkages in the image space of the mapping (see also the work of McCarthy and his co-workers for example

[25–27]). Some of the relationships between mapping of planar and spherical kinematics is discussed in [28]. It should be pointed out that the kinematic mapping for spatial kinematics developed by Ravani and Roth [23] reconfigures the so-called classical Study's Soma space parameters [15 p. 150–152, 29] into a 3D dual projective space allowing application of quaternion algebra and providing a different geometric structure for the six-dimensional space of rigid body poses. Moreover, some of the applications of kinematic mapping concepts in robotics are given in, for example, [30, 31].

In addition to its use in constrained motion synthesis for mechanical linkages, kinematic mapping has also been used for synthesis of unconstrained free motion of a rigid body. In kinematics, Ge and Ravani [32] were the first who used kinematic mapping for interpolating unconstrained motion of a rigid body through a set of desired poses considering both the rotational as well as the translation part of the motion. The method presented in [32] uses an oriented version of the 3D mapping [23] and develops interpolating spline motions with tangent, curvature and torsion continuity. Prior to [32], in computer graphics literature, Shumake [33] had studied interpolating only the rotational part using quaternions. A dual quaternion version of Shumake method is presented in [34] where in addition, the authors show how to achieve the typical subdivision properties inherent in design of Bernstein-Bezier curves which was not inherent in Shumake's algorithm. Considering interpolating rotations only, Park and Ravani [35] have developed methods for bi-invariant interpolation of rotations where the resulting motion is independent of the choice of the coordinate systems. Juttler [36] has provided discussion on coordinate independence of some of the published algorithms.

In general, the pose of a rigid body in 3D space depends on six parameters (3 for rotations and 3 for translations). The associated geometry is therefore a six-dimensional manifold. Park and Ravani [37] have generalized the De'Castelau's algorithm used in design of Bezier curves to Lie groups on Riemannian manifolds resulting from the six-dimensional space of rigid body poses. It should be noted that their algorithm results in unique solutions only for space of rigid body poses in 3D Euclidean space and not for general Lie groups or any general six-dimensional manifolds. A different approach formulating the problem as a variational problem on the Lie group of spatial rigid body displacements is presented in [38]. The methods discussed for synthesis of unconstrained 3D motions of a moving body provide for a computational geometric foundation for robot motion planning as well as computer graphics and animation.

The author and some of his co-collaborators have also extended some of the other methods used for curve design to motion design or motion synthesis. These include for example using a transfer principle and registration techniques [39], variational methods [40], and design of motions between contacting surface pairs [41].

4.2 Reuleaux's Method

Given two distinct poses of a moving body in 2D, Reuleaux's method is a geometric construction that would move the body from its initial pose to its final desired pose. It reconstructs the motion through a rotation about the Pole of the motion. Since this is a rotation only, Reuleaux's method guides the body through a minimal motion. Consider two poses of the body as depicted in Fig. 4.1. The Pole is constructed as the intersection of the perpendicular bisectors of the lines connecting two corresponding points of the body in the two given poses (homologous points). The rotation is then about the Pole with the angle of rotation being the angle between the line connecting the pole to any point of the body at the first position and the line connecting the pole to the same point after the rotation. Reuleaux's method uses perpendicular bisectors of lines connecting two pairs of homologous points.

The 3D generalization of the method presented in [19] similarly uses the perpendicular bisectors of two pairs of homologous lines (rather than points) in the two poses of the body. The two pairs of homologous lines are (L_a, L_a') and (L_b, L_b') as depicted in Fig. 4.2. S_a and S_b are the common perpendiculars between each pair of homologous lines. The perpendicular midlines (lines that are half distance and half angle between the two homologous lines) namely M_a and M_b are constructed. Two new lines C_a and C_b are then constructed that are mutually perpendicular to pairs of intersection lines M_a, S_a and M_b, S_b. The lines C_a and C_b correspond to the bisecting lines in the two-dimensional Reuleaux's method. The screw axis of the motion is then the common perpendicular between lines C_a and C_b. The angle of rotation and the translation along the screw axis are respectively the angle and distance between the common perpendicular of any line of the rigid body before motion and the screw axis and the common perpendicular of the corresponding line and the screw axis after the motion. The method in addition to its use in motion synthesis can also be used in kinematic registration in robotics applications.

Fig. 4.1 Reuleaux's construction

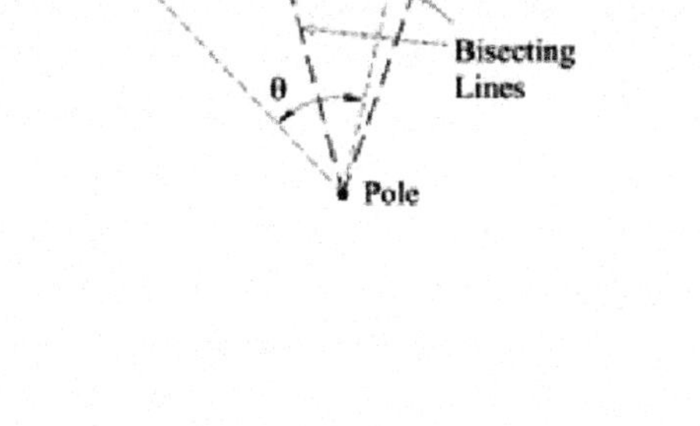

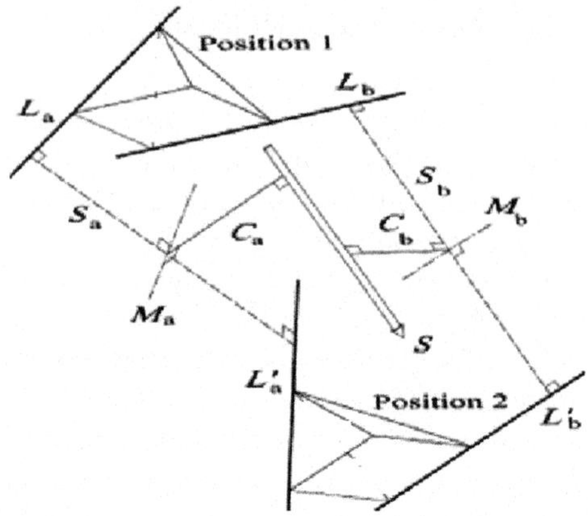

Fig. 4.2 3D-Generalization of Realeaux's method based on line geometry

4.3 Kinematic Mapping Methods

Since the pose of a body in any dimension depends on more parameters than the dimensions of the space, this has led to the idea of mapping these parameters to a higher dimensional space. This idea was first introduced by Grunwald [42] and then developed by Balschke [21] in 2D kinematics. In 3D kinematics, Study [29] used Euler parameters of rotation and a bilinear combination of them with the translation part of rigid body displacement to define Study coordinates which are then mapped into vectors in four dimensions called Study vectors. Since Study's bilinear parameters are the dual parts of a dual quaternion representing a rigid body displacement, Ravani and Roth [20] mapped these together with Euler parameters to a dual three-dimensional projective space with the understanding that only points with normalized coordinates in their mapping space correspond to rigid body poses.

For two poses of a rigid body, the associated screw axis S_{01}, rotation angle 2ϕ, and translation 2 h are depicted in Fig. 4.3 with the corresponding geodesic image curve in the dual space of the kinematic mapping.

The coordinates of an image point representing a pose of a rigid body are given by:

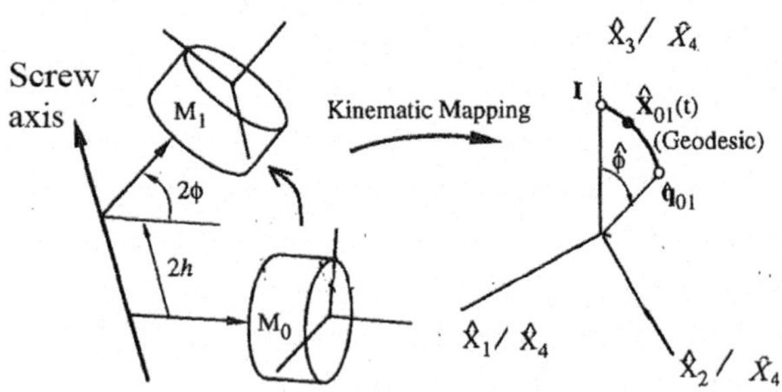

Fig. 4.3 A screw motion and the corresponding geodesic curve in the image space of the mapping

$$\hat{X}_i = \hat{s}_i \mathrm{Sin}(\hat{\phi}),\, i = 1, 2, 3;\text{ and } \hat{X}_4 = \mathrm{Cos}(\hat{\phi})$$

where $\widehat{\boldsymbol{S}} = (\hat{s}_1, \hat{s}_2, \hat{s}_3)$ is the unit dual vector representing the screw axis, and $\widehat{2\phi} = 2\phi + 2h\varepsilon \cdot (\varepsilon^2 = 0)$ is the dual angle representing the rotation and the translation about and along the screw axis. It should be pointed out that the coordinates of the image points are in fact dual Euler parameters. The geometry of the space of this mapping is a dual elliptic space and therefore well suited for the application of dual quaternion algebra.

It is clear from the above last equation that for the coordinates of an image point to represent the pose of a rigid body, they should be unit normalized which means that:

$$\sum_{i=1}^{4} \hat{X}_i^2 = 1$$

Points that are not unit normalized can be written in terms of those that are as follows:

$$\hat{R}_i = \hat{r}_i \hat{X}_i,\, i = 1, 2 \ldots, 4$$

General points represented by $\hat{R}$ can be considered as pre-images of points $\hat{X}$ which correspond to poses of a rigid body with the dual weight $\hat{r}$ representing a projection operator projecting a general point to a proper point representing a pose of the rigidbody being considered.

In motion and trajectory planning for the end-effector of a robot manipulator, the desired poses of the end effector can therefore be represented by points in the image space of the mapping and the motion design problem becomes one of finding a proper curve in the image space that interpolates or approximates these desired image points. The approach will be similar to methods used in computer aided design of curves. As an example, if four desired poses are considered represented by four pre-image points $\widehat{\mathrm{D}}_i$; $i = 0, 1, .3$ then these points can form the control polygon of a Bezier image curve (see, for example, [43]) approximating these four control points. A cubic Bezier pre-image curve can then be given by:

$$\hat{R}(t) = \left[1 t t^2 t^3\right] M \left[\hat{r}_0 \widehat{\mathrm{D}}_0 \hat{\mathrm{r}}_1 \widehat{\mathrm{D}}_1 \hat{\mathrm{r}}_2 \widehat{\mathrm{D}}_2 \hat{\mathrm{r}}_3 \widehat{\mathrm{D}}_3\right]\mathrm{T}$$

where the matrix **M** is the Bernstein-Bezier basis matrix:

$$M = \begin{bmatrix} 1 & 0 & 0 & 0 \\ -3 & 3 & 0 & 0 \\ 3 & -6 & 3 & 0 \\ -1 & 3 & -3 & 1 \end{bmatrix}.$$

Alternatively, this last equation can be written as:
$\widehat{\mathrm{R}}(t) = \sum_{i=0}^{3} \hat{r}_i \widehat{\mathrm{D}}_i B_{\mathrm{i}}^3(t)$ where B_{i}^3 is the Bernstein-Bezier basis function [43].

The actual Bezier type motion, $\hat{X}$ can then be found from $\widehat{\mathrm{R}} = \sum^{3} \hat{r}_i \hat{X}_i$ with the aid of the normalizing factor $\hat{r}_i$ which consist of arbitrary dual numbers. The real parts of it must be taken as positive numbers to make sure that the resulting displacements will be similarly oriented.

In computer aided design, a powerfull method for design of curves is the use of the De Casteljau's algorithm [43]. This algorithm uses repeated or recursive linear interpolations between the control points to generate points on the corresponding Bezier curve. In the space of the kinematic mapping, linear interpolation would mean using minimal geodesics and this does not result in control polygons which would have the subdivision property of the Bezier curves. Alternatively, if we use pre-image points then we can develop a De Casteljau type algorithm with the subdivision property as follows:

Given a set of control poses represented by image points $\widehat{\mathrm{D}}_i$; $i = 0, 1, \ldots n$ and normalizing factors $\hat{r}_i$, $i = 0, 1, \ldots, n$ then:

1. For $i = 0, 1, \ldots, n$ determine pre-image points: $\widehat{\mathrm{R}}_i^0 = \hat{r}_i \widehat{\mathrm{D}}_i$
2. For $r = 1, 2, \ldots, n$ and $i = 0, 1, \ldots, (n-r)$ calculate $\widehat{\boldsymbol{R}}_i^r(t) = (1-t)\widehat{\boldsymbol{R}}_i^{(r-1)} + (t)\widehat{\boldsymbol{R}}_{(i+1)}^{(r-1)}$

The corresponding image curve $\hat{X}(t)$ can then be computed at each step using the pre-image points generated and the normalizing factors. This algorithm does have the subdivision property for the pre-image of the image curve from which the actual image curve corresponding to the approximating motion can be generated.

4.4 Riemannian Manifolds and Lie Groups

The De Casteljau's construction used in generation of a Bezier curve provides a natural framework for its generalization to curved spaces resulting from a rigid body motion whether it is the trajectory of an end-effector in a robotic system or key-framing used in computer graphics animation. The basic concept involves using repeated linear interpolation or lines connecting control points and subdividing the control polygon of the Bezier curve.

Riemannian manifolds provide a general setting for dealing with curved spaces. In these manifolds, lines are represented by geodesics with respect to the given Riemannian metric. Bezier type curves on these manifolds can then be generated using minimal geodesics. Computing the geodesics between two points on Riemannian manifolds however requires solving a two-pint boundary value problem.

Rigid body displacements or poses represented as such form matrix Lie groups which are a special class of Riemannian manifolds. It is well known that the rotation part of a spatial displacement forms a compact Lie group. In such a case, there is a bi-invariant metric and the geodesics with respect to this metric are the one-parameter subgroup. It is also known that a Lie sub-algebra of a compact Lie group also lies on a one-parameter sub-group. For any two points on a compact Lie group there always exists a geodesic and the minimal geodesic can be found by the inverse of exponential of the logarithmic map. This means that for the Lie groups representing rigid body rotations there is no need to solve the differential equations associated with the two-point boundary value problems arising from finding the minimal geodesic and only one need to use the exponential and logarithmic functions (see [37] for details). A De Casteljeau's type algorithm can then be defined recursively by exponential of the logarithmic maps of the control points representing the desired poses.

When both rotations and translations are considered, the one parameter sub-groups are no longer geodesics with respect to any left or right invariant Riemannian metric because the Lie group representing a full spatial displacement does not admit a bi-invariant metric. In such a case, Bezier curves can be constructed, for example, by combining appropriate Bezier curves in R^3 and the rotation sub-group (SO(3)).

4.5 Conclusions

This paper has provided a summary survey of some of the work of the author and some of the other relevant works in the area of motion design synthesis. Much of the work in this area has applications in dimensional synthesis of mechanical link work, in robot motion planning, and in computer animation. In addition, they do enhance the fields of computational and theoretical kinematics and geometry.

It is shown, for example, that a three-dimensional generalization of the two-dimensional Reuleaux's method can be developed using computational line geometry where lines rather than points are used as the elements. Furthermore, it is shown that kinematic mappings as well as Lie groups can be used to transform the problem of motion design or synthesis to a mathematical problem of curve design and therefore curve design methods are reformulated to handle such motion design problems.

References

1. Burmester, L.S: Lehrbuch der Kinematik, Verlog von Arthur Felix, Leipzig (1886)
2. McCarthy, J. M.: Geometric Design of Linkages, Springer, ISBN 0-387-98983-8 (2000)
3. Roth, B.: The kinematics of motion through finitely separated positions. Trans. ASME, J. Appl. Mech. **89**(3), 591–598 (1967)
4. Roth, B.: Finite position theory applied to mechanism synthesis, Trans. ASME, J. Appl. Mech. **89**(3), 599–605 (1967)
5. Chen, P., Roth, B.: A unified theory for the finitely and infinitesimally separated position problem of kinematic synthesis. Trans. ASME J. Engr. Ind. **91**(1), 203–208 (1969)
6. Chen, P., Roth, B.: Design equations for the finitely and infinitesimally separated position synthesis of binary links and combined link chains, Trans. ASME J. Engr. Ind. **91**(1), 209–219 (1969)
7. Tsai, L.W., Roth, B.: Design of dyads with helical, cylindrical, spherical, revolute, and prismatic joints. Mech. Mach. Theory **7**(1), 85–102 (1972)

8. Tsai, L.W., Roth, B.: A note on the design of revolute-revolute cranks. Mech. Mach. Theory **8**(1), 23–31 (1973)
9. Angeles, J.: Spatial Kinematic Chains. Synthesis, Optimization, Springer-Verlag, Berlin-Heidelberg-New York, Analysis (1982)
10. Hunt, K., H.: Kinematic Geometry of Mechanisms, Clarebdon Press, Oxford (1978)
11. Reuleaux, F., ed.: Theoretische Kinematik: Grundzuge einer Theorie des Maschinenwesens (1876). English Translation: Kinematics of Machinery. Outlines of a Theory of Machine. MacMillan, London
12. Chasles, M.: Note sur les proprieties generals du systeme de deux corps semblables entre eux, places d'une manière quelconque dans l'espace; et sur le deplacement fini, ou infiniment petit d'um corps solide libre, pp. 321–336. Bulletin des Sciences Mathematiques de Ferussac, XIV (1831)
13. Euler, L.: Nova methodus motum corporum rigidorum determinandi, Novii Comentarii Academiae Scientiarum Petropolitanae, 20 (1775) 1776: 208–238, Opera Omnia (2) 9:99–125, (1776)
14. Rodrigues, O.: Des lois geometriques qui regissent les deplacements d'un systeme solide dans l'espace, et de la variation des coordonnees provenant de ces deplacements consideres independamment des causes qui peuvent les produire, Journal De Mathematiques Pures et Appliquees, **5**(1st Series), 380–440 (1840)
15. Bottema, O., Roth, B.: Theoretical Kinematics, Dover Edition, 1990, Original publication by North-Holland Publishing Co., ISBN 0-486-66346-9 (1979)
16. Horn, B.K.P.: Closed-form solution of absolute orientation using unit quaternions. J Opt Soc Am **4**(4), 629–642 (1987)
17. Ravani, B., Ge, Q.J.: Computation of spatial displacements from geometric features. Trans. ASME J. Mech. Des. **115**, 95–102 (1993)
18. Eberharter, J.K., Ravani, B.: Kinematic registration in 3D using the 2D Reuleaux method. ASME Trans. J. Mech. Des. **128**, 349–355 (2006)
19. Baroon, J., Ravani, B.: Three dimensional generalization of Reuleaux's and instant center methods based on line geometry. Trans. ASME J. Mech. Robot. **2**, 041011-1 to 041011-8 (2010)
20. Ravani, B., Roth, B.: Motion approximation and mechanism synthesis. ASME Trans. J. Mech., Transm., Autom. Des. **105**, pp. 460–466 (1983)
21. Blaschke, W.: Euklidische kinematic und nichteuklidische geometrie, Zeitschr. Math. Phys. **60**, 61–91 and 203–204 (1911)
22. Blaschke, W., Muller, H., R.: Ebene Kinematik, Munchen, p. 269 (1956)
23. Ravani, B., Roth, B.: Mappings of spatial kinematics, ASME Trans. J. Mech., Transm. Autom. Des. **106**(3), 341–347 (1984)
24. Ge, Q.J., Ravani, B.: A computational geometric approach to motion synthesis, Artificial intelligence and symbolic computing. In: Housus, E. N., Rice, J.R., (eds.): IMACS Proceedings, pp. 193–202. North Holland Pub. (1992)
25. Bodduluri, R.M.C., McCarthy, J.M.: Finite position synthesis using the image curve of spherical four-bar motion. ASME Trans. J. Mech. Des. **114**(1), 55–60 (1992)
26. Ge, Q.J., McCarthy J.M.: The constraint manifold of the coupler of an SSSS spatial four bar linkage. In: Proceedings of the Applied Mechanisms and Robotics Conference, Cincinnati, OH, (1989)
27. McCarthy, J.M.: Finite position synthesis of a 4S spatial mechanism using Kinematic Mapping. In: Proceedings of the Eighth World Congress on the Theory of Machines and Mechanisms (IFToMM), Prague, Czechoslovakia, (1991)
28. McCarthy, J.M.: On the relation between kinematic mappings of planar and spherical displacements". ASME Trans. J. Appl. Mech. **53**, 457–459 (1986)
29. Study, E.: Die Geometie der Dynamen, Leipzig, p. 437 (1903)
30. Ge, Q.J., McCarthy, J.M.: Functional constraints as algebraic manifolds in a Clifford Algebra. IEEE J. Robot. Autom. **7**(5), 670–677 (1991)

31. Perez, A., McCarthy, J.M.: Dual quaternion synthesis of constrained robotic systems". ASME Trans. J. Mech. Des. **126**(3), 425–435 (2004)
32. Ge, Q.J., Ravani, B.: Computer aided geometric design of motion interpolants. ASME Trans. J. Mech. Des. **116**, 756–762 (1994)
33. Schumake, K.: Animating rotation with quaternion curves. ACM Siggraph **19**(3), 245–254 (1985)
34. Ge, Q.J., Ravani, B.: Geometric construction of Bezier motions. ASME Trans. J. Mech. Design **116**, 749–759 (1994)
35. Park, F.C., Ravani, B.: Smooth invariant interpolation of rotation. ACM Trans. Graphics **16**(3), 77–295 (1997)
36. Juttler, B.: Visualization of moving objects using dual quaternion curves. Comput. Graph. **18**(3) (1994)
37. Park, F.C., Ravani, B.: Bezier curves on Lie groups and Riemannian manifolds with kinematic applications. ASME Trans. J. Mech. Des. **117**, 36–40 (1995)
38. Zefran, M., Kumar, V., Croke, C: On the generation of smooth three-dimensional rigid body motions. IEEE Trans. Robot. Autom. **14**(4), 576–589 (1998)
39. Hofer, M., Pottmann, H., Ravani, B.: From curve design algorithms to design of rigid body motions. Vis. Comput. **20**, 279–297 (2004)
40. H. Pottmann, H., Hoefer, M., Ravani, B.: Variational motion design, in Advances in Robot Kinematics. In: Lenarcic, J., Galletti, C. (eds.): Proceedings of ARK Conference, Kluwer, pp. 361–370, (2004)
41. Hofer, M., Pottmann, H., Ravani, B.: Geometric design of motions constrained by a contacting surface pairs. J. Comput. Aided Geom. Des. **20**, 523–547 (2003)
42. Grunwald, J.: Ein Abbildungsprinzip, welches die ebene geometrie und kinematic mit der raumlichen geomtrie verknupft, Sitzber. Ak. Wiss. Wien **120**, 677–741 (1911)
43. Farin, G.: Curves and surfaces for CAGD: A Practical Guide, 3rd ed., p. 473. Academic Press, San Diego (1993)

Chapter 5
Betancourt's Contribution to Path Generation Synthesis in Mechanisms

Juan Ignacio Cuadrado

Abstract Between the late eighteenth century and the early nineteenth century, the Spanish engineer Agustín de Betancourt y Molina (1758–1824) made two innovative contributions to the formulation and resolution of trajectory synthesis problems, specifically in the context of steam engine design. These contributions are presented in two distinct works: the Mémoire sur une machine à vapeur à double effet, which was presented at the French Académie Royale des Sciences in 1789, and the book Essai sur la composition des machines, co-authored with José Maria de Lanz y Zaldívar (1764–1839). The book was published in Paris, with its first edition dating back to 1808. This work thoroughly explores the details, advancements, limitations, and shortcomings of both works. With the accessible and studied documentation available today, it can be affirmed that both contributions can be considered the earliest formulations and resolutions of what would later be referred to as synthesis of a four-bar linkage in which the coupler point performs approximately rectilinear motion with three points of precision.

Keywords History of mechanism and machine science · Synthesis of trajectory generation · Agustin de Betancourt

5.1 Introduction

Among the various fields of interest of Professor López-Cajún, his work and contributions in the field of the History of Mechanism and Machine Science stand out. More specifically, and closely related to this work, Professor López-Cajún showed a special interest in the figure of José María de Lanz y Zaldívar [1, 1], who was born in Campeche in 1764, at that time part of the virreinato of Nueva España, which is now located in Mexico. Lanz collaborated with Betancourt as a co-author in the "Essai

J. I. Cuadrado (✉)
Polytechnic University of Valencia, Camino de Vera s/n, 46022 Valencia, Spain
e-mail: jcuadrad@mcm.upv.es

M. Ceccarelli and J. C. Jauregui-Correa (eds.), *State-of-the-Art and Innovations in Mechanism and Machine Science*, Mechanisms and Machine Science 150,
https://doi.org/10.1007/978-3-031-47040-0_5

Fig. 5.1 A portrait of Agustín de Betancourt, shown in Russian attire. 1810s portrait. St. Isaac's Cathedral Museum, Saint Petersburg, Russia. Wikipedia

sur la composition des machines" and also in the establishment in 1802 of the first Engineering School in Spain, following the model of the French Ecole Polytechnique (Fig. 5.1).

Betancourt was born on February 1, 1758, in Puerto de la Cruz (Santa Cruz de Tenerife, Spain). In 1778, he moved to the mainland and pursued scientific studies at the Reales Estudios de San Isidro and artistic studies at the Academia de Bellas Artes of San Fernando in Madrid, from 1778 to 1784.

In 1784, he received a scholarship that allowed him to participate in the activities of the École des Ponts et Chaussées in Paris to obtain a degree in Hydraulic Engineering. At the same time, he was entrusted with the task of establishing a Cabinet of Machines that would incorporate models and machines of general utility for public works and industry. To see more biographical details about Betancourt, you can refer to the following sources [3, 3, 3], these sources provide detailed biographical information about Agustín de Betancourt, his contributions to engineering, and his impact on industrial development.

During the process of establishing the Cabinet of Machines, he visited England in 1788 and became acquainted with the remarkable advancements made by James Watt (1736–1819) in the development of the steam engine. Watt introduced a series of technological innovations that significantly improved its performance [6, 6, 8].

One notable invention, presented by Watt on April 28, 1784, was a patent that included various solutions for the rectilinear guidance of the piston and its connection to the rocker arm. Among these solutions, one stood out: the use of an articulated quadrilateral, where a point on the connecting rod traced an almost straight trajectory. Later, Watt added a pantograph mechanism to the articulated quadrilateral. These solutions are respectively referred to as Watt's singular mechanism (see Fig. 5.2) and extended mechanism (see Fig. 5.3).

This innovation enabled the piston to exert its force on the rocker arm during both the upward and downward strokes. It is for this reason that Betancourt referred to the new steam engine as a double-acting machine (double effet).

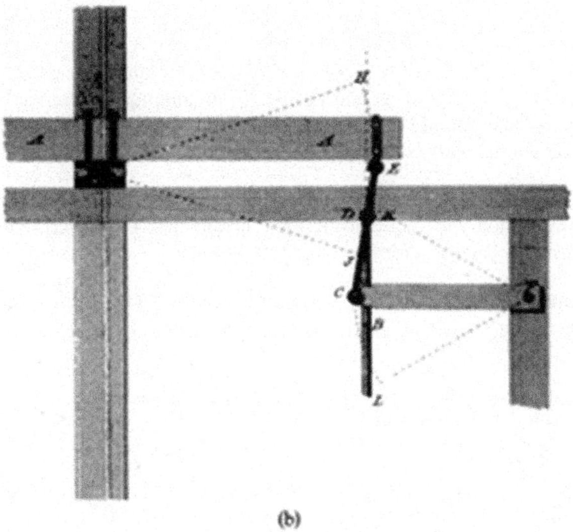

Fig. 5.2 Rectilinear guidance of the piston and its connection to the rocker: Patent submitted by Watt in 1784. The second solution (**b**) constitutes the so-called Watt's singular mechanism. In Muirhead, The origin and progress of the Mechanical Inventions of James Watt (1854), vol. III, pl. XXII and XXIII

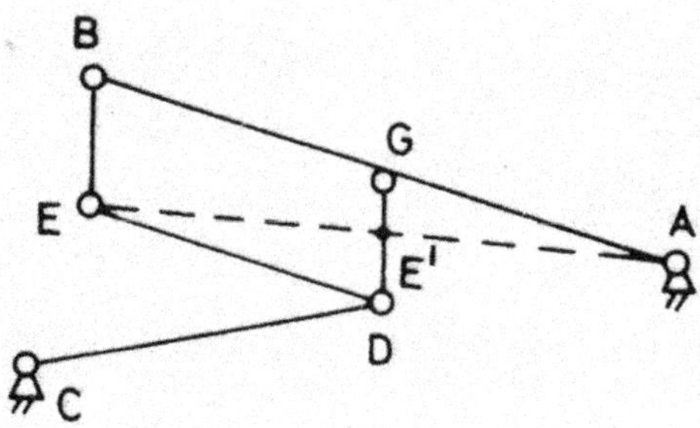

Fig. 5.3 Extended Watt mechanism: Mechanism used by Watt in his steam engines. It can be considered as a Watt's singular mechanism with a pantograph-shaped extension. Since EDGB is a parallelogram, point E describes a curve similar to the one described by the intersection of EA and GD, that is, point E′. Therefore, if AGDC form a Watt's singular mechanism, point E′ will approximately describe a straight line, and consequently, point E will also do so. In Koetsier, 1983

Despite encountering numerous challenges in accessing the new machines, he succeeded in observing a portion of the machine and grasping the implications that the new design had on enhancing performance (see Fig. 5.4).

Betancourt, among other things, points out:

> To begin with, I was surprised to see that the chain connected to the rocker arm, which suspended the piston inside the steam cylinder, had been replaced by a parallelogram. I will provide a more detailed description of this later on (...).
>
> The day after witnessing this machine, I departed for France. Upon returning home, I dedicated myself to faithfully recalling all the parts I had seen and endeavored to understand their purpose. I drew various plans and profiles in an attempt to decipher their function. Eventually, I conceived a double-acting machine. From that very moment, I embarked on constructing a model that surpassed my expectations.
>
> Recognizing the immense usefulness of this machine in mechanical arts, as well as its economical construction and fuel consumption advantages, I believed that the Academy would be pleased to receive the forthcoming description.

Upon returning to Paris, Betancourt presents the "Memoire sur une machine à vapeur à double effet" on December 15, 1789, and signs it as "Le chevalier de

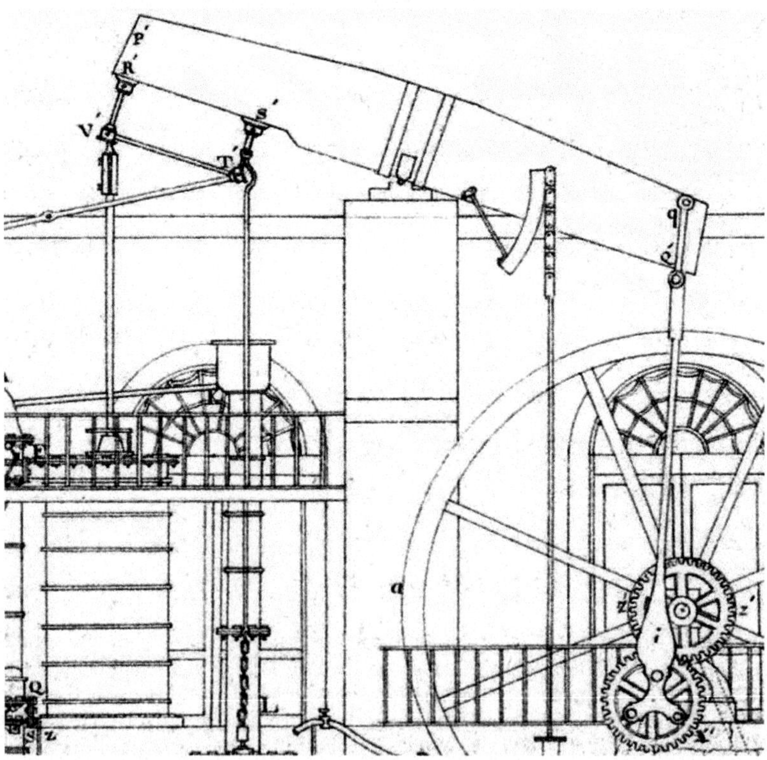

Fig. 5.4 Watt's double-acting steam engine. In Betancourt, Mémoire sur une machine à vapeur à double effet (1789), plate III

Betancourt Capitaine au service d'Espagne" (Knight of Betancourt, Captain in the service of Spain). The session records of December 16, 1789, of the Royal Academy of Sciences confirm that "Mr. Betancourt has presented a Memoire on a double-acting steam engine" and that the commissioners Jean Charles Borda (1733–1799) and Gaspard Monge (1746–1818) have been appointed to report on said Memoire. In the session on February 10, 1790, the commissioners' report concludes as follows:

> We believe the Academy should applaud Mr. Betancourt's enthusiasm and expertise for introducing to France a discovery whose knowledge would not have naturally reached him until much later. The Memoire he has presented, which deserves our approval, should be published in the collection of works by foreign scholars.

In this work, we will analyze in detail some of the drawings and mathematical developments included by Betancourt in the Memoire, particularly those related to the mechanism of rectilinear guidance.

It is crucial to highlight, as Franz Reuleaux (1829–1905) notes in the introduction to his Theoretische Kinematik [9]:

> Watt shared with us some insights into the thought process that led directly to the mentioned mechanism. 'The idea,' he wrote to his son in November 1808, 'arose as follows: Finding the double chains or racks and toothed sectors for transmitting motion from the piston shaft to the angular motion of the rocker arm highly inconvenient, I endeavored to find a means of achieving the same result through rotational movements around centers. After some time, it occurred to me that if AB and CD were two equal radii rotating around centers B and C, respectively, and connected by a rod AD moving along arcs of equal length, the deviation from a straight line would be approximately equal and opposite. Consequently, point E should trace an approximate straight line. Furthermore, if, for convenience, CD were only half the length of AB, by shifting point E closer to D, the same effect would occur. It was from this construction that the subsequently named parallel motion was derived.
>
> While we take an interest in the contents of this letter, a closer examination reveals a deficiency that he... may have also discovered. While it presents the motives and some of the final results of Watt's experiment, it lacks indications of any systematic sequence of ideas leading to the desired outcome.

What happened in the following years? Gaspard de Prony (1755–1839), in his Architecture Hydraulique [10] of 1796, developed a theoretical study for calculating deviations from the straight line of a point on the coupling rod. Jean Nicolas Pierre Hachette 1769–1834), in his Histoire des Machines à Vapeur Depuis leur Origine jusqu'à nos Jours [11] of 1830, establishes without proof that the curve described by the connection point of the piston rod to Watt's mechanism is of sixth degree.

Later, in 1897, Gino Loria (1862–1954) conducted a bibliographic study of theoretical studies on the Watt mechanism [12], including references from Henri Brocard (1845–1922) and Julien Napoléon Haton de la Goupillière (1833–1927). However, he did not include any studies prior to 1836, as indicated by Nolle [13]. The same Nolle indicates:

> It was many years after Watt's application, in 1784, of the coupler of the four-bar linkage to provide approximate straight-line motion of the piston rod on one of his steam engines, that the general usefulness of an intermediate link (not attached to the stationary link or "frame") motion was fully realized. Mathematical analysis of such intermediate link motions, usually

formulated in terms of point paths, or coupler curves, did not gather momentum until the late 1860's and investigations pertaining to questions of synthesis of coupler curves or certain aspects thereof did not find their way into published literature till some twenty years later. Possibly earlier progress was hampered by the lack of any consistent basis for the study of the kinematic properties of elements of link arrangements.

The discovery of Watt did not remain unnoticed, but despite its catalytic effect, only small contributions were made towards deeper understanding of the characteristics of coupler curves in general.

In the second half of the nineteenth century, after de appearance of the published works of Pafnuti Lvóvich Chebyshev (1821–1894) and Ludwig Burmester (1840–1927), the approximate methods of straight-line generation by means of linkages underwent rapid development.

It must be concluded that communications and dissemination of knowledge and new discoveries was still, at that time, a major isolating factor.

However, in the following two sections, we will analyze two contributions dated in 1789 and 1808 (see Fig. 5.5). The first one was signed by Betancourt, and the second one was jointly signed by Lanz and Betancourt. In our opinion, considering what was mentioned in the previous paragraphs, these contributions represent the first attempts at synthesizing straight-line generation with precision points. The first one follows a geometric-analytical approach, while the second one follows a purely geometric approach.

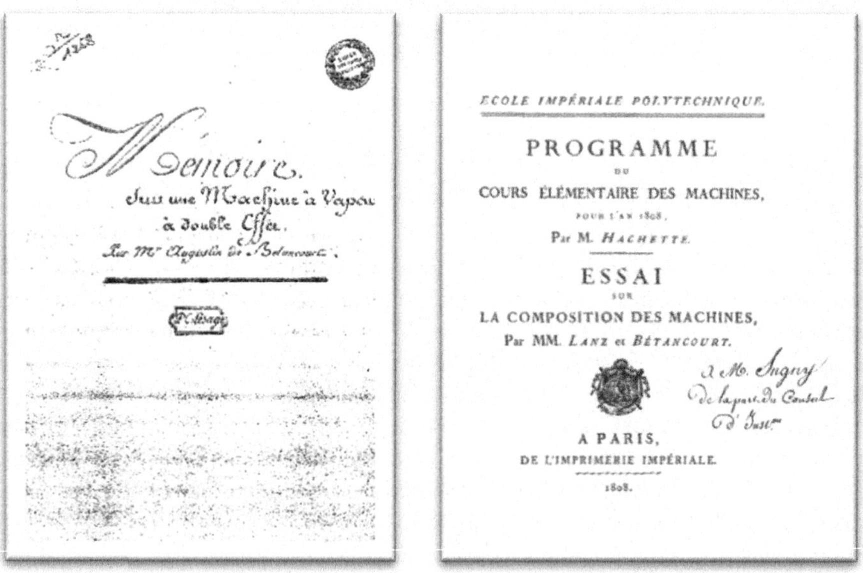
ECOLE IMPÉRIALE POLYTECHNIQUE.

PROGRAMME

DU

COURS ÉLÉMENTAIRE DES MACHINES,

POUR L'AN 1808,

Par M. HACHETTE.

ESSAI

SUR

LA COMPOSITION DES MACHINES,

Par MM. LANZ et BÉTANCOURT.

A PARIS,

DE L'IMPRIMERIE IMPÉRIALE.

1808.

Fig. 5.5 Cover pages of "Mémoire sur une machine à vapeur à double effet" and of the "Essai sur la composition des machines"

5.2 The Mémoire Sur Une Machine À Vapeur À Double Effect

The Mémoire [14] is a 31-page manuscript (see Fig. 5.7) that includes an introduction to the development of the steam engine, an account of Betancourt's visit to England focused on gathering information about advancements in steam engine design, his contacts with Watt and Boulton, and visits to facilities that already had Watt's new designs. He was only able to observe part of the machine. The manuscript provides a description of the machine's parts and its functioning, highlighting Watt's innovative mechanism. It also presents a formulation and solution for what Betancourt refers to as the specific case of the parallelogram mechanism, which allows for the calculation of the length of one of the beams. Additionally, there are seven plates included in the manuscript that reproduce parts of the steam engine, including a representation of the parallelogram mechanism in three positions. This representation serves as the basis for the calculation of the beam's length mentioned earlier.

We will now focus our attention on this last part, including the formulation presented in the manuscript, auxiliary mathematical developments that help us understand Betancourt's approach, and an analysis of the problems that arise in this formulation.

The problem he presents is as follows: given the mechanism in the three mentioned positions, the angle of the rocker arm in those positions, and the length of all the movable bars that make up the mechanism, except for the rocker arm XE, calculate the length of that rocker arm.

In modern terminology, the problem he poses is a trajectory generation synthesis problem with three points of precision.

Betancourt represents the mechanism in three chosen positions that correspond to three positions of the rocker arm: the first with an angle above the horizontal, the second in a horizontal position, and the third with an angle below the horizontal, symmetrically with respect to the first.

The represented mechanism corresponds to what we have called the extended Watt mechanism, formed by two rocker arms: in the first position, AC rotates around the fixed-point A and XE rotates around the fixed-point X. Additionally, there is a coupling bar EB, so that the bars AB, BE, and XE, all joined by kinematic pairs of rotation, correspond to a singular Watt mechanism. In this mechanism, the midpoint of the coupling bar traces an approximately straight path under certain conditions. The pantograph is added to this mechanism, formed by the extension of the bar AB to form the bar AC, and the bars CD and DE parallel to BE and BC, respectively. Point D, aligned with the center of the coupling bar BE and with point A, will also trace an approximately straight path.

The choice of the two symmetric extreme positions of the rocker arm allows Betancourt to ensure that the segments DR, EM, and BP are parallel to each other and perpendicular to the horizontal line AO. Moreover, the segments EM and BP have the same length.

Although Betancourt does not mention it in the Memoire, once the length of the crank is calculated and the perpendicular bisector of the segment EL is represented, the rotation center of the rocker arm, denoted by point X in the figure, can be located.

Betancourt's solution is a geometric resolution primarily based on trigonometric principles. As observed in the figure, the selected positions, along with a set of auxiliary lines, generate a significant number of right triangles that contribute to the resolution.

On Plate IV (see Fig. 5.6), the mechanism is depicted in three positions, and different letters are assigned to the locations of the different kinematic pairs.

Notably, points B, K, and P, on one hand, and points C, O, and Q, on the other hand, lie on two circles centered at point A, while points E, L, and M are located on a circle centered at point X. Points D, O, and R lie on a line perpendicular to segment AO, passing through point O. The location of the rocker arm in position 2 is represented by AO.

The points A, B, E, and X form an articulated quadrilateral, where points A and X indicate the positions of the pairs R with the fixed bar. Additionally, the bars CD and DE are added to this mechanism to form a parallelogram that constitutes a pantograph mechanism. Under specific conditions, point D traces an almost straight segment.

To distinguish the formulas included by Betancourt in the report from the auxiliary formulas added to facilitate Betancourt's approach, an asterisk has been added to the formula of the former.

The given starting data is:

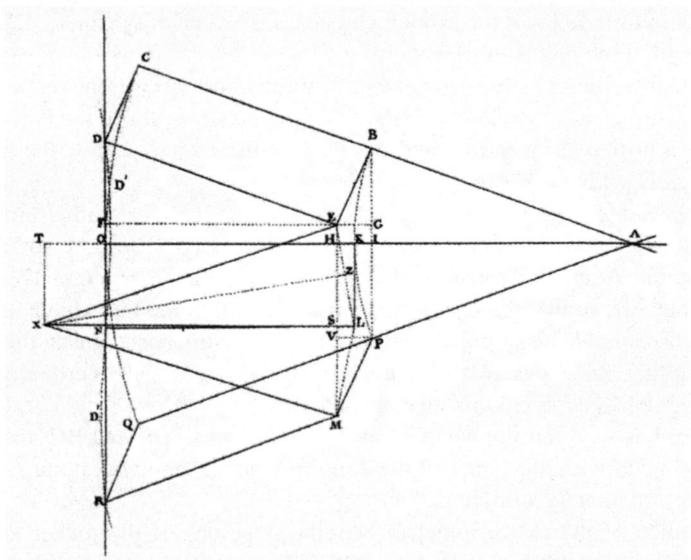

Fig. 5.6 Geometrical scheme used by Betancourt to describe the operation of the Watt's mechanism. In Betancourt, Memoire sur une machine à vapeur à double effet (1789), Plate IV

$$AB = a(*) \tag{5.1}$$

$$BC = b(*) \tag{5.2}$$

$$CD = c(*) \tag{5.3}$$

$$sinCAO = \alpha(*) \tag{5.4}$$

The question at hand is to calculate the length of the bar XE based on the lengths of segments EL and LM.

He begins by calculating the length of segment HK. H represents the vertical projection of points E and M onto the horizontal position of segment OA, which aligns with the second position of the rocker arm. K corresponds to the position occupied by point B in the second position of the mechanism:

$$HK = OK - OH \tag{5.5}$$

$$OH = FE = DE\cos\alpha \tag{5.6}$$

$$OK = DE = b \tag{5.7}$$

$$HK = b(1 - \cos\alpha) = bm \tag{5.8}$$

Continue calculating the length of segment KI. I is the vertical projection of points B and P onto the horizontal position of segment OA coinciding with the second position of the rocker:

$$KI = KA - IA \tag{5.9}$$

$$IA = AB\cos\alpha \tag{5.10}$$

$$KA = AB = a \tag{5.11}$$

$$KI = a(1 - \cos\alpha) = am(*) \tag{5.12}$$

The length of segment HI will be the sum of segments HK and KI:

$$HI = HK - KI = EG \tag{5.13}$$

$$HI = (a + b)m(*) \tag{5.14}$$

From the right triangle EGB, calculate the length of the segment BG:

$$BG = \pm\sqrt{BE^2 - EG^2} \tag{5.15}$$

$$BE = c \tag{5.16}$$

$$BG = \pm\sqrt{\left(c^2 - (a+b)^2 m^2\right)}(*) \tag{5.17}$$

Being S the vertical projection of points E and M onto the horizontal position of segment NL coinciding with the second position of bar DE, which, when the points OHSN form a parallelogram, will have a horizontal position, the length of segment ES will be:

$$ES = HS + IB - BG \tag{5.18}$$

$$HS = c \tag{5.19}$$

$$IB = ABsin\alpha = asin\alpha \tag{5.20}$$

$$ES = c + asin\alpha \mp \sqrt{\left(c^2 - (a+b)^2 m^2\right)}(*) \tag{5.21}$$

From the right triangle ESL, calculate the length of segment EL:

$$EL = \pm\sqrt{SL^2 + ES^2} \tag{5.22}$$

$$SL = HK \tag{5.23}$$

$$EL = \pm\sqrt{b^2m^2 + \left(c + asin\alpha \mp \sqrt{\left(c^2 - (b+a)^2 m^2\right)}\right)^2} = A(*) \tag{5.24}$$

In the text of the Memoire, the calculation expression for the length of segment ML is included, and it is indicated that the same steps have been followed as to obtain the previous expression. We will now proceed to follow these steps in detail.

The length of segment MS will be:

$$MS = MV - HS + VH \tag{5.25}$$

$$MV = BG \tag{5.26}$$

$$HS = c \tag{5.27}$$

$$VH = PI = -APsin\alpha \tag{5.28}$$

$$AP = a \tag{5.29}$$

$$MS = \sqrt{\left(c^2 - (b+a)^2 m^2\right)} - c - asin\alpha \tag{5.30}$$

From the right triangle MSL, we calculate the length of segment ML:

$$ML = \pm\sqrt{SL^2 + MS^2} \tag{5.31}$$

$$SL = HK \tag{5.32}$$

$$ML = \pm\sqrt{b^2m^2 + \left(\sqrt{\left(c^2 - (b+a)^2 m^2\right)} - c - asin\alpha\right)^2} = B(*) \tag{5.33}$$

Next, based on triangles MLS and LXZ, he calculates the radius of the rocker arm R, identified by the length of segment XL:

$$R = \frac{B \times A}{2bm}(*) \tag{5.34}$$

By identifying the segments depicted in the graph, we will be able to determine:

$$\frac{XL}{ZL} = \frac{ML}{SL} \tag{5.35}$$

This relationship is only true if the triangles MLS and LXZ are similar. However, that similarity does not always occur. It only occurs if:

$$ES = \frac{EM}{2} \tag{5.36}$$

In this case, the triangles MLS and ESL will be equal. Since the perpendicular bisectors have the same inclination but opposite directions, their intersection must occur at a point X that lies on the same horizontal line as point S. Therefore, the segment XL will be horizontal.

In this case, the triangles ESL and XLZ will be similar, and consequently, MLS and LXZ will also be similar:

$$ES = HS + EH =$$

$$= HS + BAsin\alpha - BG = c + asin\alpha - \sqrt{(c^2 - (a+b)^2 m^2)} \tag{5.37}$$

$$EM = BP = 2asin\alpha \tag{5.38}$$

That is to say, the formula deduced by Betancourt only holds if:

$$c + asin\alpha - \sqrt{(c^2 - (a+b)^2 m^2)} = asin\alpha \tag{5.39}$$

In other words, if:

$$c = \sqrt{(c^2 - (a+b)^2 m^2)} \tag{5.40}$$

Condition that is only fulfilled if:

$$m = (1 - cos\alpha) = 0 \tag{5.41}$$

Which, in turn, is only satisfied if:

$$\alpha = 0 \tag{5.42}$$

Which is incompatible with the design conditions; therefore, no case of similarity between said triangles can be established, and the deduced formula would not be applicable for design purposes.

In the subsequent comments to the mathematical formulation and as a consequence of the analysis of the obtained expressions, Betancourt states that the curve traced by point D will fit more closely to the straight-line DR as the sine of the rocker's rotation angle decreases, and therefore, as the rotated angle decreases.

On the other hand, in the second position of the mechanism, he points out that if a line is drawn from point N through the center of motion A, all points on this line, whose motion depends on that of the parallelogram, will deviate as little as possible from the corresponding vertical. Betancourt has established the dimensions of the extended Watt mechanism so that the end of the parallelogram passes through three points contained in a straight line, and he deduces that the midpoint of the coupler KL will also be contained in a straight line for those positions. Therefore, point D will be connected to the end of the rod piston, and the midpoint of the coupler will be connected to the air pump that supplies water to the boiler (Fig. 7).

It is surprising that Betancourt exclusively uses the trigonometric procedure, makes a mistake in formulating the similarity condition, and does not use more direct geometric procedures such as obtaining the crank radius and locating its center by applying the condition of locating points E, L, and M on a circle. However, it is undeniable that he has laid the foundations for planning a synthesis of generating rectilinear trajectories with three precision points.

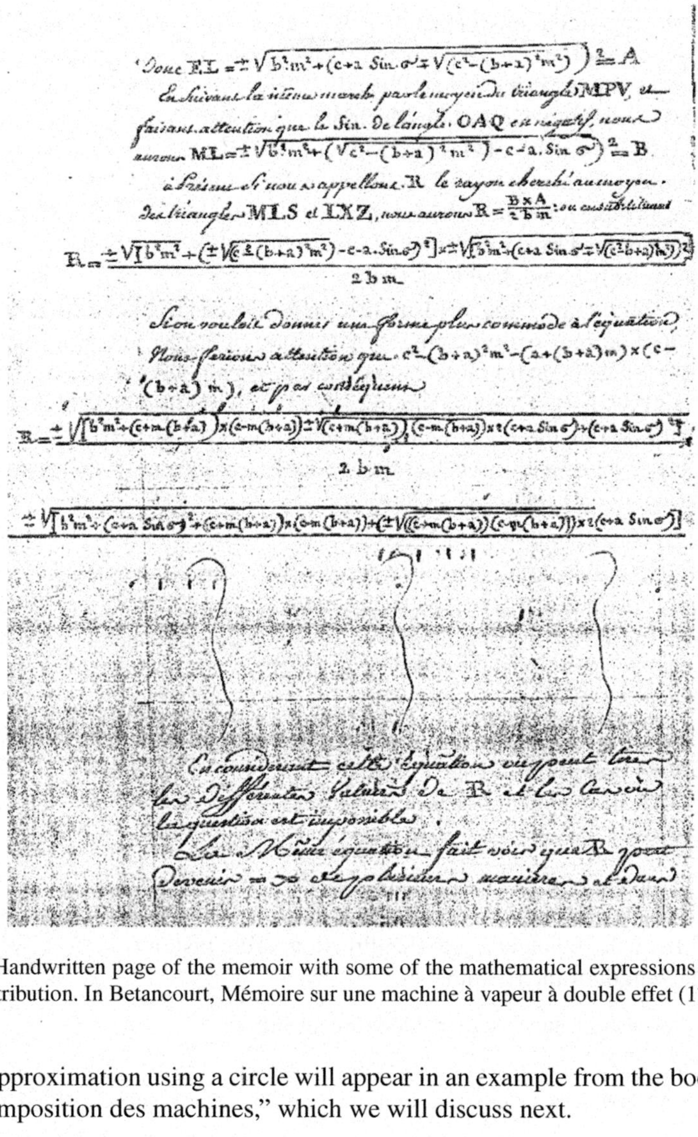

Fig. 5.7 Handwritten page of the memoir with some of the mathematical expressions referenced in the contribution. In Betancourt, Mémoire sur une machine à vapeur à double effet (1789)

The approximation using a circle will appear in an example from the book "Essai sur la composition des machines," which we will discuss next.

5.3 The "Essai Sur La Composition Des Machines"

In the year 1808, as we have already mentioned, the Essai sur la composition des machines de Lanz y Betancourt was published [15, 16]. The title of the treatise itself marked a substantial difference compared to previous works dedicated to the collection and classification of machines. It focuses on the composition of machines, that is, it analyzes not so much the machine itself but the mechanisms that constitute it.

The book was part of the material used by Jean Nicolas Pierre Hachette (1769–1834) in the Elementary Course on Machines developed at the École Polytechnique in Paris, based on the ideas of Gaspard Monge (1746–1818). The course program and the book present various types of motion transformations and classify examples of mechanisms that perform them. These examples are included in a classification table and are accompanied by descriptions of the mechanism's operation and, sometimes, by an explanation of a method to design it.

In line 17 of the table, mechanisms whose purpose is to transform reciprocating rectilinear motion into reciprocating circular motion are included. In that line, the extended Watt mechanism is listed in column H, the singular Evans mechanism in column I, and the Evans mechanism in column O. In addition to the description of the latter, a design method is provided, which we will detail next.

In this case, we directly translate the text in English from the version published by R. Ackermann in London in 1820.

It begins by explaining the structure of the mechanism. This concerns the mechanism known as Evans' mechanism (see Fig. 5.8):

> In this figure, ABB represents a side elevation of the beam of a steam engine; G its centre of rotation; nm an iron rod which is at liberty to turn freely about an axis b placed at the extremity A of the beam, and which divides the rod nm into two equal portions; the rod nm is attached by the extremity n to the piston rod f, and at the other extremity m, to the rod pq, which turns on the fixed axis q.

The starting conditions are established:

"Under this arrangement, we will suppose to be given.

- 1st. The Dimensions of the beam of the engine ABB.
- 2nd. The position of its centre of rotation G.
- 3rd. The arc bca which the extremity A of the beam will traverse at each oscillation, and which will be tangential to the direction of the piston f.
- 4th. The length of the rod nm."

The unknowns of the posed problem are specified:

> From these data it is required to determine the length of the rod pq, and the position of its centre of rotation q, so as to ensure, as nearly as possible, the rectilinear direction of the piston.

The procedure for solving the problem is described:

"The positions of the three points m, m', m", will be determined, so as to indicate the respective situation of the extremity m of the given rod nm, at the commencement-towards the middle-and the close of the oscillation of the beam; and so that in those three positions, the other extremity n shall be situated accurately in· the direction of the piston rod f. If a circle be described which shall pass through these three points; its radius will be equal to the required length of the bar pq, and its centre so determined will represent the required centre of rotation q.

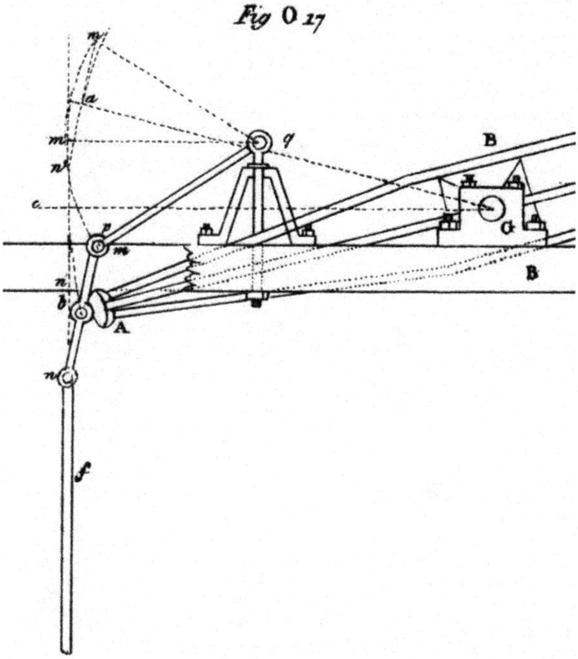

Fig. 5.8 Explanation in the Essai of the procedure for calculating the dimensions of the mechanism so that point n follows a rectilinear trajectory. In Lanz and Betancourt, Essai sur la composition des machines (1808), O 17

The curve described by the extremity n of the piston rod f, will pass through the three points n, n', n'', and will approximate to a right line, as the arc acb described by the extremity of the beam, is smaller."

The approach in this case is purely geometric. It is specified that the three successive positions of point m define a circumference, whose radius and center will provide a solution to the posed problem.

Furthermore, it is indicated that the same procedure can be applied to what is known as the extended Watt mechanism and the singular Watt mechanism (see Fig. 5.9).

"The same course of proceeding will serve to determine (in the figure H 17, plate 9.) the length of the rod f' d, and the position of the point of rotation f'; and in the figure I 17 of the same plate, will also serve to determine the length of the beam ab, and the position of the centre of rotation a."

5.4 Conclusions

Throughout the work, we have been able to detail, through two publications by Agustín de Betancourt, the Mémoire sur une machine à vapeur à double effet published in 1789 and the Essai sur la composition des machines published in 1808, a new procedure to approach the design of the extended Watt mechanism,

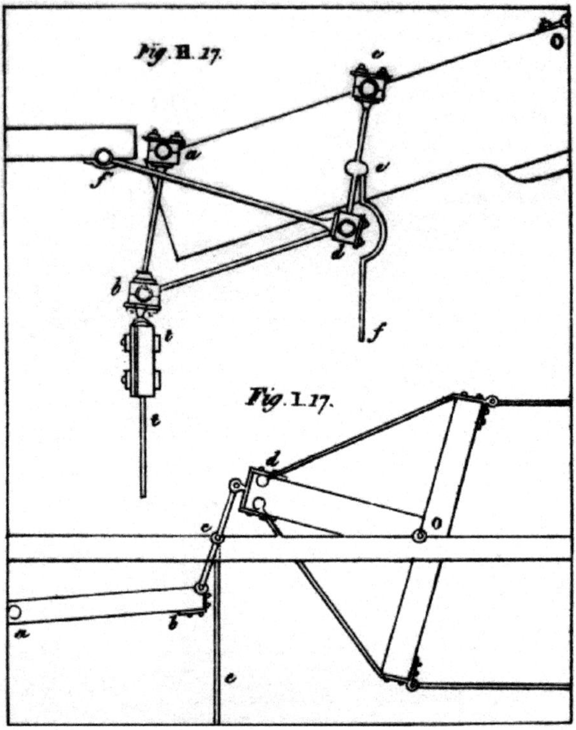

Fig. 5.9 Watt's straight-line linkages: Watt's extended linkage and Watt's singular linkage. In Lanz and Betancourt, Essai sur la composition des machines (1808), H 17 and I17

specifically the calculation of one of the rocker arms, based on the condition that a point of the mechanism be located on a straight path in three positions. Betancourt employs geometric-analytical approximations, and although one of them contains some formulation errors, we can assert that it is a completely innovative procedure for the time of publication, advancing more than half a century ahead of similar contributions and constituting, to this day, the first approach to the synthesis of generating straight-line paths with three points of precision.

In the case of the Mémoire, the work did not receive much diffusion since, although its publication was approved by the French Academy of Sciences, the revolutionary events of the time in France likely prevented its publication. In fact, only the manuscript that we have worked with and is preserved at the École des Ponts et Chaussées has survived. The case of the Essai is even more surprising because the work was highly successful throughout the first half of the nineteenth century, with versions published in French in 1808 [15], 1819 [17], and 1840 [18], in English in 1820 [19], and in German in 1829 [20]. Despite this, there are no references in later writings to this contribution. However, the immediate practical consequence of Betancourt's studies and designs was the construction of the first double-acting steam engine on the continent by the Périer brothers in 1789 in France.

References

1. López-Cajún, C.S., Cuadrado Iglesias, J.I., Ceccarelli, M.: Early modern activity on TMM by Lanz and Betancourt before 1830. In: 11th IFToMM world congress in mechanism and machine science, Tianjin, pp 939–943 (2004)
2. López-Cajún, C.S.: José María Lanz y Zaldívar (1764–1839). In: Distinguished figures in mechanism and machine science: their contributions and legacies—part 2, pp. 111–122. Springer, Dordrecht (2010)
3. Cuadrado, J.I., Ceccarelli, M.: El nacimiento de la teoría de Máquinas y Betancourt. In; Silva M. (ed.), Técnica e Ingeniería en España. Vol. III El siglo de las luces, Real Academia de Ingeniería, Zaragoza (2005)
4. Erogova, O., Ceccarelli, M., Cuadrado, J.I., López-Cajún, C.S., Pavlov, V.E.: Agustín Betancourt: An Early Modern Scientist and Engineer in TMM. In: Proceedings of ASME IDETC/CIE 2006 Mechanisms & Robotics Conference, Paper No. DETC2006–99198, Philadelphia (2006)
5. Payen J., Agustín de Betancourt y Molina, in Dictionary of Scientific Biography, Charles Scribner's Sons, New York
6. Ferguson, E.S.: Kinematics of mechanisms from the time of watt. In: Contributions from the Museum of History and Technology, Paper 27, pp. 186–230, Washington (1962)
7. Muirhead, J.P.: The Origin and Progress of the Mechanical and Inventions of James Watt, 3 volumes. Murray, London (1854)
8. Koetsier, T.: A contribution to the history of kinematics – I. watt straight-line linkages and the early french contributions to the theory of the planar 4-bar coupler curve. Mech. Mach. Theory **18**, 37–42 (1983)
9. Reuleaux, F.: Theoretische Kinematik. Druck und Verlag von Friedrich vieweg und Sohn, Braunschweig, Grundzüge einer Theorie des Maschinenwesens (1875)
10. Prony, R.: (1796) Nouvelle architecture hydraulique. Firmin Didot, Paris
11. Hachette, J.N.P.: Histoire des Machines à Vapeur Depuis leur Origine jusqu'à nos Jours. Corby, Paris (1830)
12. Loria, G.: Indications Bibliographiques au sujet de la courbe de Watt. L'Intermèdiaire des Mathématiciens, Tome IV, pp. 184–185, Paris (1897)
13. Nolle, H.: Linkage coupler curve synthesis. a historical review – I. Developments up to 1875. Mech. Mach. Theory **9**, 147–168 (1974)
14. Betancourt, A.: (1789) Mémoire sur une machine à vapeur à double effet, Bibliothèque de l'École National des ponts et chaussés, Paris
15. Lanz, J.M., Betancourt, A.: Essai sur la composition des machines: Programme du cours élémentaire des machines pour l'an 1808 par M. Hachette, Imprimerie Impériale, Paris (1808)
16. Ceccarelli M, Cuadrado JI (1997) Sobre el Essai sur la composition des machines por José María de Lanz y Agustín de Betancourt en 1808, 3er Congreso Iberoamericano de Ingeniería Mecánica, La Habana.
17. Lanz, J.M., Betancourt, A.: Essai sur la composition des machines. 2 éd, rev., corr. y augm., Bachelier Libraire, Paris (1819)
18. Lanz, J.M., Betancourt. A.: Essai sur la composition des machines. 3 éd, rev., corr. y augm., Bachelier Imprimeur-Libraire de l'Ecole Polytechnique, Paris (1840)
19. Lanz, J.M., Betancourt, A.: Analytical Essay on the Construction of Machines. R. Ackerman, London (1820)
20. Lanz, J.M., Betancourt, A.: Versuch über die Zusammenstzung der Machinens. A. Rücker, Berlin (1829)
21. Lanz, J.M., Betancourt, A.: Ensayo sobre la composición de las Máquinas. Colegio de Ingenieros de Caminos, Canales y Puertos, Madrid (1990)

Chapter 6
A Brief History of Robotics Development of CAS

Xiao Dou, Yanyu Wang, and Yibing Fang

Abstract The key accomplishments and contributions of the Chinese Academy of Sciences in the three stages of China's robotics development from 1970 to 2020s are outlined, through which this paper presents that the Chinese Academy of Sciences has played a pivotal role in China's robotics industry.

Keywords Robotics · Chinese academy of sciences · History

Recognizing the interest of prof Carlos Lopez-Cajun in investigating and keeping track of technical developments of communities, this contribution refers to his interest in the history of Robotics, but with focus on Chinese evolution. China's robotics research dates back to 1975, when the Shenyang Institute of Automation submitted a robot project application to the Chinese Academy of Sciences that was subsequently approved. Since then, China's robot development has undergone three stages, namely the initial period (1975–1985), the growth period (1986–2005), and the rapid development period (2006–2020), during which CAS has consistently played a pivotal role.

X. Dou · Y. Wang · Y. Fang (✉)
Institute for the History of Natural Sciences, Chinese Academy of Sciences, Beijing, China
e-mail: Yibing@ihns.ac.cn

X. Dou
e-mail: jinghu_yaye@163.com

Y. Wang
e-mail: wangyanyu1982@163.com

M. Ceccarelli and J. C. Jauregui-Correa (eds.), *State-of-the-Art and Innovations in Mechanism and Machine Science*, Mechanisms and Machine Science 150,
https://doi.org/10.1007/978-3-031-47040-0_6

6.1 The Initial Period (1975–1985)

Prior to the 1970s, China lacked a cohesive stance on pursuing robotics research due to a dearth of relevant scientific research conditions and experiences. In the initial stages, several pertinent ministries, research institutions and universities embarked on preliminary investigations into robotics, with the Chinese Academy of Sciences assuming a prominent role in this endeavor.

In 1975, the Shenyang Institute of Automation (SIA) of CAS proposed China's first robot research project application, which was ultimately approved by CAS and marked the beginning of robotics research in China. However, the application process was initiated in 1972 by Ye Qiang, the deputy director of the Shenyang Institute of Automation, who focused on establishing a project for intelligent systems. Three years later, the research project on "artificial intelligence and robotics" proposed by the Shenyang Institute of Automation was included in the "Automation Discipline Plan (1976–1985)" of the Chinese Academy of Sciences for the first time. This marks a significant milestone in the history of China's robotics development, signifying the official commencement of robotics research in China.

Subsequently, Professor Jiang Xinsong (see Fig. 6.1), who was elected as a member of Chinese Academy of Engineering, made significant contributions to promoting robotics research in SIA through three aspects.

Firstly, the research directions were determined through the formulation of the "Shenyang Institute of Automation 1981–1990 Ten-year Development Plan" by an academic committee consisting of 22 members led by Jiang Xinsong. The plan identified three key areas for research: artificial intelligence and robotics, information

Fig. 6.1 Jiang Xinsong, member of Chinese academy of engineering [1]

systems and control engineering, as well as image processing and pattern recognition. Secondly, professional research teams were established with the formation of two teams by Jiang: a basic research team and an engineering team, both of which underwent specialized training in robotics. Thirdly, the implementation of study abroad programs was pursued by CAS. In 1979, Jiang led a delegation from the robotics and artificial intelligence field to Japan for investigative purposes. Through this investigation, the research direction of China's robotics was further clarified. The implementation of these reform measures by Jiang Xinsong propelled SIA to become the leading force in China's robotics research at that time, effectively steering China's robotics research activities onto a path of progress and innovation. [1].

In the initial period, research on robots of the Chinese Academy of Sciences were mainly carried out by three institutes: Shenyang Institute of Automation, Changchun Institute of Optical Mechanics (CIOM) and Hefei Institute of Artificial Intelligence (HIAI). 3 projects were carried out, mainly focused on 4 aspects: robot basic technology, robot and general controller, marine and underwater robot and the construction of robot demonstration engineering center. A series of representative results have been achieved, which are shown as follows.

In the field of the basic technology of robots, research works included the basic technology of underwater robot, machine vision system, speech recognition, and mobile robot environment model. Meanwhile, on the basis of investigating the development of robots in the United States, Japan, the United Kingdom, the Soviet Union and other countries, the SIA compiled a research report of 100,000 words called "Foreign robot technology development background, model, current situation analysis and prospect", which has played an important reference role in the development of China's robot technology. In addition, the key technologies of the second and third generation robots, such as image recognition and speech recognition, were developed by the SIA in this period.

The development of industrial robotics constituted the second facet of accomplishments. The SIA developed China's first "teaching and reproduction" robot SZT-1 (Fig. 6.2), which entered the debugging stage in February of 1982. This robot was controlled by a computer, driven by electro-hydraulic servo system, it contained 5 degrees of freedom and two control modes which were point control and trajectory control. The relevant technical indicators of the robot SZT-1 were close to the UNIMATE-2000 industrial robot which were widely used abroad in the 1970s, SZT-1 also won the second prize of the Chinese Academy of Sciences Scientific Achievement Award. After that, two units of SZT-1 were produced by the SIA and Shenyang No.2 machine tool factory in 1986, one was used in the No. 2 Automobile Factory in Hubei, mainly for water tank spot welding operations, and the other was used in Changchun No.1 Automobile Manufacturing Plant for gear heat treatment and blanking. SZT-1 was the first modern industrial robot of the true sense in the history of Chinese robot development, which was a real milestone [1]. At the same time, the Changchun Institute of Optics and Mechanics of CAS successfully developed an encoder [2].

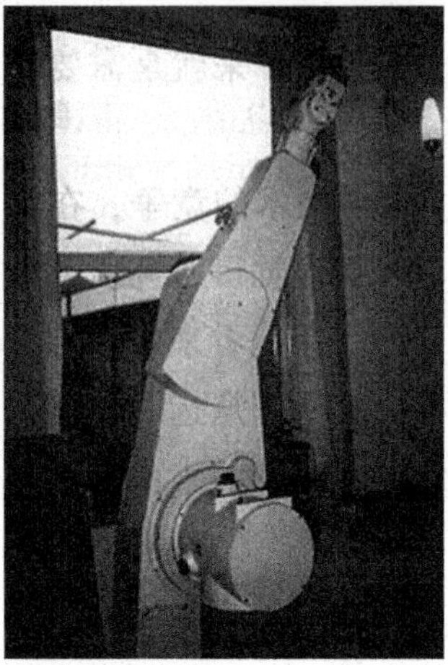

Fig. 6.2 SZT-1 playback robot manipulator [1]

The third achievement is the development of underwater vehicles. In this period, The Shengyang Institute of Automation of CAS led the development of two underwater vehicles, one was HR-01, the first underwater vehicle in China (see Fig. 6.3), and the other was the prototype of underwater vehicle Goldfish No. 1. The development of the HR-01 began in 1981 and was completed in 1985[3]. It is noteworthy that the HR-01's function and control system have been recognized as reaching a global standard, indicating China's capability to develop and utilize unmanned remote-controlled submersibles at the time.

In summary, during the initial phase of Chinese robot research and development, the role of the Chinese Academy of Sciences was primarily manifested in two aspects: firstly, conducting cutting-edge fundamental research and accumulating knowledge for second- and third-generation robot technology based on the introduction and assimilation of advanced foreign robotic products; The second achievement was the

Fig. 6.3 HR-01 underwater vehicle [1]

initiation of research and development in underwater robotics, which proved to be a highly strategic move due to its emphasis on independent research from the outset. This direction yielded achievements that reached advanced levels globally at that time. Since then, this research direction has emerged as a pivotal area of focus within China's growing robot industry.

6.2 The Growth Period (1986–2005)

Based on accumulation in early stage, Chinese robot cause experienced its growth age until 2005. Since 1986 China launched a series of major science and technology program successively, such as Key Technologies R&D Program, 863 program, and Scientific and Technological strategy of national strengthening. Chinese robot cause headed down to an organized development path under national plans, with the industrialization process also undergoing initial stages of growth. Among the national scientific and technological programs, the Key Technologies R&D Program during the Seventh Five-Year Plan (1986–1990), and the 863 Programs during the Eighth, Ninth, and Tenth Five-Year Plans (1991–2005) were of paramount importance. Through the successfully development of underwater vehicle in these programs, CAS has gradually emerged as a leader in the field of Chinese robotics.

In the Seventh Five-Year Key Technologies R&D Program, item 75 focused on "robot development" research, specifically targeting industrial robots as the primary field of Chinese robot technology research sponsored by the Ministry of Machinery Industry, among which CAS was mainly responsible for one project—the development of underwater robots [2]. However, among all the 19 robot technological achievements made under the Seventh Five-Year Key Technologies R&D Program, all six of those relevant to underwater robots accomplished by CAS have reached an international advanced level, which are listed in Table 6.1 [4].

Table 6.1 Six achievements of underwater robots accomplished by CAS

Year of identification	Achievement	R&D organization
1990	Prototype of "Sea Crab" the six -legged undersea locomotive robot	Shenyang Institute of Automation, CAS
1990	300 m middle-sized underwater vehicle——RECON-IV-300-SIA-X	Shenyang Institute of Automation, CAS
1990	6 function underwater manipulator	Shenyang Institute of Automation, CAS
1991	M851 high resolution color acoustic image and sonar	The Institute of Acoustics, CAS
1991	Technical development of high strength seawater resistant polymer materials	Changchun Institute of Applied Chemistry, CAS
1989	Research on application of underwater optics and camera system	Xi'an Institute of Optics and Precision Mechanics, CAS

Fig. 6.4 300 m middle-sized underwater vehicle—RECON-IV-300-SIA-X [5]

300 m middle-sized underwater vehicle (see Fig. 6.4)—RECON-IV-300-SIA-X developed by SIA was the most representative among the six achievements above. It provided technical supports for works in extreme environments, such as the exploration and exploitation of submarine oil and gas, monitoring and maintenance of hydroelectric dams, port construction and underwater salvage. These contributions were crucial to the development of marine energy, coastal defense construction and the development of underwater vehicle itself. This accomplishment has also yielded certain economic benefits, including $370,000 in rent and $600,000 from the sale of two sets of equipment to the United States.

Following the outstanding performance in the Seventh Five-Year Key Technologies R&D Program, the 863 program has officially become a leading force in China's robotics industry, with CAS at the forefront of this movement. CAS has established a research framework for robot technology, with SIA under the leadership of Jiang Xinsong as the main leading unit, and special robots as the primary focus, while industrial robots serve as a subsidiary field. During this period, two technological achievements had a significant impact on the development of robotics in China.

Firstly, CAS successfully developed EXPLORER underwater vehicle (see Fig. 6.5) and CR-01 underwater vehicle (see Fig. 6.6). In 1992, The Shenyang Institute of Automation collaborated with the Institute of Oceanology, Russian Academy of Sciences, with responsibility for the robot control system falling to the Chinese team and structural design being handled by their Russian counterparts. The mid-term objective of the collaboration was to develop an autonomous underwater vehicle named EXPLORER, with a depth capacity of 1000 m. This goal was achieved in 1994, marking a significant transition from wired to wireless technology. In 1995, China achieved a significant milestone with the completion of the CR-01 autonomous underwater vehicle, capable of reaching depths up to 6000 m. This achievement propelled Chinese underwater technology to the forefront of global advancements and marked a substantial leap from its previous capabilities limited to depths of only 300 m. This implies that China possesses the capability to catch up with and even

Fig. 6.5 EXPLORER underwater vehicle [1]

Fig. 6.6 CR-01 underwater vehicle [6]

surpass foreign advanced technology in a specific field, thereby making underwater vehicles an unwavering focus of Chinese robotics development.

Secondly, CAS successfully developed AGV (Automated Guided Vehicle) (see Fig. 6.7). During this period, CAS was responsible for two projects: the development of practical technology for stereoscopic warehouse AGV and development of AGV series products. The omni-directional AGV prototype, developed in the former project, was integrated with relevant technology for stereoscopic warehouse logistics systems, 3 sets of which were provided to Shanghai cigarette plant cutting workshop stereoscopic warehouse. 30 sets of AGV product in 4 types, developed by the latter project, has been provided to 8–10 production line. Simultaneously, a proprietary technology transfer agreement was executed with Samsung Corporation of Korea, marking China's inaugural robot technology transfer to another nation. Furthermore, during that time, China's robot market at risk of being completely dominated by foreign countries, it was in the AGV field where Jiang Xinsong found an opportunity to establish a foothold for China's self-developed products and system design. This is China's initial attempt to industrialize robot. In 2000, Shenyang Siasun Robot

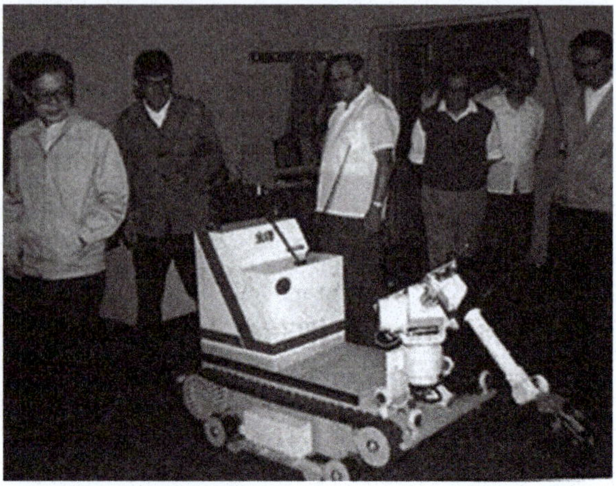

Fig. 6.7 Mobile robot [1]

Automation Co., Ltd. was officially established and has since maintained its position as the leading company in China's robotics industry.

6.3 The Rapid Development Period (2006–2020)

In 2006, China issued the Outline of the National Medium and Long Term Scientific and Technological Development Plan (2006–2020). Since then until 2020, China's robotics entered a rapid development period. The main feature of this period is that robot technology research and development has achieved full coverage of industrial robots, service robots and special robots, and the independent innovation of robot technology has made remarkable progress. During this period, the Chinese Academy of Sciences continued to leverage its advantages in the field of underwater robots and achieved three breakthrough achievements.

Firstly, "Jiaolong" manned underwater robot was a landmark achievement of this period. (see Fig. 6.8). The development of the "Jiaolong" was led by the Office of China Ocean Mineral Resources Exploration and Development Association, with over 100 scientific research units and enterprises including the CAS, China Shipbuilding Industry Group Co. (CSIS), Ministry of Land and Resources, and State Oceanic Administration collaborating on its completion. In 2009, "Jiaolong" manned underwater robot completed the 1,000-m trial, with a total of 20 dives and a maximum depth of 1,109 m, making China the fifth country with 1,000-m manned deep-diving capability after the United States, Russia, Japan and France, which is a milestone in the history of China's robot development [7]. In 2010, the "Jiaolong" underwater robot, which was the first independently designed and independently integrated manned submersible in China, completed the 3,000-m sea trial, with the deepest diving depth of 3,759 m, and the longest bottoming time of 72 min, which is represent a new milestone in China's robotics development.

Fig. 6.8 The Jiaolong deep-sea manned submersible [8]

On July 26, 2011, the "Jiaolong" deep-sea manned submersible completed a 5,000-m dive trial, with a total of five dives, eight divers, and a maximum dive depth of 5,188 m. At 11:47 p.m. on June 27, 2012, in the Mariana Trench of the Western Pacific Ocean, the "Jiaolong" deep-sea manned submersible set a world record for diving 7,062 m, opening the door for the Chinese to enter the deep-sea world. In 2013, the "Jiaolong" deep-sea manned submersible passed the acceptance inspection, marking that China has systematically mastered the design, construction and trialling technology of large-depth manned deep submarines, and has become one of the advanced countries in manned deep diving.

The "Shenhaiyongshi" (see Fig. 6.9) and the "Fendouzhe" (see Fig. 6.10) manned underwater robot represent two significant accomplishments in this field. The development of "Shenhaiyongshi" was spearheaded by CSIC and collaboratively executed by ninety-four domestic entities. In 2018, the "Shenhaiyongshi" achieved a series of domestic "firsts" in South China Sea scientific exploration, including continuous diving, night diving and multi-submersible joint operations. It also pioneered new applications such as undersea salvage and rescue and deep-sea archaeology, resulting in significant breakthroughs. In 2016, China launched the "Fendouzhe" project to develop a manned submersible capable of reaching depths of up to 10,000 m, with significant contributions from the "Jiaolong" and "Shenhaiyongshi" teams. In 2020, the "Fendouzhe" achieved a successful dive to a depth of 10,058 m in the Marianas Trench, setting a record for manned deep diving in China. The submersible was officially delivered in Sanya in 2021. The deployment of "Fendouzhe" has significantly bolstered China's capacity for full-sea depth operations, attesting to the country's comprehensive prowess in the realm of advanced marine technology.

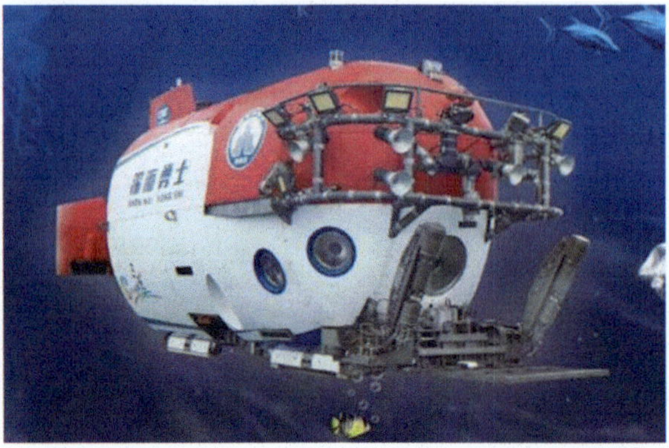

Fig. 6.9 "Shenhaiyongshi" deep-sea manned submersible [9]

Fig. 6.10 "Fendouzhe" deep-sea manned submersible [10]

6.4 Conclusion

In summary, the Chinese Academy of Sciences plays a pivotal role in China's robotics industry through three key aspects.

Firstly, as China's academic hub in the 1960s and 1970s, the Chinese Academy of Sciences possessed a robust foundation in automation, which paved the way for the research and development of China's robotics technology through the efforts of Shenyang Automation Research Institute after the 1970s.

Secondly, through the guidance of national science and technology plans, Jiang Xinsong played a pivotal role as a strategic scientist by strategically selecting underwater robotics as the primary research direction, thereby propelling the Chinese Academy of Sciences to ascend as one of the preeminent leaders in China's burgeoning robot industry.

Thirdly, CAS has been at the forefront of exploring the industrialization of robotics technology in China, exemplified by its establishment of Shenyang Siasun Robot Automation Co., Ltd., which remains the preeminent manufacturer of robots in China to this day.

After 2006, China began to comprehensively develop industrial robots, special robots, and service robots. As a result, the center of robotics technology research gradually shifted from scientific research institutes and universities to enterprises. Despite this shift towards industry involvement in robotics development, the Chinese Academy of Sciences still maintains a leading role in the field of special robots such as underwater robots.

Acknowledgements The author acknowledges thankfully the support of the Strategic Planning Bureau of CAS for the funding of the special project "History of Discipline Dvelopment in CAS".

References

1. Guangrong, X.: Biographies of academicians of Chinese academy of engineering: Biography of Jiang Xinsong. People's Publishing House, Aviation Industry Press, Beijing (2016). (In Chinese)
2. Robot modular architecture research group: Robot development strategy research report—Process, technology, industry, standards and policy. Ordnance Industry Press, Beijing. (In Chinese) (2009)
3. Xisheng, F., Yiping, L.: Thirty years of marine robotics. Sci. Bull. **58**(S2) (2013)
4. SPC: Selected introduction of national seventh five-year key technologies R&D program major achievements, vol. 2. Chemical Industry Press, Beijing (1993). (In Chinese)
5. Xinsong, J.: Unmanned underwater vehicles. Liaoning Science and Technology Press, Shenyang (2016). (In Chinese)
6. Dalong, T: Chinese robots towards the new century—retrospect and prospect of national 863 program intelligent robot. Liaoning Science and Technology Press, Shenyang (2001). (In Chinese)
7. MSTPRC: 2009 China science and technology development report. Science and Technology Academic Press, Beijing (2010). (In Chinese)
8. Qinan, X., Cong, Y., Shuai, W.: Development review and prospect of Jiaolong manned submersible in ocean exploration. Chin. J. Nonferrous Metals **31**(10), 2739 (2021)
9. Hong, Y.: Shenhaiyongshi manned submersible. Modern Physics **32**(01), 51 (2020). (In Chinese)
10. IACAS: Scientific research results: "Fendouzhe" full-sea depth manned submersible. Sci. Soc. **11**(01) (2021) (In Chinese)

Chapter 7
Education in Mechanism and Machine Science

Cristina Castejón and Eduard Krylov

Abstract The objective of this chapter is to highlight the value of educational activities in the mechanism and machine science (MMS) in engineering education. It includes a description of the general contents of MMS that every engineer should take in order to have a broad knowledge of this discipline. It also highlights the work of the teacher in the learning process inspired by the tasks carried out by Professor Lopez-Cajun, a teacher very close to his students who knew how to transmit the passion for this science. Finally, the activities to be carried out to attract the talent of young people to engineering careers are highlighted, as this is a great demand of society. In the chapter we also highlight the experiences of the Student International Olympiads in MMS, one of the activities promoted by the IFToMM.

Keywords Education in MMS · Engineering curricula · MMS contents · Olympiads

7.1 Introduction

MMS covers a discipline of vital importance for the engineer that includes the general study of the kinematics and dynamics of machines, as well as the knowledge of the mechanical elements that are part of it. In this line, Professor Carlos Lopez-Cajun was a reference in the transmission of knowledge of this discipline to his undergraduate and postgraduate students in the different universities in which he taught.

Mechanism and Machine Science (MMS) is a very important discipline in the engineering curricula and plays a crucial role in the future world. It is the base for the design, analysis, and optimization of different mechanical systems needed for the industries such as automotive, aerospace, robotics, and manufacturing. In all

C. Castejón (✉)
University Carlos III de Madrid, 28911 Leganes, Spain
e-mail: castejon@ing.uc3m.es

E. Krylov
Kalashnikov Izhevsk State Technical University, 649526 Izhevsk, Russia

M. Ceccarelli and J. C. Jauregui-Correa (eds.), *State-of-the-Art and Innovations in Mechanism and Machine Science*, Mechanisms and Machine Science 150,
https://doi.org/10.1007/978-3-031-47040-0_7

this cases, Mechanism and Machine Science provides the theoretical and practical knowledge necessary to understand, create, and innovate in these fields. All these issues are referred in the definition of MMS included by the IFToMM Permanent Commission for the Standardization of Terminology [4].

Among the competences required of an engineer are the ability to have and apply scientific and technological knowledge of mathematical, analytical, and numerical methods in engineering, as well as the ability to design, analyse and test machines. These competences enable them to project, calculate and design products, processes, installations, and plants, manage, plan, and supervise multidisciplinary teams, carry out research, development and innovation in products, processes, and methods, among others.

The MMS discipline is covered by different subjects inside the mechanical engineering curricula, which differ from university and country, but academically cover all three parts of an engineer's education: fundamentals of science, basic engineering, and technology for engineering [5]. In the first years, students train in fundamentals of mechanics, dynamics, and kinematics, that it provides with a solid understanding of fundamental concepts such as Newton's laws and the principles of work and energy. In the upper courses, students apply the basic concepts to delve into more specialized areas as mechanism design, robotics, or machine elements (gear systems, linkages, cam mechanisms).

Another important part of an engineer's training in the MMS field is laboratory practice and design projects. These practical exercises allow students to apply their theoretical knowledge to real-world situations, achieving skills such as critical thinking, problem solving and teamwork (Fig. 7.1).

The construction and testing of prototypes, with the help of new digital technologies, guarantee the comprehensive training of the engineer (see Fig. 7.2). The inclusion of digital technology in the last decades has changed the way of teaching (master class on blackboard) especially in the fields of engineering and, in particular, in MMS where the use of applications, videos and simulations helps students to understand some complex concepts faster.

7.2 MMS Contents

By way of summary, this chapter includes the contents covered by the MMS discipline, according to the classic books [6, 7], including the contributions of Professor Lopez-Cajun about mechanisms [2] and cams [3].

Fig. 7.1 Group of students working to participate in motostudent competition

Fig. 7.2 Books edited by springer with the results of the ISEMMS congress (Since 2013)

7.2.1 Introduction to Mechanism and Machine Science

The chapter covers an introduction to MMS, overview of basic principles and engineering applications. This are topics that introduce the student to the study of machines from all points of view.

The machine can be presented as an energy, power or motion transmission system. A Mechanism system can be presented as a whole, thus justifying the existence of a number of auxiliary systems, such as lubrication, braking, regulation, flywheels, etc., for its correct operation. Finally, the mechanisms are classified according to their type and application.

7.2.2 Engineering Mechanics Fundamentals

The contents of this subject are oriented to the knowledge of the kinematics and dynamics of the rigid body (Newton's laws of motion, forces and moments), to understand the fundamentals of balancing systems, as well as the concepts of work, power, energy and performance, with particular emphasis on friction and its effects.

Students acquire the knowledge related to the rotational velocity and acceleration as kinematic properties of a rotational rigid body and they learn how to describe the kinematic state of an object calculating the velocity and acceleration field related to the time. Moreover, students learn the concepts of sliding velocity and instantaneous axis of rotation and minimum slip, and they are able to identify and describe the particular motions of the rigid solid.

On the other hand, the fundamental laws of dynamics, the calculation of the inertia tensor and the angular momentum of a solid with respect to a point are studied.

The application of the fundamental concepts of rigid solid dynamics with general motion allows the study of particular solutions such as rigid solid motion with a fixed point, gyroscopic motion and torque-free motion. The application allows the study of support reactions, static and dynamic equilibrium conditions, as well as shaft balancing procedures.

Finally, from the study of passive resistances due to relative movements between two bodies, wear and friction are studied, as well as the concepts of work, power, and energy.

In this chapter it is very important that the student is familiar with the use of vector mathematics, trigonometry, and graphical resolution methods.

7.2.3 Kinematics of Mechanisms

The study of the kinematics of articulated mechanisms is addressed, presenting the analytical and graphical methods of analysis. Kinematics is the study of motion without considering the causes that produce it. Therefore, its study focuses on the calculation of the positions, velocities and accelerations of the points belonging to the mechanism.

Prior to the kinematic study it is necessary to know the concept of mobility, i.e. the number of parameters that define the position of a kinematic chain or a mechanism (number of degrees of freedom, DOF). The systems are classified according to the number of DOF by applying the Grübler-Kutzbach formula and/or the constraint criterion, considering the limitations of both formulas. Within this analysis, the concept of desmodromy is presented. Finally, a first approach to the numerical synthesis will be made by applying the Grübler's criterion to obtain two-, three-, and four-bar mechanisms with kinematic pairs of one degree of freedom.

The topological study is the first step towards the analysis and synthesis of the mechanisms.

7.2.4 Dynamics of Machinery

This section deals with the study of the dynamics of planar mechanisms, presenting the analytical and graphical methods of analysis. In this part, the problem of inverse dynamics is first presented to students, which consists of knowing the driven and resistant forces to be considered in the mechanism, knowing the rest of the actions and the kinematics of the mechanism.

On the other hand, applications to the dynamic study of one DOF mechanisms and their systems are presented. Applications to shaft balancing (balancing of rotating and reciprocating masses), governors and flywheels are studied inside this subject.

7.2.5 Machine Elements

This chapter covers the study of the so-called machine elements, transmission systems other than linkages. This group includes cam mechanisms, gears and gear trains, belt and chain drives as transmission systems. On the other hand, auxiliary elements as brakes, clutches, and couplings are also important to study.

7.2.6 Advanced MMS

In addition to these classic themes, found in the traditional literature [6, 7], new applications, needs and the advancement of technology open up a much wider field for the study of MMS in different fields. Thus, it becomes interesting for the engineer to continue applying the fundamental concepts of MMS in fields such as:

- Robotics and Automation (robot design, control and programming, Industrial Automation Systems)
- Micro/Nano Mechanisms and Systems
- Biomechanics and Human–Machine Interaction
- Computer-aided design and simulation (for modelling, analysis and virtual/augmented reality).

From the general concepts, the great motivation of the subject is the number of applications and possibilities opens for an engineer. In this sense Prof. Lopez-Cajun worked from the teaching, researching and professional point of view with their transfer to industry and society. Applications in the design of robots/mechatronics [10], cam transmissions [3], rotodynamic, among others [8, 9]. All these topics are covered by the IFToMM's Technical and Permanent commissions (TC/PC).

7.3 MMS Teaching

Today's society needs many engineers to address this paradigm of digitization in which we are immersed. In that sense, the importance of transmitting the MMS concepts from the first years of life must be considered to train the future engineers in the different disciplines (telecommunications, data science, mechanical, electronic, energy, materials, etc.). This makes it necessary for MMS education to make an effort to attract as much talent as possible to these fields, which in recent years the interest of students have been lost. So that the transmission of knowledge from schools to universities should be gradual and seeking to combine such knowledge with the interests and culture of young people. Hence, activities such as international competitions have a fast visibility in the world of social networks. As well as certain machines (motorcycles, cars, robots) that attract the attention of young people and encourage them to undertake studies that allow them to design, understand, use and develop new technology for the improvement of society.

On the other hand, and as part of MMS teaching, the need to transmit and apply the knowledge of physics and mathematics science to specific machine applications is a priority. This knowledge is very well stated and detailed in the teaching publications used as textbooks in different universities (see as classical reference [6, 7]).

In the MMS teaching activity, Professor Lopez-Cajun stands out for his generosity, an aspect of great importance that the PC of Education defends through the ISEMMS congresses that are organized every 4 years, where teachers can share and pool their experiences and methodologies that allow them to transmit the theoretical and practical concepts of the MMS (see Fig. 7.3) [11–13]. It is in these environments where the dialogue and the knowledge and use of different tools, more and more oriented to digitalization, can be used, transformed and improved by the whole IFToMM community. Some of the advances to highlight are:

- Applications of digital tools such as Geogebra or others.
- Activities carried out in different universities, related to collaborative work, such as the international competitions SIOMMS (Student International Olympiad on MMS), Moto Student (International competition where student teams apply all the knowledge in a real industrial project, by designing, developing and manufacturing a real racing motorbike prototype), Formula Student (international engineering race car design competition) among other. These activities allow

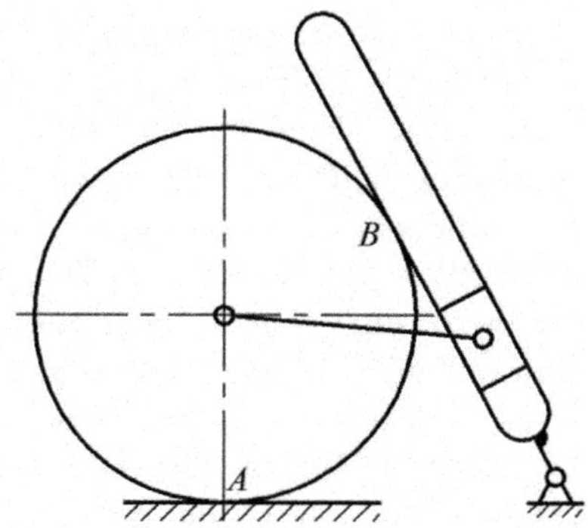

Fig. 7.3 Planar mechanism for calculating mobility

students and teachers to share very positive experiences from a training and internationalization point of view.

– Finally, an activity of great interest that Prof. Lopez-Cajun maintained not only from the point of view of research but also teaching is the dissemination about the history of machines. This subject allows the engineers to understand the why of some elements and machines, being able to advance towards the future with the basis of the past and the present [14, 15].

7.4 Talent Attraction to MMS and New Prospectives

The priorities of today's youth have changed with respect to previous generations, being aware that they must care for, protect and heal the planet, and in that sense, sustainable development objectives must be present in a transversal way in the collective mind of society. The engineering profession has always been oriented to the service of society, trying to develop systems that make our lives easier.

So the best way to attract talent and the best students to engineering careers and in particular to the MMS is to focus on the strong social impact that an engineer is able to contribute, from the development of sustainable transport systems, mechanisms that make life easier for disabled people, low-cost devices for use in undeveloped territories, in addition to the mechanical devices and systems used in the clinical sector that facilitate the work of healthcare workers.

MMS is a science rather difficult to study, because it contains large amount of knowledge related to many types of mechanisms and machine parts. Students often find it difficult to get through a large number of drawings and formulas. Therefore, for students to successfully master this discipline, three conditions must be satisfied:

- *motivation* towards studying;
- *endurance* and the ability to perform a large amount of computing and graphic work;
- have the opportunity to develop *creative and critical thinking* when solving non-standard problems.

MMS competitions, called Student Olympiads contribute to all free points. The *motivation* is ensured, firstly, by the competitive atmosphere, and, secondly, that students are given the opportunity to prove themselves as an analyst, designer, computing engineer [16]. During the entire period of preparation for the Olympiad, the participants have to solve a large number of problems of varying complexity and study textbooks in depth. This provides experience and promotes *endurance.* In the twenty-first century, engineers face the challenges of complexity, uncertainty and ambiguity of post-industrial technology, therefore well-developed cognitive skills; critical thinking and creativity become no less important than professional knowledge [17]. This should give engineering education a new focus on student creativity and innovation. As a rule, contest problems are non-standard, in order to solve them, one need to do critical data analysis and think creatively. Therefore, the Olympiads contribute to the development of *critical intelligence* and *creative potential* of students.

Themes of the contest problems follow main topics of MMS study course. They are: structural analysis & structural synthesis; kinematical analysis (mainly of planar mechanisms); kinetostatics for finding reactions in supports; analysis of dynamics, including fluctuation of rotational speed and determining moment of inertia of a flywheel; gearing (kinematics, geometry and efficiency); synthesis, kinematics and dynamics of cam mechanisms; balancing. These topics are not necessarily included in every Olympiad, but students are prepared for them all.

Kalashnikov Izhevsk State Technical University holds regional Student Olympiad every year. In order to give an idea of the themes and the level of complexity of contest problems, we present here a full set of problems that were proposed at the recent Olympiad in the spring semester, 2023.

Problem 1. Determine the mobility and the number of idle degrees of freedom (local motilities) associated with the mechanism (see Fig. 7.4). Show the equations used and identify any assumptions made when determining your answers. Assume cam contact at points *A* and *B*.

Fig. 7.4 Planar mechanism for structural analysis

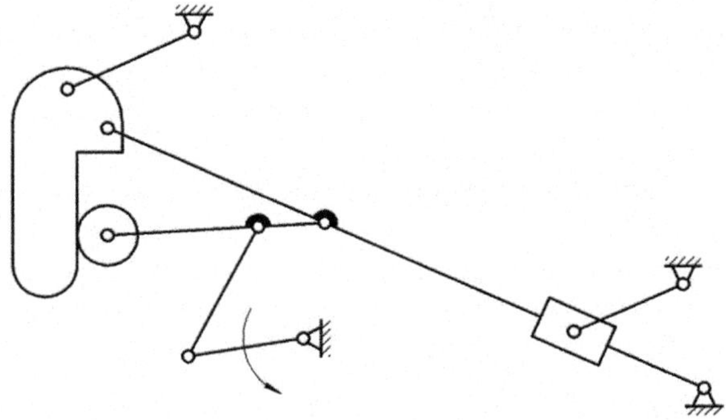

Problem 2. Make structural analysis of the planar mechanism shown in Fig. 7.5, divide it in Assur's groups, and write down a structural formula. Take the link marked with a curved arrow as an initial one.

Problem 3. The six-bar linkage is shown in an arbitrary position in Fig. 7.6. The slider is the output link. The sizes are given: $l_{AB} = 50\ mm$; $a = 120\ mm$; $b = 80\ mm$; $BC = 3.8AB$; $CD = 2AB$; $CE = 4AB$.

Questions:

(a) determine geometrically or analytically the crank angles corresponding to the working and idle strokes of the mechanism;

(b) find time the ratio Q;

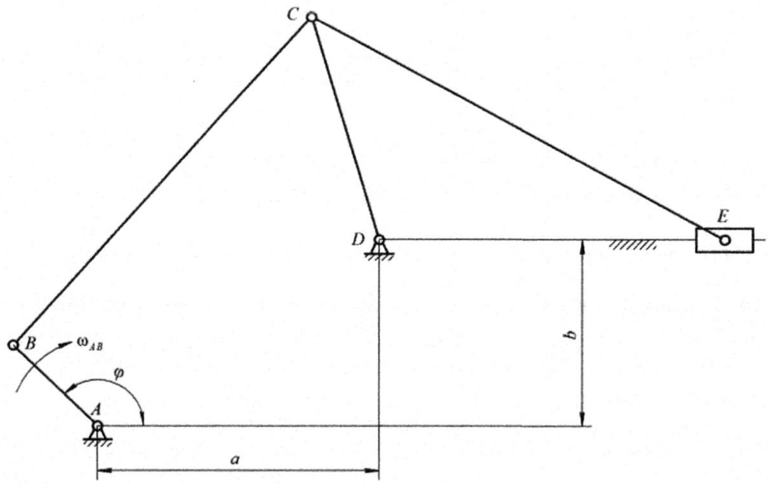

Fig. 7.5 Six-bar linkage for the structural synthesis

Fig. 7.6 The mechanism for kinematical analysis

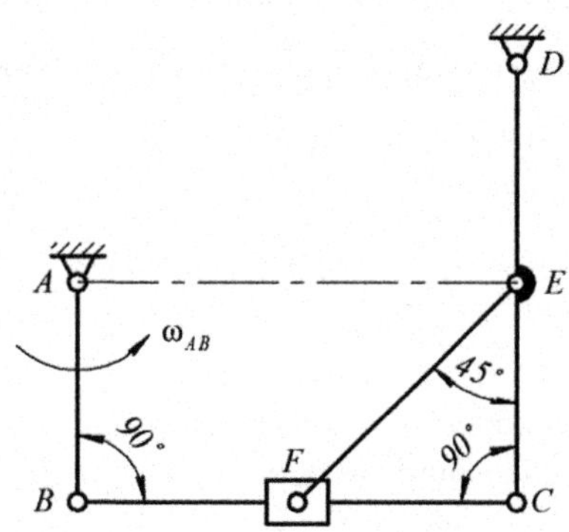

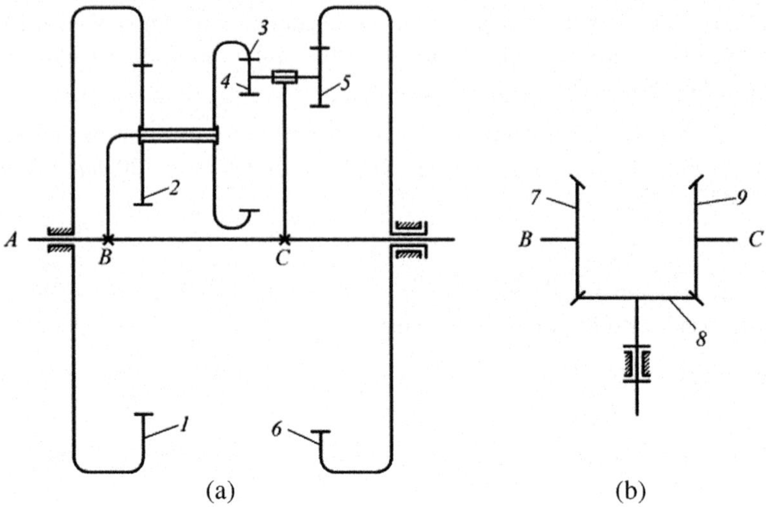

Fig. 7.7 The epicyclic gear train

(c) determine the maximum acceleration of the crank if link AB rotates with $\omega_{AB} = 2\ rad/s$.

For graphical solution use worksheet with the mechanism on a scale.

Problem 4. Find the angular velocity and angular acceleration of link EF at the position illustrated in Fig. 7.7, if $\omega_{AB} = 10$ rad/s = const, and lengths are $AB = \frac{1}{2}BC = \frac{1}{2}CD = 50$ mm. The problem can be solved either analytically or graphically. For graphical solution use worksheet with the mechanism on a scale.

Problem 5. For the epicyclic gear train in Fig. 7.8, *a* the numbers of teeth are $N_1 = 100$; $N_2 = 40$; $N_3 = 60$; $N_4 = 15$; $N_5 = 25$; $N_6 = 130$. Shaft A makes $n_A = 100\ rpm$.

Questions:

(a) what is the mobility m of the mechanism?

(b) what are angular velocities (expressed in rpm) of gears 4 and 6?

(c) how will the solution change if part of the shaft A between points B and C is removed?

(d) how will the solution change if the additional gear box consisting of 3 identical conical wheels ($N_7 = N_8 = N_9$, see Fig. 7.8, *b*) is introduced between points B and C?

Problem 6. The car wheel has a mass of 12 *kg*. During the wheel rotation on the balancing workbench it was found to be unbalanced (Fig. 7.9*a*). In order to balance it, two identical loads with the mass of $m_1 = m_2 = 60$ g were fixed to the wheel disc at a radius of 70 *mm*, as shown in Fig. 7.9b.

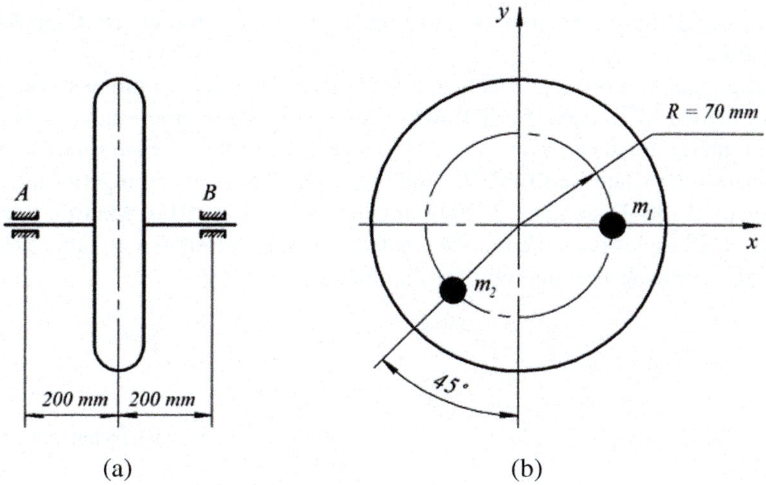

Fig. 7.8 The balancing of a wheel

Fig. 7.9 Poster presented in the IFToMM World Congress 2019 in Krakow related the first events SIOMMS

Find the maximum reaction forces R_A and R_B that acted in supports A and B ***before balancing***.

Related the promotion of Machine and Mechanism Science among young people all over the world, in 2009 the IFToMM decided to hold international competitions, Student International Olympiad on MMS (SIOMMS) [18, 19]. Five such Olympiads have been held to date (see Table 7.1 and Fig. 7.10). The recent competition was held on November, 30–December, 2, 2022 in University Carlos III of Madrid, Spain.

The MMS topics presented in the contest problems are shown in Table 7.2 for all SIOMMS (+ means that the topic was included).

Table 7.1 SIOMMS in progress

Title	Date, year	University, country
SIOMMS 2011	April, 19–21, 2011	Kalashnikov Izhevsk State Technical University (ISTU), Russia
SIOMMS 2013	October, 12–13, 2013	Shanghai Jiao Tong University, Shanghai, China
SIOMMS 2016	October, 20–21, 2016	University Carlos III of Madrid, Spain
SIOMMS 2018	October, 24–26, 2018	Pontifical Catholica University of Peru, Lima, Peru
SIOMMS 2022	November, 11–December, 1, 2022	University Carlos III of Madrid, Spain

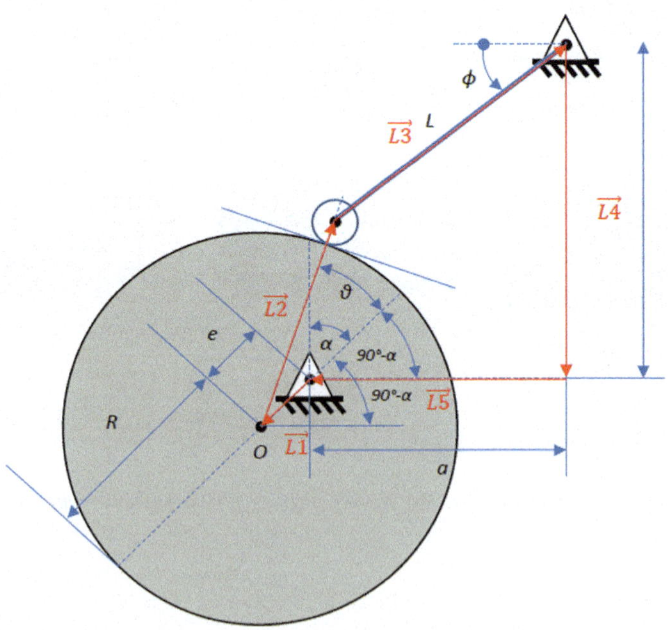

Fig. 7.10 Problem of dimensional synthesis of a cam mechanism

Table 7.2 Problem topics

Topic	SIOMMS 2011	SIOMMS 2013	SIOMMS 2016	SIOMMS 2018	SIOMMS 2021/2022
Structural analysis and dimensional synthesis of planar mechanisms	+	+	+	+	+
Kinematics of planar mechanisms	+	+	+	+	+
Cams	+	+	–	–	+
Geometry of toothed gearing	+	+	+	–	–
Gear trains	+	+	+	+	+
Balancing	+	+	–	+	–
Mechanism dynamics, differential equation of motion, fluctuation of energy, flywheel	+	+	+	+	+

The theory of cam mechanisms was in the focus of Professor Carlos Lopez-Cajun's researches [20–27]. We are therefore particularly pleased to note that the contest problem on the synthesis of cam mechanisms has been included in the recent SIOMMS 21/22. To pay tribute to his memory, we present here the complete text of this problem.

Problem B3 (SIOMMS 22). Given an eccentric with a radius $R = 35$ *mm* and with a roller rotation follower, as shown in the schematic in Fig. 7.10 (merely illustrative).

Knowing the location of the supports ($a = 45$ *mm*, $b = 60$ *mm*), we intend to design the cam for angle coordination with three precision points as follows:

$$\alpha 1 = 0^\circ \rightarrow \phi 1 = 40^\circ$$

$$\alpha 2 = 0^\circ \rightarrow \phi 2 = 20^\circ$$

$$\alpha 3 = 180^\circ \rightarrow \phi 3 = 0^\circ$$

It is asked:

a. determine the value of the eccentricity *e*, and the length *L* so that the angle coordination is as required. It is recommended to use the Freudenstein method and the angles indicated in the figure;
b. if the cam rotates clockwise with constant speed of $\omega = 0.5$ *rad/s*, plot the follower displacement diagram (φ) as a function of cam rotation angle (α); assume that the displacement diagram is symmetric and can be approximated by a harmonic function;

c. determine the value of the maximum velocity of the follower (in °/s), and specify, for which angular position(s) it occurs;
d. indicate, as a function of the angles in the figure, the value of the pressure angle.

We are very pleased that after a troublesome period of the coronavirus pandemic, SIOMMS resumed in 2022. We hope that the IFToMM will continue to support SIOMMS, because the best memory for outstanding scientists, including Professor Lopez-Cajun, is the involvement of young people in the scientific process.

7.5 Conclusions

In this chapter we have tried to collect the activities related to education in the field of MMS, as a tribute to Professor Lopez-Cajun. The importance of the basic sciences as formative for its application in engineering has been highlighted, with particular emphasis on machines and mechanisms, as well as the detail of the theoretical concepts covered by MMS, trying to be as general as possible and being aware of the wide range that allows the development of this science. Finally, the new trends and the need to promote vocations in the STEM field have been presented, with special mention of the MMS Student Olympiads (SIOMMS) as valuable activity that enriches engineering students from different points of view.

References

1. Ceccarelli, M., Correa, J.C.J.: Carlos López-Cajún (1948–2020). In: Ceccarelli, M., Gasparetto, A. (eds.), Distinguished figures in mechanism and machine science. history of mechanism and machine science, vol. 41. Springer, Cham (2023). https://doi.org/10.1007/978-3-031-18288-4_4
2. López-Cajún, C.S., Ceccarelli, M.: Mecanismos/Mechanism: Fundamentos Cinematicos Para El Diseno Y Optimización. Editorial Trillas Sa De Cv. ISBN 10: 9682481813. (2008)
3. Angeles, J., López-Cajún, C. S.: Optimization of cam mechanisms. Springer Dordrecht, (1991). https://doi.org/10.1007/978-94-011-3572-6
4. Segla, S.: IFToMM terminology commission and standardization of mechanism and machine science terminology. FME Transactions **42**(3), 177–180 (2014). https://doi.org/10.5937/fmet1403177s(2014)
5. Ceccarelli, M. Bragastini, R.: Historical accounts on the figure of engineers and academic mission for their formation In: García-Prada, J., Castejón, C. (eds.) New trends in educational activity in the field of mechanism and machine theory. Mechanisms and Machine Science, vol. 19. Springer, Cham (2014). https://doi.org/10.1007/978-3-319-01836-2_1
6. Norton, R. L.: Design of machinery. McGraw-Hill Education, (2019)
7. Shigley, J.E., Uicker, J.J.: Theory of machines and mechanisms. Oxford University Press (2020)
8. González-Cruz, C.A., Jáuregui-Correa, J.C., López-Cajún, C., Sen, M.: Domínguez-González A, Experimental analysis of synchronization and dynamics in an automobile as a complex system. Mech. Syst. Signal Process. **60**, 472–4841 (2015)
9. Jáuregui-Correa J.C., López Cajún C.S., Sen, M.: Analysis of experimental data from complex multibody system. Mech. Mach. Sci. **25**, 211–218. In: 2015 5th international symposium on multibody systems and mechatronics. MUSME, Huatulco 2014, Code 117419

10. Jáuregui-Correa, J.C., López-Cajún, C.S., García-Arredondo, A., Hernández-Martínez, E.E., Ceccarelli, M.: Validation process of pose accuracy estimation in parallel robots. J. Dyn. Syst., Meas. Control., Trans. ASME **137**(61) Article number 064503 (2015)
11. García-Prada, J.C., Castejón, C. (eds.).: New trends in educational activity in the field of mechanism and machine theory, part of the book series: mechanisms and machine science (Mechan. Machine Science, vol. 19), Springer, (2014)
12. García-Prada, J.C., Castejón, C. (eds.).: New trends in educational activity in the field of mechanism and machine theory (2014–2017), Part of the book series: Mechanisms and Machine Science (Mechan. Machine Science, vol. 64), Springer, (2019)
13. García-Prada J.C., Castejón C., Pedrero J.I. (eds.).: New trends in educational activity in the field of mechanism and machine theory (2018–2022), Part of the book series: Mechanisms and Machine Science (Mechan. Machine Science, vol. 128), Springer, (2023)
14. Egorova, O., Ceccarelli, M., Cuadrado Iglesias J.I., Lopez-Cajun C.S, Pavlov V.E.: Agustin Betancourt: An early modern scientist and engineer in TMM. In: ASME 2006 international design engineering technical conferences and computers and information in engineering conference. (2006). https://doi.org/10.1115/DETC2006-99198
15. López-Cajún, C., Ceccarelli, M. (eds.).: Explorations in the history of machines and mechanisms. In: Proceedings of the Fifth IFToMM symposium on the history of machines and mechanisms, part of the book series: history of mechanism and machine science (HMMS, vol. 32). (2016)
16. Balyakin, V.B., Krylov, E.G.: Cultural and educational significance of MMS competitions for future engineers. In: García-Prada, J.C., Castejón, C (eds.), New trends in educational activity in the field of mechanism and machine theory, MMS 64, pp. 38–48. Springer Nature Switzerland AG, (2019). https://doi.org/10.1007/978-3-030-00108-7_5
17. Krylov, E.G., Devyaterikov, S.A.: Developing students' cognitive skills in MMS classes. STEM Educ. **3**(1), 28–42. https://doi.org/10.3934/steme.2023003. EDN UUYIRQ. https://www.aimspress.com/article/doi/10.3934/steme.2023003
18. Krylov, E.G., Goldfarb, V.I., Serova, T.S.: IFToMM contribution to attraction of youth to MMS development and promotion. Mech. Mach. Theory **100**, 3–7 (2016)
19. Krylov, E., Gubert, A., Devyaterikov, S., Egorova, O.: SIOMMS: Evolution and development. Mech. Mach. Theory. **153**, 104029. https://doi.org/10.1016/j.mechmachtheory.2020.104029. EDN DCQYPO. (2020)
20. Angeles, J., López-Cajún, C.S.: The displacement program. In book: Optimization of cam mechanisms (pp.28–60). (1991). https://doi.org/10.1007/978-94-011-3572-6_2
21. Angeles, J., López-Cajún, C.S.: Kinematic analysis of cam mechanisms. (1991). https://doi.org/10.1007/978-94-011-3572-6_7
22. Angeles, J., López-Cajún, C.S.: Cam profile geometry. (1991). https://doi.org/10.1007/978-94-011-3572-6_3
23. Angeles, J., López-Cajún, C.S.: Optimization of planar cam mechanisms with roller-followers. (1991). https://doi.org/10.1007/978-94-011-3572-6_5
24. Angeles, J., López-Cajún, C.S.: Optimization of planar cam mechanisms with flat-face followers. (1991). https://doi.org/10.1007/978-94-011-3572-6_49
25. Angeles, J., López-Cajún, C.S.: The computer-aided drafting and manufacture of cams. (1991). https://doi.org/10.1007/978-94-011-3572-6_9
26. Carbone, G., Lanni, C., Ceccarelli, M., Lopez-Cajùn, C.: Dynamic effects of curvature change in the profile of two circular-arc cams. V Congreso Iberoamericano de Ingegneria Mecanica CIDIM2001, Mérida. pp. 1247–1252 (2001)
27. Lanni, C., Ceccarelli, M., Cajun, C.L.: An experimental validation of three circular-arc cams with offset followers. Mech. Based Des. Struct. Machines **34**, 261–276 (2006). https://doi.org/10.1080/15397730600830351

Chapter 8
Cam Mechanisms in the MMS Study Course

E. Krylov and N. Barmina

Abstract Cam mechanisms are widely used in engineering, so they are comprehensively analyzed in many researches, papers, and books. Cam-related topics are included in all MMS study courses and other courses devoted to the machinery. Cams are interesting not only from a purely scientific point of view, but also as a subject of study at the university within Machine and Mechanism Science (MMS) course. They are involved in many other MMS topics since they provide illustrative material for many ideas, concepts, and methods in Structural Analysis & Synthesis, Kinematics, and Dynamics. The paper shows how studying the topic of cam can positively affect the intellectual development of students. If this development is in the focus, then one should monitor the mental operations associated with active learning and other types of students' activities. Revealing these operations along with having feedback from students via surveys and examining their difficulties can make cam-related topic interesting and important for students. The paper also demonstrates how cams are presented in MMS Students Olympiads.

Keywords Cam · MMS · Teaching · Intellectual development · Problems · Olympiads

8.1 Introduction

The methods used in the machines and mechanisms science (MMS) are very close to those suggested by the mechanics of a solid body, since both areas focus on motions (kinematics) and forces associated with these motions (statics, kinetostatics, and dynamics). But MMS is an applied science, so all general methods and rules are implored to design some structure that should be able to perform the prescribed

E. Krylov (✉)
Kalashnikov Izhevsk State Technical University, Izhevsk, Russia
e-mail: 649526@mail.ru

N. Barmina
MIP Mechanic Ltd, Izhevsk, Russia

M. Ceccarelli and J. C. Jauregui-Correa (eds.), *State-of-the-Art and Innovations in Mechanism and Machine Science*, Mechanisms and Machine Science 150,
https://doi.org/10.1007/978-3-031-47040-0_8

motion and operate efficiently. The process of design of a new mechanism or machine is often very complex and influenced by such factors as the number of parts, shapes of bodies, geometry of surfaces, materials, limitations imposed by working conditions of the object of design, and many others. A standard MMS course suggests comprehensive analysis of three types of mechanisms—linkages, gears, and cam mechanisms.

Cam section is included in any complete MMS textbooks: Artobolevsky [1], Frolov et al. [2], Khurmi and Gupta [3], Uicker et al. [4] and others. Fundamentals of cam mechanisms design can be found in books and monographs by Rothbart and Klipp [5], Williams [6] and others. Over the last 30 years, a large number of articles can be found in the scientific literature on various aspects of design and production of cam mechanisms.

Here we have a pleasure to pay tribute to the **Professor Carlos Lopez-Cajun's** valuable contribution to the theory of cam mechanisms, namely:

- the novel approach for generating the displacement program of follower [7];
- thorough kinematic analysis of cam mechanisms that allowed to get the methods for the digitization of profile points on the base of Cartesian coordinates and their derivatives, obtained by a spline approximation of the cam profile [8];
- the optimum-design algorithms for cam profile geometry [9–11];
- the design of spatial cams [12];
- the methods of computer-aided drafting and manufacture of cams [13];
- experimental investigation of cam mechanisms dynamics, including prototyping and testing for experimental validation [14, 15].

With great number of publications devoted to various issues of research, production and operation of cam mechanisms, the matter of teaching of cams in universities is not so popular in literature. Some of researches are focused on teaching students techniques for synthesizing cam profiles. Shooter and West introduce two simple concepts of conjugate geometry to derive the governing vector equation for the cam profile [16], Kosenok et al. suggest to use the method of closed vector contours for the same purpose [17, 18]. Chen et al. claim that the cam workbench aims at sustaining the learned design process and visualizing learned effects, so they show the conception of a cam workbench for educational purposes [19]. Shiue recommends students to use Visual Basic application for designing the disk cam with a reciprocating follower to analyze the required motions of the follower and plot the cam profile, displacement, velocity, acceleration, and jerk diagrams. This application helps students in designing disk cams with the required motions and in visualizing the cam profile for the corresponding motion of the follower [20]. So, the articles on the teaching of cam mechanisms are focused mainly on three issues: visualization, computation, and testing.

In this manuscript we would like to raise the question of the effect of studying the topic of cams on the *intellectual development of students*. In the modern post-industrial society meta-subjective cognitive skills, critical thinking and creativity become no less important than professional knowledge, especially if the latter is interpreted narrowly [21]. Better conditions for the development of these skills can be found if students are involved in active learning activities while studying MMS in general and cam topic in particular. One possible means of doing so would be to monitor the mental operations associated with active learning and other types of activities.

8.2 Aspects of Active Learning in Section *Cams* in the MMS Study Course

There is no need to consider here the whole content of the section Cams in the MMS study course. Our purpose is to reveal something special that can contribute to the mental development and help students to master this topic. Table 8.1 shows how cams are included in typical MMS curriculum.

The investigation of cam mechanisms gives the excellent opportunity for application of variety of engineering methods. Besides, what is especially interesting about cams, they provide illustrative material for many MMS topics and concepts. Several aspects of active learning are listed below.

Analysis of a mechanism structure. The cam mechanisms appear at the very first MMS topics. Figure 8.1a shows a planar mechanism which has mobility $m = 1$, since the position of the output vertical rod is completely determined by the angle of the cam rotation. But the Chebyshev formula (Kutzbach criterion) gives incorrect result because the rotation of the roller does not affect the overall motion of the mechanism. For the sake of ongoing investigation of the higher kinematical pair the roller as a separate moving part can be removed in such a way as *to keep the circular shape*. Now the rocking follower has a circular shape at the place of contact with the cam,

Table 8.1 Cams in MMS topics

No	MMS topic	Knowledge and skills
1	Structural analysis	1. Idle degrees of freedom (local mobility) 2. Classification of kinematical pairs (higher pairs, j_2) 3. Equivalent linkages
2	Structural synthesis	1. Phases of a mechanism motion 2. Angle of pressure 3. Angle of motion transmission 4. Inversion of motion
3	Kinematics	1. Concept of relative motion 2. Differentiation of the follower displacement, first-order and second-order kinematic coefficients 3. Graphical differentiation and integration
4	Kinetostatics	1. Forces in kinematical pairs 2. Friction
5	Dynamics	3. Impacts and the choice of the follower motion 2. Reduced moments and forces, dynamical model of a cam mechanism

see Fig. 8.1b. The contact between surfaces provides both rotation and sliding; hence it is classified as a two-degree-of-freedom kinematical pair j_2. The example given by this kinematical pair is so typical that it gives a generic name of ***cam contact*** for some conjugation of surfaces, see Fig. 8.2 [22].

If a mechanism containing cam-follower elements should be analyzed as a combination of Assur structural groups, it is necessary to replace the higher pair with

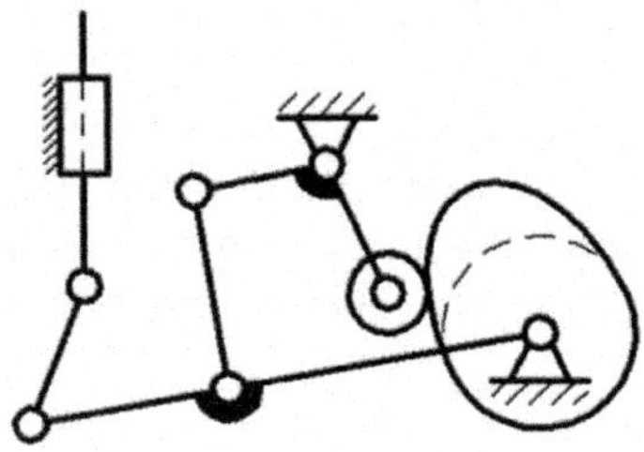

(a) roller gives idle DOF

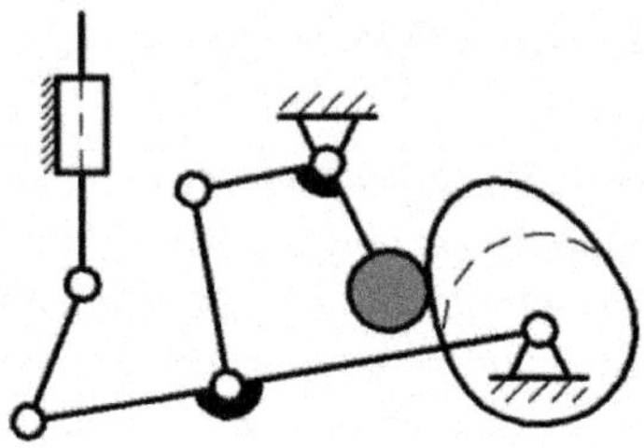

(b) roller is removed, rocking follower and cam make the higher pair

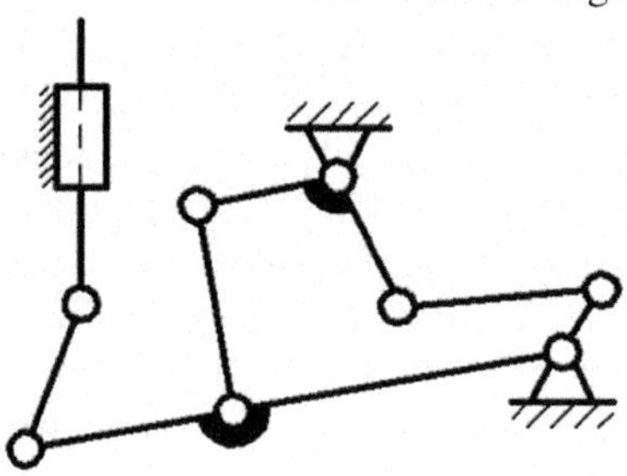

(c) equivalent linkage

Fig. 8.1 Cam illustrates basic concepts of structural analysis **a** roller gives idle DOF **b** roller is removed, rocking follower and cam make the higher pair **c** equivalent linkage

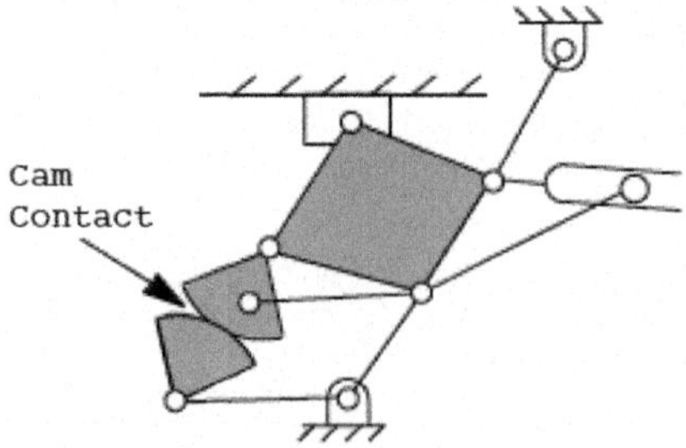

(a) cam contact

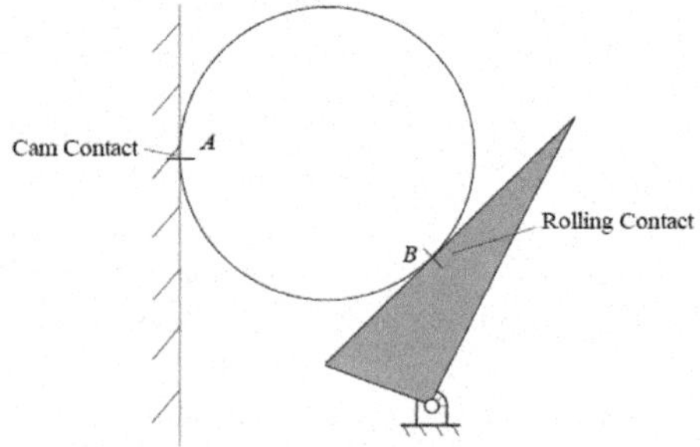

(b) cam contact vs. rolling contact

Fig. 8.2 Cam contact in the planar mechanism **a** cam contact **b** cam contact vs. rolling contact

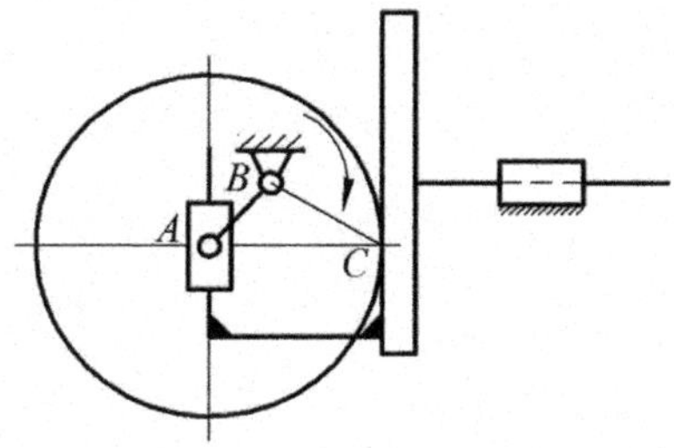

Fig. 8.3 Equivalent linkage for the cam mechanism with the flat follower

the equivalent linkage. Figure 8.1c shows an example where both contact surfaces are convex, the equivalent linkage for the cam mechanism with a flat reciprocating follower is shown in Fig. 8.3 [22]. Equivalent linkages find also an application for kinematical analysis of cam mechanisms, for example they can provide reference data for validation of experimental researches [15].

Easy to explain. A concept of angle of pressure is common for linkages, gears, and cam mechanisms. This angle plays an important role in design, since it characterizes the force transmission between bodies and should be taken into account to grant smooth operation without jam and collapse of links. Cam mechanism suggest simple and evident idea of the pressure angle as the angle between the common normal at the cam -follower contact point and the velocity of this point that belongs to the follower [23], as shown in Fig. 8.4, where the pressure angle is denoted by α. From empirical considerations the pressure angle is recommended to lie within 30°. The transmission angle γ completes the pressure angle up to 90°, $\alpha + \gamma = 90°$.

Generalization. Cam mechanisms along with gears include the higher kinematical pairs, which have a great importance for the operation and performance of these mechanisms. Typically, students consider the theory of a higher kinematic as a topic only related to gears. Indeed, the concept of the instant center of relative rotation allows deriving the definition of the pitch point and gear ratio, the most important notions in the theory of in-plane toothed gearing. However the analysis of the relative motion of the points belonging to the conjugate profiles is also valid for the cam-follower contact. Figure 8.5 [24] demonstrates the technique of finding the instant center P for two rotating profiles (cam-rocking follower mechanism).

For the case of reciprocating follower the instant center P is considered as a point in the cam the velocity vector of which is equal to the velocity of follower. The location of the instant center is based on the consideration of a velocity diagram, as for the mechanism with the rocking follower; see Fig. 8.6.

Bringing Ideas Together, Comparison. One of the most important ideas in the topic of structural synthesis is *how the shape of part and geometry of surface are related to the motion of the output link*. Again, the cam mechanism gives a good example. The geometry of the cam as a separate part can be described by profile angles β (rise, dwell, return) between lines that separate segments of different curvature. In Fig. 8.7 the cam mechanism with the offset knife-edge reciprocating follower is shown at the

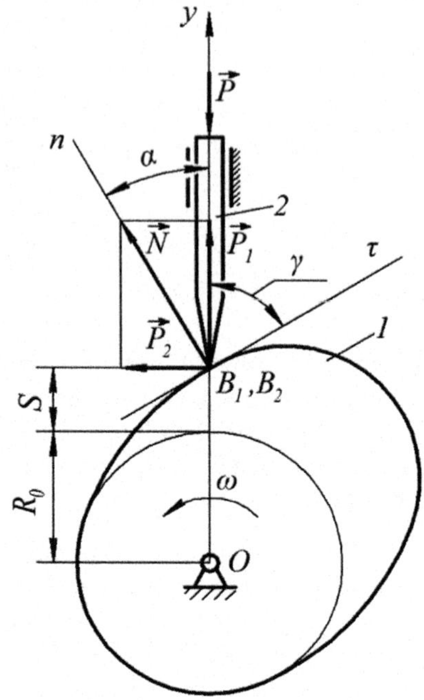

Fig. 8.4 Angle of pressure α and transmission angle γ

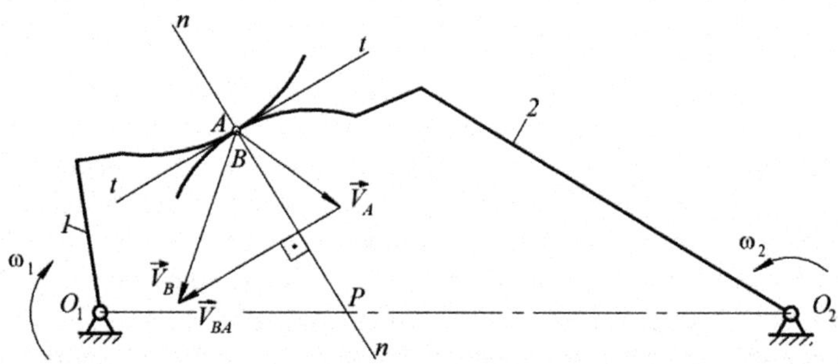

Fig. 8.5 Velocity diagram and location of pitch point for cam-rocking follower mechanism

beginning of rise. It can be seen that when the cam turns through the profile angle of rise β_r, the follower will not reach the upper extreme position. In order that the follower could take this position, the cam should be rotated further for an extra angle β_e.

So, the phase angle of rise is combined as $\varphi_r = \beta_r + \beta_e$.

The relation between geometry and design is quite normal and typical in MMS, but the cam topics gives one of the most evident and clear examples of the profile design based on kinematics.

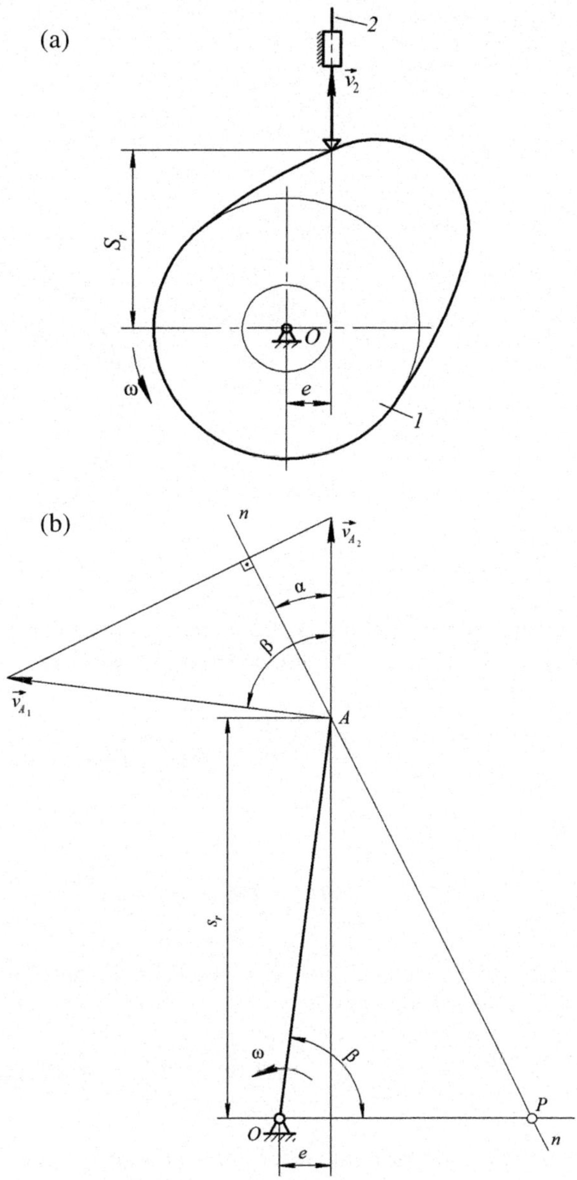

Fig. 8.6 Velocity diagram and location of the instant center for the cam mechanism with a reciprocating follower

Figure 8.8 illustrates the problem of layout of a cam in a mechanism with flat-face follower (Fig. 8.8 a). The follower rises with a constant speed (Fig. 8.8b), its maximum rise is $S_{max} = H = 40mm$; the angle of rise is $\varphi = 2\pi rad$; the cam rotates with a constant speed. It is required to layout a profile of the cam provided the prime radius is $r_{Omin} = 40mm$. For an arbitrary position of the mechanism the velocity diagram is presented by vector triangle Bb_2b_1 (Fig. 8.8c)—that is *kinematics*.

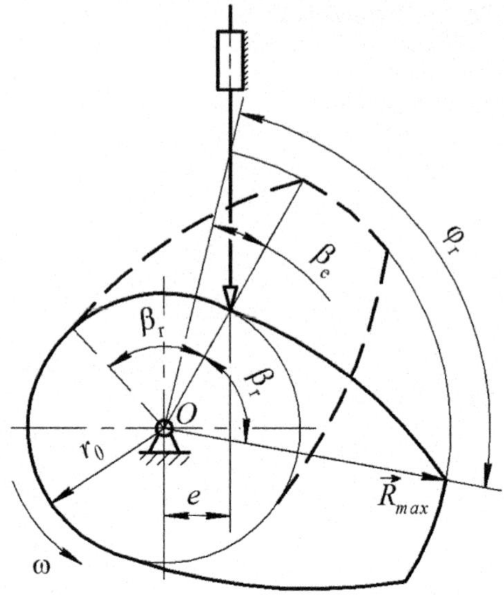

Fig. 8.7 Profile and phase angles for the offset knife-edge follower

Triangle *ObB* (nothing with kinematics, pure *geometry*) is similar to the triangle Bb_2b_1. From similarity of these triangles we have:

$$\frac{\mathrm{b_2b_1}}{\mathrm{Bb}} = \frac{\mathrm{Bb_2}}{\mathrm{Ob}} = \frac{\mathrm{Bb_1}}{\mathrm{OB}} = \frac{\mathrm{V}_{B_1}}{\mathrm{r}} = \omega$$

when

$$BC = Ob = \frac{Bb_2}{\omega} = \frac{V_{B_2}}{\omega} = \frac{dS_2}{d\phi} = S_2'$$

where S'_2 is the analog of velocity. With the given motion of the follower $S = a\phi$, this analog is found as.

$$S_2' = a = \frac{H}{\phi} = \frac{40}{\pi} = 12.73\ mm = const$$

Hence, the position vector in an arbitrary point is given by

$$r_i = \sqrt{(S_i + r_{O\,\min})^2 + (S_2')^2}$$

On the base of this formula a graphical image of the cam profile can be easily obtained by the method of inversion of motion (Fig. 8.9).

The set of steps during analyzing the similarities between a kinematical triangle and geometrical triangle is not trivial but quite popular in MMS. So, this example

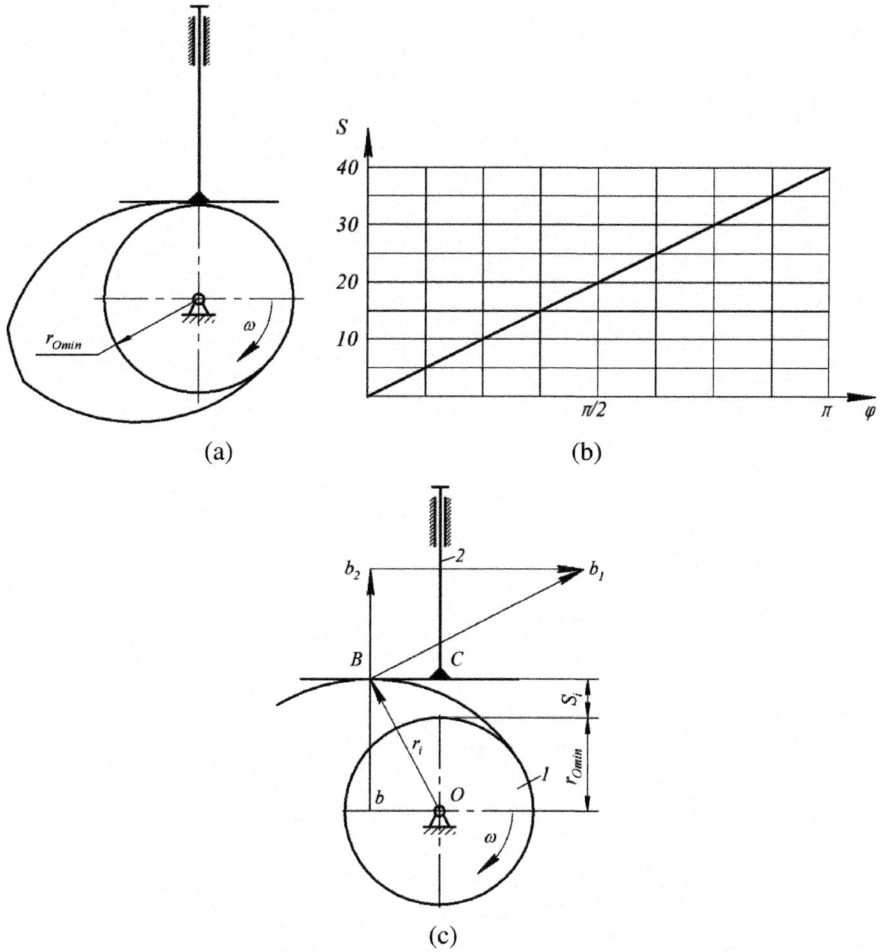

Fig. 8.8 Design of the cam profile in a mechanism with a flat-face follower

shows students the importance of analogy and how *bringing things together* can provide the key to solution.

8.3 Cams by Students' Eyes

It is no secret that teachers and students think of and feel the same educational material differently. We tried to find out which topics were the most interesting for students (Fig. 8.10) and which were the most difficult for them (Fig. 8.11). Fourteen participants of the regional Olympiad on MMS, which took place in May, 2023 in Izhevsk, Russia, were interviewed for this purpose.

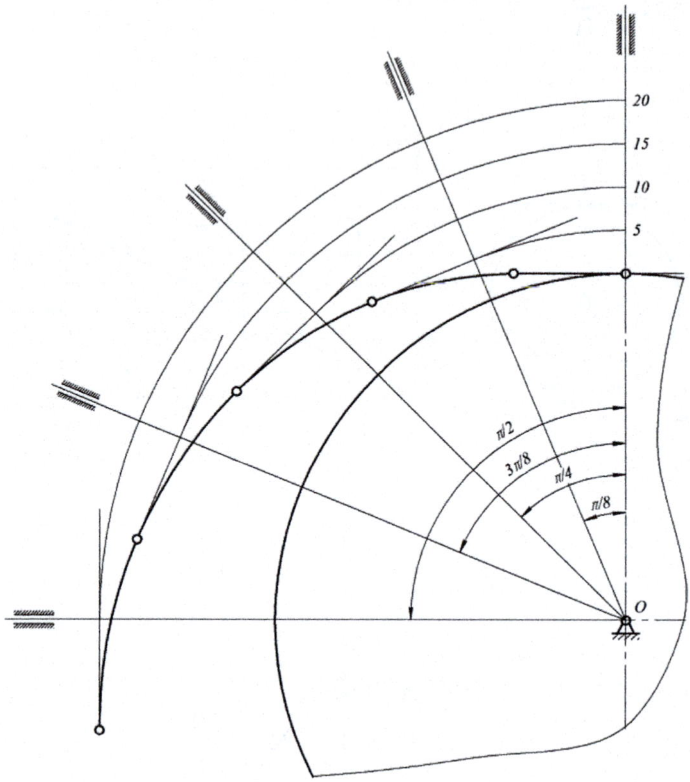

Fig. 8.9 Geometry of the cam profile by the inversion of motion

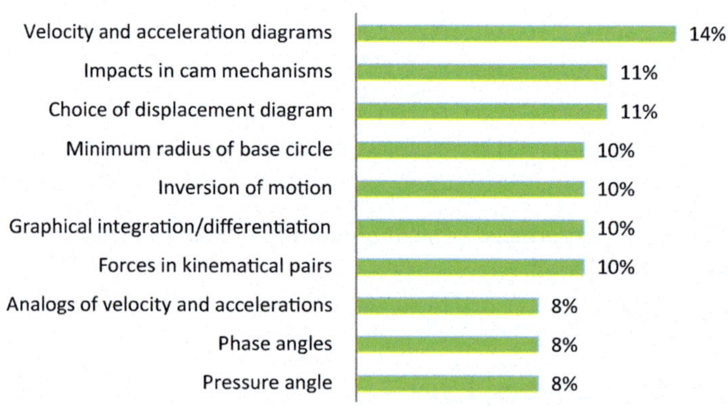

Fig. 8.10 Which cam-related topic did you find most interesting?

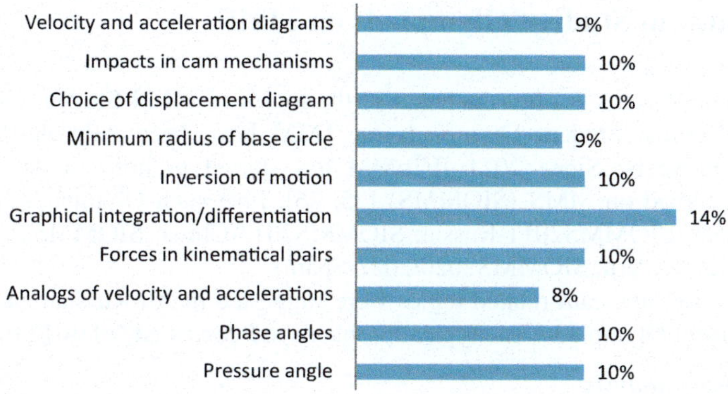

Fig. 8.11 Which cam-related topic did you find most difficult?

The survey showed that the topics, which are not of much interest to students, felt by them subjectively as more difficult. Hence we see the importance of involvement of students in studying on the base of clear explanations, accompanied by examples, demonstrations, and visualization (as it was mentioned in the Introduction). This facilitates active learning.

While Fig. 8.11 illustrates the difficulties encountered, Fig. 8.12 explains why, from the students' point of view, it was difficult for them to master the teacher's material.

It seems that students are ready to comprehend, but the problem of big data, usual for modern educational systems, makes the learning process quite difficult. Therefore educators are advised not to overload the teaching process with pictures and formulas, and try to stimulate mental activity on the base of techniques mentioned in Sect. 8.2.

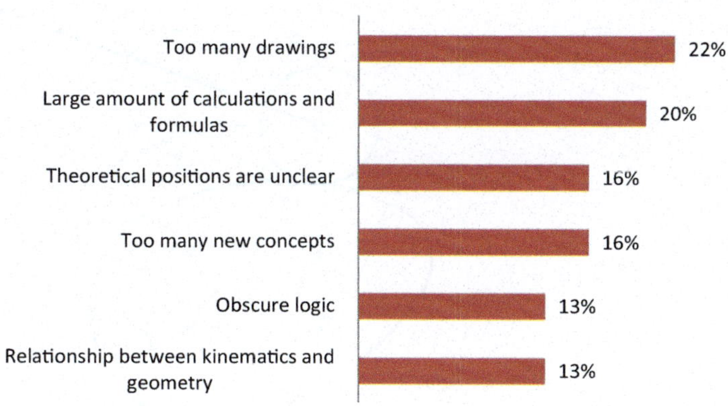

Fig. 8.12 Which was the main obstacle to the effective comprehension while studying cams?

8.4 Cams in Student Olympiads on MMS

The section of cams is often included in the topics of MMS Student Olympiads. The All-Russian Student Olympiads on TMM have been held in Russia for the last 30 years. Since 2011 IFToMM has started to hold Student International Olympiad on MMS (SIOMMS) [25, 26]. Five such Olympiads have been held to date (SIOMMS2011-Russia; SIOMMS2013-China; SIOMMS2016-Spain; SIOMMS2018-Peru; SIOMMS2020/2022-Spain).

To this day, five cam-related topics were suggested in All-Russian and International Student Olympiads. The typical problems are described below [23].

Kinematical Analysis

All sizes of the cam mechanism shown in Fig. 8.13 are given, including the radius of curvature of a pitch curve at the point of contact with the roller. For the given position make a complete kinematical analysis of the rocker O_2B, and find the angular velocity of roller 3 provided it rolls without sliding on the cam profile.

It is expected to use the method of relative motion, *no equavalent linkage is allowed.*

Forces in Kinematical Pairs

For the mechanism with the reciprocated knife-edge follower (Fig. 8.14) we know all sizes, moment of friction in the cam shaft, and the coefficient of friction in the follower guides. The cam rotates with uniform velocity. It is asked for *(a)* forces of reaction in all kinematical pairs, *(b)* value and direction of a driving torque, *(c)* set up a condition of the mechanism's self-locking.

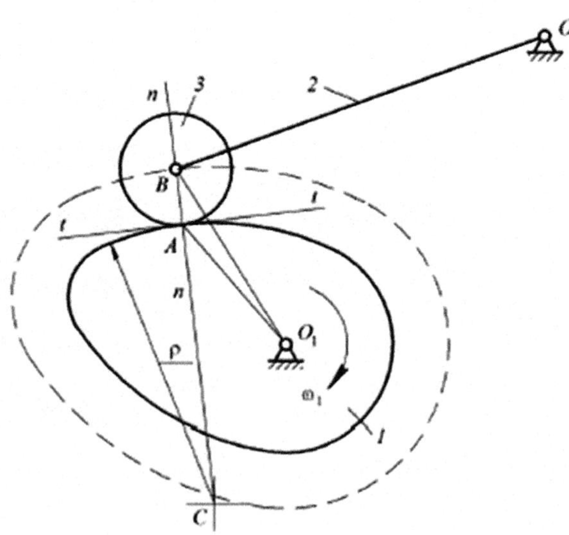

Fig. 8.13 Kinematical analysis of the cam mechanism

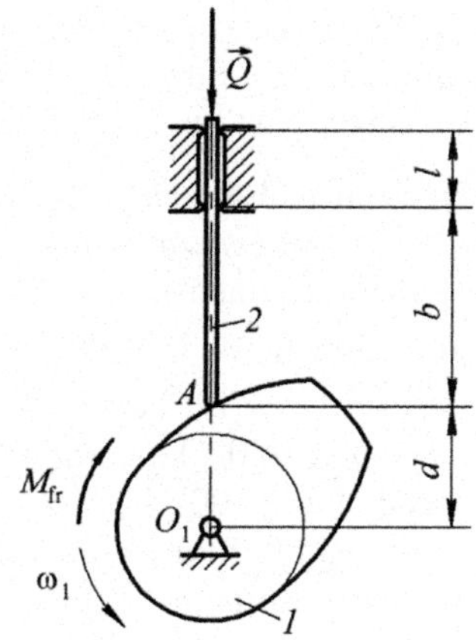

Fig. 8.14 Forces in kinematical pairs

Dynamical Model of a Cam Mechanism, Reduced Forces and Moments

It is considered a steady state running of the cam mechanism shown in the Fig. 8.15. The radius of cam r and eccentricity e are given. The coefficient of sliding friction in the pair cam-follower is f. Driving moment M_d is constant, but its value is unknown. The constant resistance force Q acts only when the follower rises. Inertia moment and inertia force are reduced and replaced by a constant moment of inertia calculated with respect to the cam shaft. All forces and moments except M_d, Q, and F_{fr} are neglected.

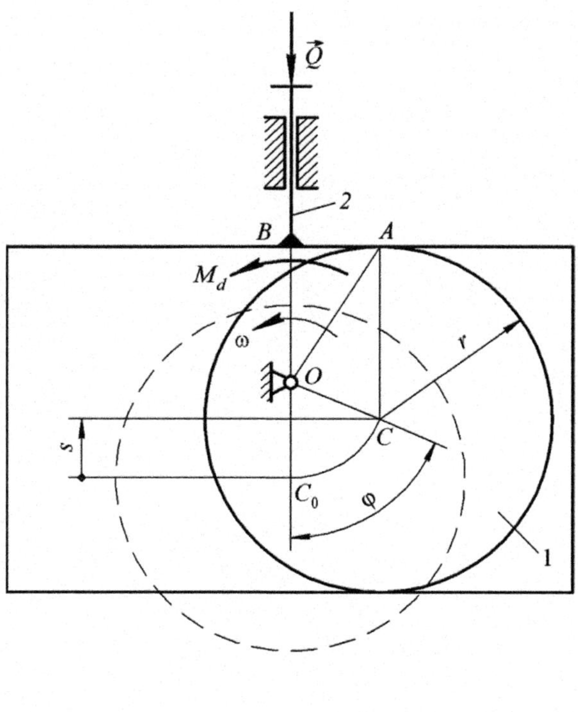

Fig. 8.15 Dynamical model of a cam mechanism, reduced forces and moments

Determine: *(a)* the moment of resistance (depending on Q and F_{fr}) reduced to the camshaft, as a function of angle φ, *(b)* value of driving moment M_d reduced to the camshaft. Don't use the equivalent linkage.

Dynamics of the cam mechanism investigated in generalized coordinates. *Here* $q = \varphi$ *is the generalized coordinate specifying the angular position of the cam.*

In the mechanism (Fig. 8.16) the reciprocated knife-edge follower rises according to the function $s = 0.5S_{max}\left(1 - \cos\frac{\pi q}{q_r}\right)$, where q is a variable angular position of the cam, $q_r = 120°$ corresponds to the maximum rise of the follower. The follower has a mass of 0.1 kg, its stroke is $S_{max} = 30$ mm. The camshaft rotates with constant speed of $\dot{q} = 100^{rad}$.

Set up the formula for driving moment M_d and find its maximum value. What angular position q^* of the cam corresponds to the maximum value of the driving moment ($M_d = M_{dmax}$)?

Use Lagrange equations to solve this problem.

Layout of cam profile and analytical solution (SIOMMS-2013, Shanghai, Shanghai Jiao Tong University).

The mechanism is shown to scale (see Fig. 8.17). There are three questions: *(a)* plot the pressure angle α and follower displacement *s* for cam rotation of $\theta = 60°$ from the position shown in figure; *(b)* complete the diagram of the follower displacement $s = f(\theta)$; *(c)* find the follower travel *H*, rise interval *Ø*, inner dwell interval $Ø_S$ and follower rise at 75° of the cam rotation angle; *(d)* derive the analytical expression of

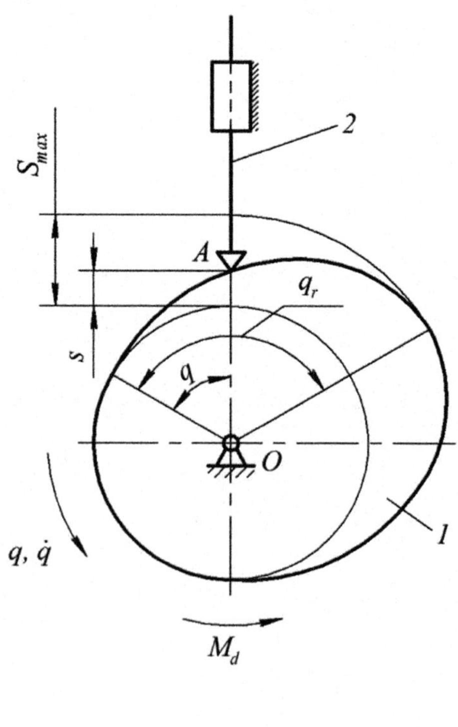

Fig. 8.16 Dynamics of the cam mechanism investigated in generalized coordinates

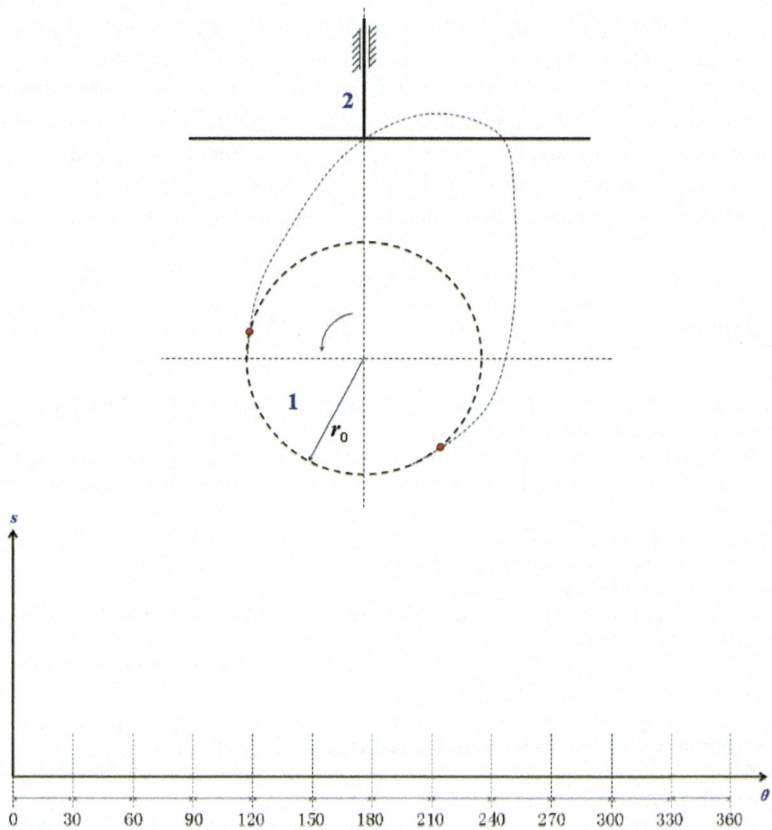

Fig. 8.17 Layout of cam profile and analytical solution

the distance from the contact point between the follower and the cam to the centerline of the follower; *(e)* determine the condition where the maximum distance appears; *(f)* derive the analytical expression of the cam contour.

Analysis of participants' decisions shows that although the cam mechanisms themselves are relatively simple, solving problems on this topic is not easy for students.

8.5 Conclusion

Cam mechanisms are widely used in engineering, so they are comprehensively analyzed in many researches, manuscripts, and books. Cam-related topics are included in all MMS study courses and other courses devoted to the machinery. Cams are interesting not only from a purely scientific point of view, but also as a subject of study at the university. They are involved in many other topics since they provide illustrative material for many ideas, concepts, and methods in Structural Analysis & Synthesis, Kinematics, and Dynamics. The investigation of cam mechanisms gives the excellent opportunity for application of variety of engineering methods. Studying cams suggests a plenty of opportunities to enhance critical and creative thinking. Therefore, the analysis of cam-related topics needs to be continued in order to find ways to effectively educate students and help their intellectual development.

References

1. Artobolevsky, I.I.: Teoriya mekhanizmov i mashin [Mechaisn and machine theory], 4th ed Science. Moscow. (in Russian) (1988)
2. Frolov, K.V. (ed.).: Teoriya mekhanizmov i mekhanika mashin [Theory of mechanisms and machine mechanics]. 5th ed. Publishing house of Bauman Moscow State Technical University, Moscow. (in Russian) (2004)
3. Khurmi, R.S., Gupta, J.K.: Theory of machines, 14th ed. Chand & Co Ltd (2005)
4. Uicker, J., Pennock, G., Shigley, J.: Theory of machines and mechanisms, 4th ed. Oxford university press, New York–Oxford (2011)
5. Rothbart, H.A., Klipp, D.L.: Cam Design Handbook. J. Mech. Des. **126** (2004). https://doi.org/10.1115/1.1723466
6. Williams, J.J.: Introduction to analytical methods for internal combustion engine cam mechanisms. (2013). https://doi.org/10.1007/978-1-4471-4564-6
7. Angeles, J., López-Cajún, C.S.: The displacement program. In book: Optimization of Cam Mechanisms, pp. 28–60 (1991). https://doi.org/10.1007/978-94-011-3572-6_2
8. Angeles, J., López-Cajún, C.S.: Kinematic analysis of Cam mechanisms (1991). https://doi.org/10.1007/978-94-011-3572-6_7
9. Angeles, J., López-Cajún, C.S.: Cam profile geometry (1991). https://doi.org/10.1007/978-94-011-3572-6_3
10. Angeles, J., López-Cajún, C.S: Optimization of planar Cam mechanisms with roller-followers (1991) https://doi.org/10.1007/978-94-011-3572-6_5
11. Angeles, J., López-Cajún, C.S.: Optimization of planar Cam mechanisms with flat-face followers. (1991). https://doi.org/10.1007/978-94-011-3572-6_4 9
12. Angeles, J., López-Cajún, C.S.: An introduction to spatial cams. (1991) https://doi.org/10.1007/978-94-011-3572-6_6.9
13. Angeles, J., López-Cajún, C.S.: The computer-aided drafting and manufacture of Cams. (1991) https://doi.org/10.1007/978-94-011-3572-6_9
14. Carbone, G., Lanni, C., Ceccarelli, M., López-Cajún, C.: Dynamic effects of curvature change in the profile of two circular-arc cams. In: V Congreso Iberoamericano de Ingegneria Mecanica CIDIM2001, Mérida, pp. 1247–1252 (2001)
15. Lanni, C., Ceccarelli, M., López-Cajún, C.: An experimental validation of three circular-arc cams with offset followers. Mech. Based Des. Struct. Mach. **34**, 261–276 (2006). https://doi.org/10.1080/15397730600830351
16. Shooter, S., West, R., Reinholtz, C.: A unified approach to teaching analytical cam design using conjugate geometry. Int. J. Mech. Eng. Educ. **23**, 145–156 (1995). https://doi.org/10.1177/030641909502300209
17. Kosenok, B., Balyakin, V., Krylov, E.: Dimensional synthesis of a cam profile using the method of closed vector contours. In the theory of machine and mechanism study. Mechanisms and machine science (book series) vol. 73, pp. 753–763 (2019) EDN BUIYHB. https://doi.org/10.1007/978-3-030-20131-9_75

18. Kosenok, B., Balyakin, V., Krylov, E.: Method of vector closed contours in design problems of study course "Internal combustion engines: kinematics and dynamics". In: mechanisms and machine science (book series). vol. 73, pp. 775–784 (2019) EDN QDOSIG. https://doi.org/10.1007/978-3-030-20131-9_77
19. Chen, H., Nguyen, T.T.N., Müller, M., Kurtenbach, S., Pan, C., Huesing, M., Corves, B.: Application of a Cam workbench for education in mechanical engineering. (2016). https://doi.org/10.1007/978-3-319-45450-4_18
20. Shiue, Y.-S.: Visual basic as a development tool for the design of disk cam with reciprocating followers. **8**, 34–35 (1998)
21. Krylov, E.G., Devyaterikov, S.A.: Developing students' cognitive skills in MMS classes. STEM Educ. **3**(1), 28–42 (2023). EDN UUYIRQ. https://doi.org/10.3934/steme.2023003
22. Waldron, K.J., Kinzel, G.L.: Kinematics, danamics, and design of machinery. Wiley, USA (2003)
23. Gazizova, Z.S. et al.: Problems on the theory of mechanisms and machines: All-Russian and international olympiads on the theory of mechanisms and machines. Competitive problems with answers and solutions. Izhevsk: Kalashnikov Izhevsk State Technical University named after M.T. Kalashnikov. ISBN 978–5–7526–0720–2. EDN UVDMLH. [Zadachi po teorii mekhanizmov i mashin : praktikum (Vserossijskie i mezhdunarodnye olimpiady po teorii mekhanizmov i mashin. Konkursnye zadachi s otvetami i resheniyami)/Z. S. Gazizova, S. V. Ezerskaya, S. A. Devyaterikov [i dr.].—Izhevsk : Izhevskij gosudarstvennyj tekhnicheskij universitet im. M.T. Kalashnikova] (in Russian) (2016)
24. Krylov, E.G., Barmina, N.A.: Gears in Russian university courses on mechanism and machine science. In: mechanisms and machine science (book series), vol. 101, pp. 225–241. EDN WQSFXM (2012). https://doi.org/10.1007/978-3-030-73022-2_10
25. Goldfarb, V.I., Krylov, E.G., Serova, T.S.: IFTOMM contribution to attraction of youth to MMS development and promotion. Mech. Mach. Theory **100**, iii–vii (2016). https://doi.org/10.1016/j.mechmachtheory.2016.02.008
26. Krylov, E.G., Devyaterikov, S.A., Gubert, A.V., Egorova, O.V.: (2020) SIOMMS: evolution and development. Mech. Mach. Theory **153**, 104029 (2020). https://doi.org/10.1016/j.mechmachtheory.2020.104029

Chapter 9
Examples of a Learning-By-Doing Approach for Bachelor and Master Students Approaching Robot Design

Elio Matteo Curcio, Francesco Lago, Stefano Rodino, and Giuseppe Carbone

Abstract This chapter introduces a learning-by-doing approach tailored for bachelor and master students in robot design, driven by the intricacies involved in designing new robots. The complexity of teaching young students arises from the necessity for robust foundational backgrounds in multidisciplinary fields, resulting in an arduous and lengthy learning process. In light of these challenges, we propose a learning-by-doing learning approach that draws inspiration also from discussions with Prof. Carlos Lopez Cajun, who was well aware of the challenges in the educational process. The proposed approach seeks to augment practical skills and enhance comprehension of robotics by prioritizing hands-on experiences. Case studies illustrate the efficacy of this method in stimulating students to tackle the complexities of robotics engineering. The proposed approach provides a potential remedy for the uncanny valley learning obstacles faced by the younger generations in the learning process for robot design.

Keywords Robotics · Design · Learning-by-doing

9.1 Introduction

Robots are widely used to help human beings and/or to execute various tasks in a range of service applications requiring direct human–robot interaction. While robotics technology has made great strides, researchers are continuing to investigate robotics with the aim of improving performance and expanding applications, also addressing the safety implications of human–robot interactions. The fast-paced evolution of robotic technology has sparked interest in creative solutions and has created a growing market in cutting-edge industries. In this context, there is a need for fresh ideas, approaches, and applications to address the challenges of these new

E. M. Curcio · F. Lago · S. Rodino · G. Carbone (✉)
DIMEG, University of Calabria, Arcavacata di Rende, Italy
e-mail: giuseppe.carbone@unical.it

M. Ceccarelli and J. C. Jauregui-Correa (eds.), *State-of-the-Art and Innovations in Mechanism and Machine Science*, Mechanisms and Machine Science 150,
https://doi.org/10.1007/978-3-031-47040-0_9

emerging technologies. These cutting-edge application domains, such as assistive robotics, service robotics, surgical and rehabilitative robotics, are becoming increasingly important, not just from a technological and financial perspective but also in terms of their impact on society, [1].

Unfortunately, the design of new robots is quite complex and requiring multidisciplinary expertise as well as, possibly, a large team with various backgrounds. This is also reflected at educational levels where it becomes quite complex to motivate and teach young students as they need to be covering all the necessary knowledge with strong foundational backgrounds, whose learning becomes more and more difficult to accomplish for young generations. Accordingly, new educational approaches should be attempted also to overcome the above-mentioned uncanny valley learning barriers. It is worth mentioning that these aspects were extensively discussed also with Prof. Carlos Lopez Cajun in the past 20 years, giving inspiration to keep searching with no rush for solutions to stimulate and attract the interest and motivation of students.

The concept of learning-by-doing can be a very promising learning approach that can be conveniently implemented in robot design. This chapter provides some general concepts and examples to inspire the new generations for learning and loving robot design.

This chapter is organized as follows: Sect. 9.2 gives a background and discussion on the learning-by-doing concepts; Sect. 9.3 describes the proposed design learning approach; Sect. 9.4 gives some insight on the following step of optimizing a design concept; Sect. 9.5 provides some examples as based on experiences at University of Calabria for both bachelor and master students.

9.2 Learning by Doing Concepts

The concept was first introduced by John Dewey in his book "Experience and Education" published in 1938, [2]. Dewey argued that learning should not only be theoretical but also practical and that students should learn by doing. This approach has been applied in STEM education and has been widely discussed in various journals and books, including, for example, [3–9] that provide a wide range of examples as applied, in particular, to mechanical [3–7] and to mechatronic education [8, 9].

Overall, the learning-by-doing approach is a valuable method for teaching mechanical engineering students by providing them with practical experience and hands-on learning opportunities. The learning-by-doing approach in mechanical engineering is a valuable educational method that emphasizes hands-on learning and practical experience in addition to theoretical concepts. It involves creating and building real-world mechanical systems, testing them, and analysing the results. This approach is effective in enhancing students' understanding of complex mechanical concepts and theories by providing them with a practical understanding of how those concepts apply in real-world scenarios.

Recent research has shown that the learning-by-doing approach in mechanical engineering has a significant impact on students' learning outcomes, including improved critical thinking skills, problem-solving abilities, and enhanced creativity. The approach also promotes teamwork and collaboration skills, as students often work in teams to design and build complex mechanical systems. Several recent references have focused on the learning-by-doing approach in mechanical engineering. For example, a study by Zhang et al. [6], explored the effectiveness of project-based learning in teaching mechanical engineering students. Another recent study by Jang and Kim [7], evaluated the effectiveness of the learning-by-doing approach in developing students' design thinking skills in mechanical engineering. The authors found that the approach significantly improved students' design thinking skills, leading to better design solutions.

Based on the literature outcomes, advantages of a learning-by-doing approach can be summarized as

- Active Learning: Learning by doing is an active learning process that engages students in the development of a mechatronic system. This approach encourages students to participate in hands-on activities, which results in a deeper understanding of the subject matter.
- Practical Skills: Learning by doing enables students to develop practical skills in designing, building, and testing a mechatronic system. This approach allows students to apply the theoretical concepts they have learned in the classroom to real-world situations.
- Improved Retention: Learning by doing helps to improve retention as students are more likely to remember information that they have discovered on their own. This approach enables students to develop a deeper understanding of the subject matter, leading to improved retention.
- Enhanced Creativity: Learning by doing encourages creativity and innovation as students are given the freedom to explore and experiment. This approach allows students to develop their problem-solving skills, leading to enhanced creativity.

Similarly, drawbacks of a learning-by-doing approach can be summarized as

- Time-Consuming: Learning by doing can be time-consuming as it requires students to spend a significant amount of time designing, building, and testing a mechatronic system. This approach can be challenging for students who have a heavy workload or other responsibilities.
- Resource-Intensive: Learning by doing can be resource-intensive as it requires access to specialized equipment, materials, and software. This approach can be challenging for institutions with limited resources.
- Lack of Standardization: Learning by doing can lack standardization as it relies on the creativity and innovation of the students. This approach can result in variations in the quality of the mechatronic systems developed.
- Safety Concerns: Learning by doing can pose safety concerns as students may be working with dangerous equipment or materials. This approach requires proper safety measures and supervision to ensure the safety of the students.

The above drawbacks should be carefully considered when identifying the specific applications and implementations of the learning approach.

9.3 The Proposed Robot Design Learning Approach

Designing a robotic or mechatronic system using a learning-by-doing approach involves several steps that we summarize in the scheme of Fig. 9.1 as based on the following main steps.

Problem Definition: The first step in designing a robotic or mechatronic system is to define the problem. This involves identifying the task or function that the robot or mechatronic system needs to perform.

Literature Review: The next step is to conduct a thorough literature review. This involves searching for relevant publications, research papers, and technical reports that describe similar systems or solutions to the problem. The literature review provides important background information and helps the designer gain an understanding of the state-of-the-art in the field.

Product Design Specification: Based on the literature review, a product design specification (PDS) is created. The PDS lists the functional requirements, design constraints, and specifications for the robotic or mechatronic system.

Concept Generation: In this step, several concepts are generated for the robotic or mechatronic system. This can be done through brainstorming sessions or by using creative problem-solving techniques. The concepts are evaluated based on their ability to meet the PDS requirements.

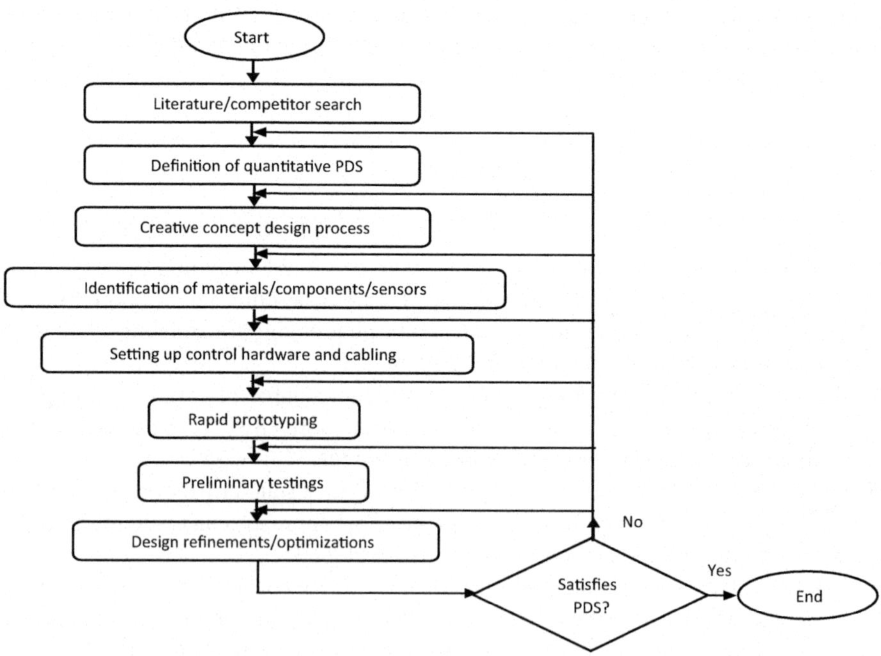

Fig. 9.1 Flowchart of the proposed design procedure

Concept Selection: The best concept is selected from the generated concepts. This involves evaluating the concepts based on factors such as feasibility, effectiveness, cost, and ease of implementation.

Detail Design: Once a concept has been selected, the detailed design process begins. This involves creating detailed design drawings and specifications for the various components of the system. The design should be optimized for manufacturability, cost-effectiveness, and reliability.

Prototyping: A prototype of the robotic or mechatronic system is built. The prototype can be used to validate the design, identify potential issues, and make improvements.

Testing and Evaluation: The prototype is tested and evaluated to determine if it meets the design requirements and specifications. Any issues or problems are identified, and modifications are made to the design as necessary.

Final Design: Based on the results of the testing and evaluation, the final design of the robotic or mechatronic system is created.

Implementation: The final design is implemented by manufacturing and assembling the various components of the robotic or mechatronic system.

Operation and Maintenance: The robotic or mechatronic system is operated and maintained according to the specifications and instructions provided by the designer.

The scheme in Fig. 9.2 provides a further graphical representation of the design loop with representation of the just the key steps that can be followed with an iterative logic.

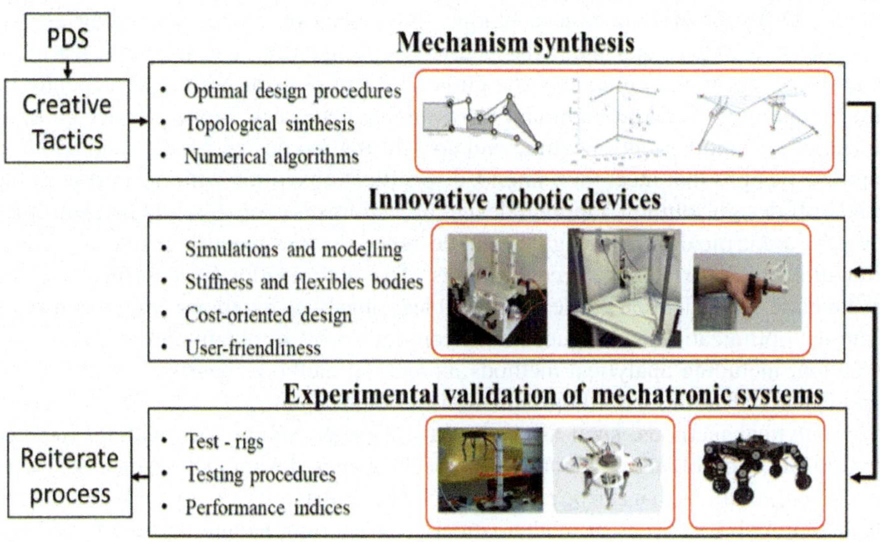

Fig. 9.2 Scheme of the proposed logical procedure

9.4 The Need of Optimal Design Criteria

The design of robots is a complex process that requires careful consideration of multiple design criteria to ensure optimal performance. A systematic approach is necessary to ensure that the robot's performance is not compromised while it operates safely and efficiently within its intended application domain. The proposed design procedure cannot be limited just at trial error approach. Accordingly, one should systematize the design also through a multi-objective approach that considers several design criteria simultaneously. One of the most critical aspects of the proposed design procedure is the need to define specific optimal design criteria. These criteria must be tailored to the specific application domain, and they should include quantitative measures for safety. Safety is an essential consideration in the design of robots, as they often interact with humans and operate in environments that pose potential hazards. The design criteria should be established with input from all stakeholders, including designers, manufacturers, and users. The proposed design procedure considers multiple design criteria to ensure that the robot's performance is optimal. The design criteria must be integrated into the design process to ensure that they are met. This integration requires careful consideration of the specific needs of the application domain. For example, in the case of rehabilitation robotics, the design criteria would need to consider the precision and accuracy required for the procedure and the safety measures needed to protect the patient.

Optimal design procedures play a significant role in ensuring that mechatronics and robotics devices meet their intended functionalities and performance requirements. Different optimization techniques have been developed, including analytical methods, numerical methods, heuristic optimization, and machine learning-based optimization techniques. The choice of optimization technique depends on the complexity of the problem and the available resources. Properly selected optimization techniques can save time and cost during the design process and result in optimal designs that meet their intended specifications, often with the best possible trade-off design solution. Optimal design procedures for robotics and mechatronics devices are critical in ensuring that the designed devices meet their intended functionalities and performance requirements. The optimization process involves the systematic identification of design variables, objective functions, and constraints. Various optimization techniques have been developed to obtain an optimal design solution, including analytical methods, numerical methods, heuristic optimization, and machine learning-based optimization techniques.

Analytical methods, such as mathematical programming, are widely used for optimization problems in robotics and mechatronics design. These methods offer analytical solutions that can be easily interpreted, and the solutions can be guaranteed to be optimal. Examples of mathematical programming techniques that have been applied in mechatronics and robotics design include linear programming, nonlinear programming, and mixed-integer programming. However, these methods are limited to simple models with few design variables and constraints, [10–12]. Numerical methods are also commonly used in robotics and mechatronics design. Finite element analysis (FEA) is one of the widely used numerical methods for analyzing and optimizing the mechanical components of mechatronics systems. The FEA technique involves dividing the physical system into small finite elements, and then analyzing the behavior of each element under specific conditions. Other numerical techniques used in mechatronics and robotics design include boundary element analysis (BEA), and multibody dynamics analysis (MDA).

After a proper optimal problem formulation. One has to select a solving algorithm that can efficiently search for the optimal solutions. Some of the commonly used optimization solving algorithms are:

- Gradient-based optimization: This method uses the gradient of the performance function to iteratively improve the design. Examples of gradient-based optimization methods include gradient descent, conjugate gradient, and quasi-Newton methods. These methods are computationally efficient for small-scale problems but may suffer from convergence issues for large-scale problems.
- Evolutionary algorithms: Evolutionary algorithms mimic the process of natural selection to search for the optimal design. Examples of evolutionary algorithms include genetic algorithms, particle swarm optimization, and differential evolution. These methods are effective for large-scale problems and can handle non-linear and non-convex design spaces but may require a large number of evaluations to converge to the optimal solution.
- Surrogate modeling: Surrogate modeling involves building a predictive model of the performance function based on a set of sampled design points. The model is then used to guide the search for the optimal design. Examples of surrogate modeling techniques include response surface methodology, kriging, and radial basis functions. These methods are computationally efficient and can handle noisy and expensive performance functions but may suffer from accuracy issues if the model is not representative of the true performance function.
- Heuristic optimization algorithms, such as genetic algorithms, particle swarm optimization, and simulated annealing, have been widely applied in mechatronics and robotics design. These algorithms provide solutions that are close to the global optimum, and they can handle complex models with numerous variables and constraints. Machine learning-based optimization techniques, such as artificial neural networks, support vector machines, and fuzzy logic systems, are becoming popular in mechatronics and robotics design, particularly for complex and high-dimensional problems.
- Multi-objective optimization: multi-objective optimization involves optimizing multiple conflicting objectives simultaneously, such as maximizing performance while minimizing cost. Examples of multi-objective optimization techniques include Pareto optimization, weighted sum, and ε-constraint methods. These methods can handle complex trade-offs between objectives but may require a large number of evaluations to identify the optimal trade-off solution.

One of the main advantages of optimal design procedures is the ability to improve the performance and efficiency of robotic and mechatronic systems. For example, optimal design of the robot manipulator can reduce the energy consumption and increase the accuracy of the robot's movements, resulting in a more efficient and reliable system. Another advantage is the ability to reduce the time and cost of the design process. By automating the design process, optimal design procedures can help to reduce the amount of time and resources required to design a robotic or mechatronic system. However, there are also some drawbacks to optimal design procedures. One of the main drawbacks is the complexity of the design process. The use of optimization techniques requires a good understanding of the underlying principles and methods, which can be challenging for some designers. Additionally, the use of optimization techniques can result in a solution that is difficult to implement in practice due to manufacturing constraints that need to be properly considered from the early design stage. Despite these drawbacks, optimal design procedures have proven to be an effective approach to improve the performance and efficiency of robotic and mechatronic systems. As such, they continue to be an area of active research in the field of robotics and mechatronics.

The literature on optimal design is extremely rich. Few examples are reported just for example in [13–28].

9.5 Exemplificative Case Studies

The following examples are based on the experiences of the author during the interdisciplinary course "RobotSumo competition" that is coordinated by the author for the bachelor students of mechanical engineering and the course "Mechatronics" delivered to master students of mechanical engineering at university of Calabria with quite a few cases of study as also reported, for example, in [29]. One should note that clearly the level and implementation of the learning by doing approach varies according to the levels and backgrounds of the students.

9.5.1 The RobotSumo Competition for Bachelor Students

Robot competitions have been utilized as an effective educational approach in various contexts to expedite robotics research and foster interest in science, technology, engineering, and mathematics (STEM) education. These competitions offer a cost-effective means of engaging students and encouraging their pursuit of careers in science and technology. One such engaging format is the robot-sumo competition, where two robots face off in a head-to-head match, akin to traditional human sumo matches. These robots are not allowed to use weapons or flip each other, and the goal is to push the opponent out of the arena. The competitions accommodate different weight classes and control systems, with autonomous robots competing against each other. Each match involves two teams. The robots used in the matches are constructed by the teams themselves, adhering to specific provided specifications. The matches commence upon the judge's command and continue until a contestant earns two points, with the judge determining the ultimate winner. Overall, these competitions serve as a versatile and engaging method to achieve a range of educational objectives across various courses.

The University of Calabria (Fig. 9.3) initiated a project aimed at experimenting with a novel teaching methodology to evaluate the skills of bachelor's degree students in dealing with multidisciplinary problems. This assessment was conducted through a multidisciplinary project called "RobotSumo competition", which allowed the students to demonstrate their acquired skills during the program. Rather than relying on commercial robotics kits, the project involved designing, modifying, and integrating components and programming control software. The course was limited to 30–40 students, organized in groups of a maximum of 5, with priority given to those with the highest number of CFUs (course credits). It featured a strong interdisciplinary approach, encompassing mechanical component selection (motors, gearboxes, sensors), CAD design/rapid prototyping, electronics, control, and programming. Students took charge of their activities, engaging in weekly discussions with tutors and lecturers, while also attending seminars on specific aspects. To aid autonomous learning, reference manuals, tutorials, and video lectures were provided. The course assessment comprised writing a short report, constructing a functional robotic prototype, and participating in a competition. Overall, this initiative fostered a deeper awareness of the students' abilities and practical experience across different disciplines.

The students participated in a series of activities, including well-structured theoretical and practical lessons, conducted in person. Initially, there were two months of weekly theoretical lessons and specific seminars. Competent teachers from various

Fig. 9.3 Pictures taken at the robot-sumo competition in May 2023 at University of Calabria

disciplines essential for the project's implementation led these sessions. The theoretical courses began with a review of mechanics concepts previously covered in other courses and then delved into robotics, starting with an overview of existing robots in industrial and non-industrial applications. This enabled students to grasp the key components to incorporate into their projects, acquainting them with different movement systems, actuators, and sensors. Subsequently, electronics lessons guided the students in selecting appropriate sensors and/or actuators by exploring their technical characteristics and operating principles. As each group conceptualized their robots, practical lessons were conducted in the laboratory, where they iteratively refined their prototypes through a process of 'trial and error.' This hands-on approach improved their practical comprehension of robot design, pushing them to further develop their academic knowledge.

9.5.2 Examples of Mechatronics' Group Projects at Master Level

This section delves into the design and implementation of an experiential education program in mechatronics, specifically tailored for master students in mechanical engineering at the prestigious University of Calabria. The course spans 6 credits and aims to bridge the gap between theoretical knowledge and real-world applications, empowering students with practical skills and problem-solving capabilities.

Mechatronics inherently demands the integration of mechanical, electrical, and computer engineering principles. This interdisciplinary approach is fostered through experiential learning, providing students with a holistic understanding of mechatronics beyond traditional classroom boundaries.

The "mechatronics" course revolves around team-based learning, with 4 or 5 students forming cohesive groups. Each team is assigned a unique case study topic, carefully selected to encompass the multidisciplinary nature of mechatronics. Students embark on a comprehensive design process, from conceptualization to the development and preliminary testing of proof-of-concept prototypes. Central to this section is the emphasis on real-world problem-solving. As students delve into the investigation of all design phases, they gain valuable insights into how theoretical knowledge translates into practical applications. With continuous guidance from specialized instructors, students are encouraged to employ critical thinking and collaboration to optimize their proposed solutions. Multiple examples have resulted even in conference and journal publications such as reported, for example, in [29–44].

In the next subsections, examples of team works will be reported, showcasing how experiential education enhances problem-solving abilities, instills a profound appreciation for teamwork, and cultivates adaptability to complex challenges. These skills are essential for shaping future leaders in the field of mechatronics.

9.5.2.1 In Pipe Inspection

White-water pipelines often face blockages caused by debris from various sources. These pipelines typically consist of straight pipes with diameters ranging from 200 to 600 mm, separated by 500 × 500 mm square inspector manholes. This design simplifies the mechanical solution and reduces costs, making it suitable for small businesses and local administrations, who are the intended users of the proposed mechatronic device. The system should be user-friendly to eliminate the need for highly specialized personnel and should navigate sloped pipes with different diameters.

After thorough literature analysis and specific application considerations, the main design requirements have been identified as follows: Suitable for 200–600 mm diameter pipelines, device length below 400 mm to fit into the 500 × 500 mm manholes easily, power supply below 60 W for autonomous driving and power autonomy, weight below 3 kg for transportability and pipeline protection, holding force of 20–30 N to prevent slippage and pipeline damage, capable of moving along a straight pipeline, as inspector manholes are present at direction changes, and a speed not lower than 5 cm/s for efficient task execution.

The device must be user-friendly, enabling semi-automatic operation without requiring specialized skills. The design process involves morphological charts and results in a topology with three modules: front and back "grasping modules" with telescopic elements that hold the robot to the pipe walls, and a central body module housing a crank-shaft mechanism for worm-like movement Fig. 9.4. The locomotion module displaces the front module relative to the back module, allowing forward or backward movement. The following sections elaborate on the proposed mechanism design based on the outlined conceptual design.

After conducting an extensive array of simulations and preliminary tests, a first prototype of the robot was constructed. The robot is equipped with multiple sensors,

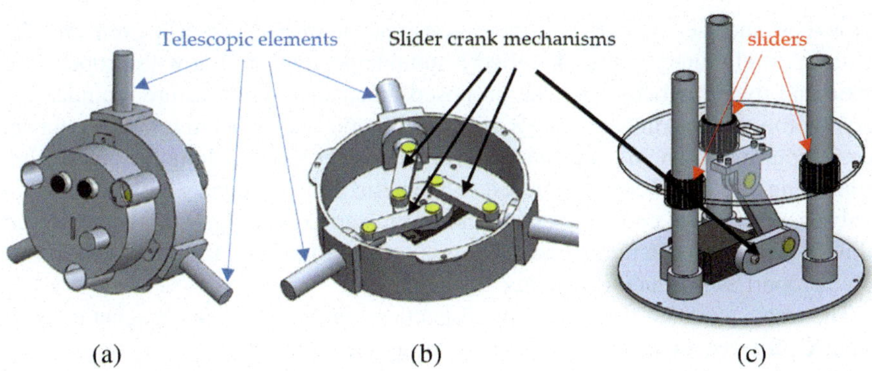

Fig. 9.4 Proposed PEIS design comprised of three main modules: **a** the front module, **b** the pipeline grasping module utilizing three slider-crank mechanisms to actuate the telescopic elements using a single motor, and **c** the central body driving module employing one slider-crank mechanism and three sliders

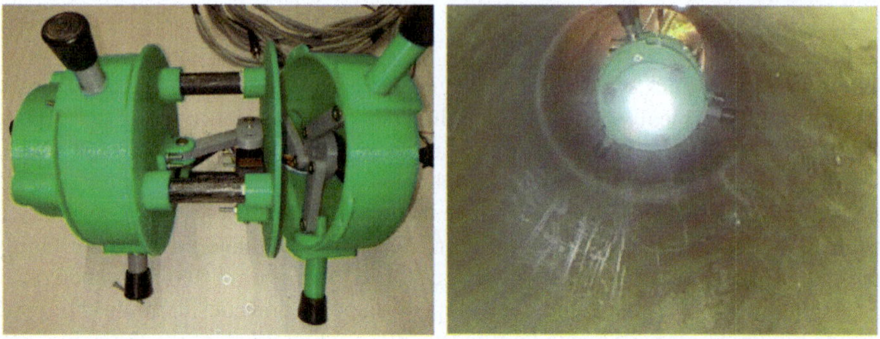

Fig. 9.5 Details of PEIS at DIMEG, University of Calabria: **a** view of the back module assembly; **b** view with the onboard lighting in dark conditions

which draw power from a battery located in the rear module. On the other hand, the motors are externally powered. A control board, featuring an alert LED and manual command buttons, is positioned outside the pipe, as depicted in Fig. 9.5.

Operating the device is straightforward. Users simply place it at the pipeline's starting point and activate a latch switch. A second switch enables the selection of the operation direction. The robot automatically halts when it encounters an obstruction, and the camera provides a live view for inspection. The sensor feedback aids in determining the obstruction's position relative to the robot's initial placement.

Further details on this case study can be found at [45].

9.5.2.2 High Powerline Inspection

A low-cost and user-friendly system was designed to inspect powerline transmission cables. A preliminary study identified a suitable design solution with proper kinematic and dynamic behavior. The proposed mechanical architecture includes two servomotors and one linear actuator with a telescopic leg and a tail for balancing and sensing purposes. The device can perform a constant motion along the cable while monitoring the parameter of interest. Kinematic and dynamic simulations, static analysis, and control code implementation were carried out Fig. 9.6. The maximum allowed weight for the powerline cable was evaluated using a formula based on the cable properties and factor of safety.

The tests simulated the powerline inspection operation using a magnetic sensor to detect defects, an accelerometer to monitor body orientation and motion quality, and an ultrasonic sensor for frontal obstacles Fig. 9.7. A scaled model with a frame and an aluminum beam was created for testing. A step size of 60 mm was found to be the best compromise between stability and performance. The pace analysis and magnetic field detection were performed, and a warning signal appeared on the user interface and a LED started blinking when a magnetic field was detected along the

Fig. 9.6 The ADAMS view model for calculation of the torque and force required for each motor as well as the contact force between arm and the powerline cable

cable Fig. 9.8. The same behavior was observed when an obstacle was detected, and the motion automatically stopped. Accelerometer data was also shown in a graph with three directions. Future experimental tests will be carried out to assess human–robot safety interactions.

Further details on this case of study can be found at [46].

Fig. 9.7 A picture of a simulated powerline monitoring test experiment

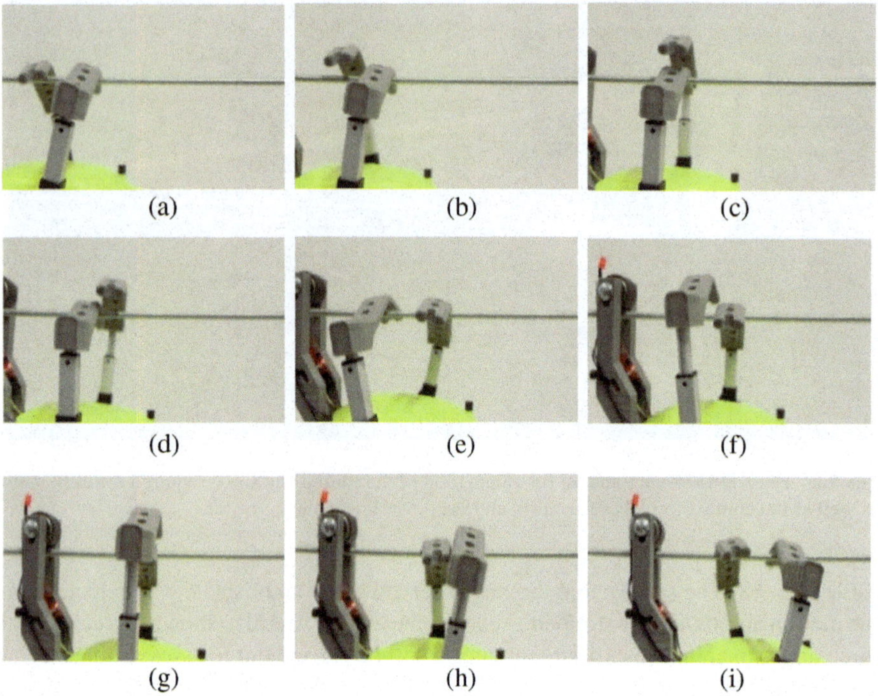

Fig. 9.8 Sequence of pace in forward direction

9.5.2.3 Four-Legged Locomotion

Designing for inspection and rescue tasks involves various requirements and constraints, which must be systematically approached to meet specific application needs. A design procedure has been defined through key steps outlined in Fig. 9.1, starting with an in-depth literature analysis to collect quantitative data for a product design specification (PDS) table. Next, a team brainstorming session can identify potential design solutions, which are compared and evaluated using quantitative calculations and software-aided simulations. This proposed design approach can serve as a valuable learning tool for mechanical engineering students to gain theoretical, simulation, operational, and experimental skills. The proposed design solution offers a 2-dof kinematic architecture for each leg, a crank and rope driving mechanism, and a balancing spring, resulting in an effective and cost-efficient design for dynamic walking and trotting gaits Fig. 9.9.

The prototype has shown its capability to achieve complex dynamic gaits in the preliminary experiments Fig. 9.10, indicating the success of this challenging topic in motivating students to learn complex theoretical, numerical, and practical skills.

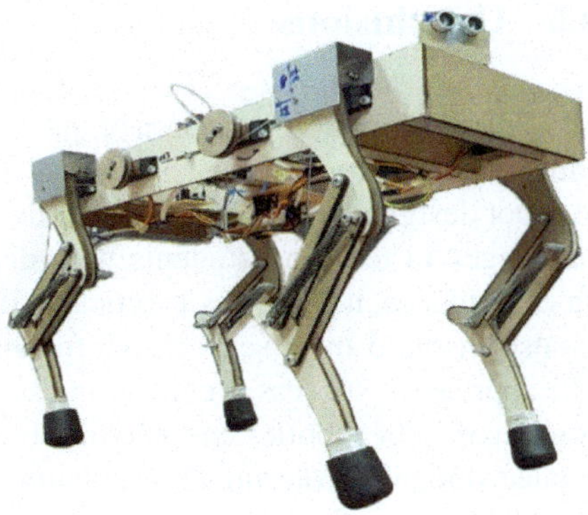

Fig. 9.9 Pictures of the built prototype

Further research will be conducted in the future to investigate the potential applications of the built prototype, such as exploration, patrolling, and surveillance of unstructured environments, as proposed for example in [48, 49].

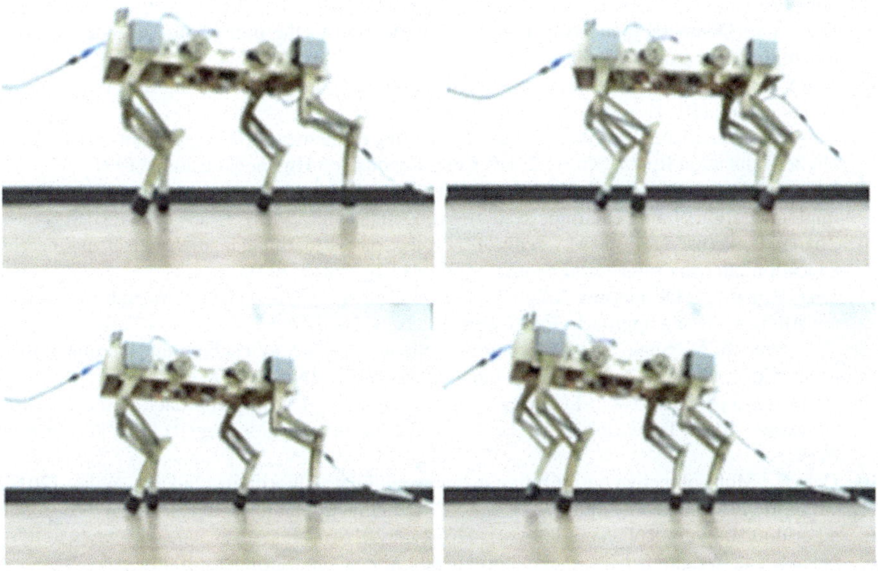

Fig. 9.10 Snapshots taken during a trotting experiment

9.6 Conclusions

The learning-by-doing approach presented in this chapter proves effective in educating bachelor and master students in robot design. Motivated by the complexity of robot design and the need for multidisciplinary expertise, the approach addresses challenges in teaching students with diverse backgrounds. By prioritizing hands-on experiences, it enhances practical skills and deepens comprehension of robotics, as also inspired by several insightful discussions with Prof. Carlos Lopez Cajun. The approach's success is evident in case studies, preparing students to navigate the complexities of robotics engineering. It offers a promising solution to motivate and engage students, fostering their passion for robotics and equipping them with skills for this evolving field, nurturing the next generation of competent engineers.

References

1. International Federation of Robotics. IFR2023. Available from https://ifr.org. Accessed 23 Aug 2023
2. Dewey, J.: Democracy and education: an introduction to the philosophy of education. The Macmillan Company, (1916)
3. Kolb, D.A.: Experiential learning: experience as the source of learning and development. Prentice Hall, (1984)
4. Bonwell, C.C., Eison, J.A.: Active learning: creating excitement in the classroom (ASHE-ERIC Higher Education Report No. 1). ERIC Clearinghouse on Higher Education (1991)
5. Almeida, D.A., Ferreira, D.C.: Learning by doing in mechanical engineering: design and implementation of a mechatronic system. Int. J. Eng. Educ. **34**(5), 1535–1543 (2018)
6. Zhang, J., Wang, L., Liu, M., Peng, T.: Project-based learning in mechanical engineering education: an exploratory study. Sustainability **13**(6), 3306 (2021)
7. Jang, Y., Kim, J.: Developing design thinking skills in mechanical engineering education using a learning-by-doing approach. Eur. J. Eng. Educ., 1–16 (2020)
8. McClamroch, N., Shah, S.L.: A learning-by-doing approach to teaching mechatronics. IEEE Trans. Educ. **62**(3), 231–240 (2019). https://doi.org/10.1109/TE.2018.2861659
9. Howell, B.J., Brown, R.G.: Learning by doing: an active-learning approach to mechatronics education. IEEE Trans. Educ. **63**(2), 184–193 (2020). https://doi.org/10.1109/TE.2019.2968299
10. Deb, K.: Multi-objective optimization using evolutionary algorithms, 2nd edn. Wiley, (2014)
11. Ertugrul, S., Inanc, T.: Optimal design of robot manipulators using artificial bee colony algorithm. J. Intell. Rob. Syst. **83**(3–4), 423–440 (2016)
12. Nocedal, J., Wright, S.J.: Numerical optimization, 2nd edn. Springer, (2006)
13. Carbone, G., Ceccarelli, M., Oliveira, P.J., Saramago, S.F.P., Carvalho, J.C.M.: An optimum path planning for Cassino parallel manipulator by using inverse dynamics. Robotica **26**(2), 229–239 (2008)
14. Carbone, G., Di Nuovo, A.: A hybrid multi-objective evolutionary approach for optimal path planning of a hexapod robot: a preliminary study. Comput. Sci. (Incl. Subser. Lect. Notes Artif. Intell. Lect. Notes Bioinform.) **9668**, 131–144 (2016)
15. Carbone, G., Laribi, M.A., Zeghloul, S.: On the optimal design of cable driven parallel robot with a prescribed workspace for upper limb rehabilitation tasks. J. Bionic Eng. **16**(3), 503–513 (2019)

16. Malyshev, D., Rybak, L., Carbone, G., Semenenko, T., Nozdracheva, A.: Optimal design of a parallel manipulator for aliquoting of biomaterials considering workspace and singularity zones. Appl. Sci. **12**(4), 2070 (2022). https://doi.org/10.3390/app12042070
17. Castejón, C., Carbone, G., Prada, J.C.G., Ceccarelli, M.: A multi-objective optimization of a robotic arm for service tasks. Stroj. Vestn./J. Mech. Eng., **56**(5), (2010)
18. Hernandez, E., Valdez, S.I., Carbone, G., Ceccarelli, M.: Design optimization of a cable-driven parallel robot in upper arm training-rehabilitation processes. Mech. Mach. Sci. **54**, 413–423 (2018)
19. Hu, J., Wang, L.: Optimization design of a parallel robot based on improved genetic algorithm. J. Mech. Eng. **50**(15), 146–151 (2014)
20. Jin, H., Hu, J., Gao, H.: Optimization of bipedal robot gait parameters based on genetic algorithm. J. Mech. Sci. Technol. **30**(9), 4071–4077 (2016)
21. Jones, D.R., Schonlau, M., Welch, W.J.: Efficient global optimization of expensive black-box functions. J. Global Optim. **13**(4), 455–492 (1998)
22. Kaneko, M., Takenaka, H., Sekiguchi, T.: Multi-objective optimization of robot manipulability and energy efficiency using nonlinear programming. J. Robot. Mechatron.S **26**(3), 372–382 (2014)
23. Laribi, M.A., Carbone, G., Zeghloul, S.: On the optimal design of cable driven parallel robot with a prescribed workspace for upper limb rehabilitation tasks. J. Bionic Eng. **16**(3), 503–513 (2019)
24. Oliveira, G.P., Ferreira, P.R.: Optimal design of robot manipulators: a review. Robot. Comput.-Integr. Manuf. **31**, 51–63 (2015)
25. Rodríguez, N.E.N., Carbone, G., Ceccarelli, M.: Optimal design of driving mechanism in a 1-DOF anthropomorphic finger. Mech. Mach. Theory **41**(8), 897–911 (2006)
26. Tavakoli, S., Haeri, M., Khajehoddin, M.V.: Optimal design of parallel robots: state-of-the-art and challenges. Robot. Comput.-Integr. Manuf. **27**(2), 193–202 (2011)
27. Tucan, P., Vaida, C., Ulinici, I., Carbone, G., Pisla, D.: Optimization of the aspire spherical parallel rehabilitation robot based on its clinical evaluation. Int. J. Environ. Res. Public Health **18**(6), 1–20, 3281 (2021)
28. Yao, S., Ceccarelli, M., Carbone, G., Zhan, Q., Lu, Z.: Analysis and optimal design of an underactuated finger mechanism for LARM hand. Front. Mech. Eng. **6**(3), 332–343 (2011)
29. Carbone, G., Curcio, E.M., Rodinò, S., Lago, F.: A robot-sumo student competition at UNICAL as a learning-by-doing strategy for STEM education. STEM Educ. **2**(3), 262–274 (2022). https://doi.org/10.3934/steme.2022016
30. Gonçalves, R.S., Rodrigues, L.A.O., Humbert, R., Carbone, G.: A user-friendly nonmotorized device for ankle rehabilitation. Robotics **12**(2), 32 (2023)
31. Rodinò, S., Lago, F., Malyshev, D., Carbone, G.: Design of a movable palm for a 3-fingers robotic hand. Int. J. Mech. Control. **24**(1), 177–218 (2023)
32. Voloshkin, A., Rybak, L., Cherkasov, V., Carbone, G.: Design of gripping devices based on a globoid transmission for a robotic biomaterial aliquoting system. Robotica **40**(12), 4570–4585 (2022)
33. Gonçalves, R.S., de Souza, M.R.S.B., Carbone, G.: Analysis of the leap motion controller's performance in measuring wrist rehabilitation tasks using an industrial robot arm reference. Sensors **22**(13), 4880 (2022)
34. Rodríguez-León, J.F., Castillo-Castañeda, E., Aguilar-Pereyra, J.F., Carbone, G.: Experimental characterization of A-AFiM, an adaptable assistive device for finger motions. Machines **10**(4), 280 (2022)
35. Alves, T., Gonçalves, R.S., Carbone, G.: Serious games strategies with cable-driven robots for bimanual rehabilitation: a randomized controlled trial with post-stroke patients. Front. Robot. AI **9**, 739088 (2022)
36. Malyshev, D., Rybak, L., Carbone, G., Semenenko, T., Nozdracheva, A.: Optimal design of a parallel manipulator for aliquoting of biomaterials considering workspace and singularity zones. Appl. Sci. (Switzerland) **12**(4), 2070 (2022)

37. Pisla, D., Tarnita, D., Tucan, P., ...Carbone, G., Plitea, N.: A parallel robot with torque monitoring for brachial monoparesis rehabilitation tasks. Appl. Sci. (Switzerland) **11**(21), 9932 (2021)
38. Curcio, E.M., Carbone, G.: Mechatronic design of a robot for upper limb rehabilitation at home. J. Bionic Eng. **18**(4), 857–871 (2021)
39. Fomin, A., Antonov, A., Glazunov, V., Carbone, G.: Dimensional (Parametric) synthesis of the hexapod-type parallel mechanism with reconfigurable design. Machines **9**(6), 117 (2021)
40. Antonov, A., Fomin, A., Glazunov, V., Kiselev, S., Carbone, G.: Inverse and forward kinematics and workspace analysis of a novel 5-DOF (3T2R) parallel–serial (hybrid) manipulator. Int. J. Adv. Robot. Syst. **18**(2), (2021)
41. Orozco-Magdaleno, E.C., Cafolla, D., Castillo-Castaneda, E., Carbone, G.: Static balancing of wheeled-legged hexapod robots. Robotics **9**(2), 23 (2020)
42. Carbone, G., Gerding, E.C., Corves, B., ...Russo, M., Ceccarelli, M.: Design of a Two-DOFs driving mechanism for a motion-assisted finger exoskeleton. Appl. Sci. (Switzerland) **10**(7), 2619 (2020)
43. Tucan, P., Gherman, B., Major, K., Carbone, G., Pisla, D.: Fuzzy logic-based risk assessment of a parallel robot for elbow and wrist rehabilitation. Int. J. Environ. Res. Public Health **17**(2), 654 (2020)
44. Görgülü, İ, Carbone, G., Dede, M.İC.: Time efficient stiffness model computation for a parallel haptic mechanism via the virtual joint method. Mech. Mach. Theory **143**, 103614 (2020)
45. Salvatore, M.M., Galloro, A., Muzzi, L., Pullano, G., Odry, P., Carbone, G.: Design of PEIS: a low-cost pipe inspector robot. Robotics **10**, 74 (2021). https://doi.org/10.3390/robotics10020074
46. Carbone, G., Forciniti, G., Fuoco, F., Scarfone, L., Viapiana, D.: A mechatronic system for automatized inspection of powerline transmission. Mech. Mach. Sci. **88**, 457–467 (2021)
47. Rodinò, S., Curcio, E. M., Di Bella, A., Funaro, M., Carbone, G.: Design, simulation, and preliminary validation of a four-legged robot. Machines **8**(4), 1–16, 82 (2020)
48. Carbone, G., Iannone, S., Ceccarelli, M.: Regulation and control of LARM hand III. Robot. Comput.-Integr. Manuf. **26**(2), 202–211 (2009). https://doi.org/10.1016/j.rcim.2009.05.002
49. Orozco-Magdaleno, E.C., Gómez-Bravo, F., Castillo-Castañeda, E., Carbone, G.: Evaluation of locomotion performances for a Mecanum-Wheeled hybrid hexapod robot. IEEE/ASME Trans. Mechatron.Mechatron. **26**(3), 1657–1667 (2021). https://doi.org/10.1109/TMECH.2020.3027259

Chapter 10
Innovations in Design of Worm-type Gears in the Last Two Decades

Evgeniy S. Trubachev and Natalya A. Barmina

Abstract The manuscript presents a review of innovations in the design of worm-type gears which are made at the Research and Production Center "Institute of Mechanics named after Prof. Veniamin I. Goldfarb", Izhevsk, Russia. The main innovations considered are spiroid gears with special cases of low speeds and small gear ratios, worm cylindrical gears with steel hardened gearwheels, and planetary spiroid gears. The features of low-speed spiroid and worm gears in multi-nomenclature series production are extremely heavy loads, the need to take into account layout limitations, unification and simplification of gear cutting tools, implementation of several gear ratios in one design. For small gear ratios, the spiroid gear can be an effective alternative to hypoid and bevel gears due to its favorable meshing characteristics and economical and affordable production technique. The use of hardened steel for worm gearwheels is related with a favorable arrangement of the conjugate contact lines for lubrication and scuffing resistance—at large angles with respect to the vector of relative velocity. Two versions of such worm gears are considered. The use of worm and spiroid gears in a planetary mechanism provides an extremely high gear ratio, compact design and smooth operation.

Keywords Worm gear · Spiroid gear · Innovative solutions

E. S. Trubachev (✉)
Institute of Mechanics N.a. Prof. V. I. Goldfarb, Kalashnikov ISTU, 7 Studencheskaya Str, 426069 Izhevsk, Russia
e-mail: truba@istu.ru

E. S. Trubachev · N. A. Barmina
MIP Mechanik Ltd, Izhevsk, Russia

M. Ceccarelli and J. C. Jauregui-Correa (eds.), *State-of-the-Art and Innovations in Mechanism and Machine Science*, Mechanisms and Machine Science 150,
https://doi.org/10.1007/978-3-031-47040-0_10

10.1 Introduction

Despite being well known and well studied, improvements of relatively simple mechanisms are the subject of efforts of modern specialists (in particular, **Professor Carlos Lopez-Cajun** [1–3]) and have a significant impact, making the work of engineers more efficient.

In 2024 the Institute of Mechanics of Kalashnikov Izhevsk State Technical University created by Prof. Veniamin I. Goldfarb and now bearing his name celebrates its 30th anniversary. The activity of this department together with the small innovative enterprise "Mechanik" is an example of integration of education, science, and production [4]. This integration provided the shortest path from initiation of a scientific or engineering idea to practical testing and cost-effective implementation.

This manuscript presents a review of some innovations made and implemented in the field of worm-type gears which seem to be simple mechanisms and, in many respects, conventional both in terms of varieties and in terms of production techniques.

10.2 The Main Object of Development–spiroid Gear

The history of invention of spiroid gear varieties and development of their production methods is well described in works of Prof. V.I. Goldfarb [5], and V. Bolosh [6]. Being, on the one hand, related to the worm gear by design and production procedures, the spiroid gear is similar to the hypoid one by layout features (Fig. 10.1) [5]. From the moment of its invention and, in many respects, nowadays it is an innovation in itself. Its innovativeness—implementation instead of traditional solutions—is conditioned by aspiration to use its advantages in comparison with analogues:

- greater load and overload capacity and static strength resulting from multipair contact and other favorable properties of meshing, or (the other side of the same question) smaller sizes and lower cost of a drive;
- greater smoothness of operation;
- lower sensitivity to the action of errors;

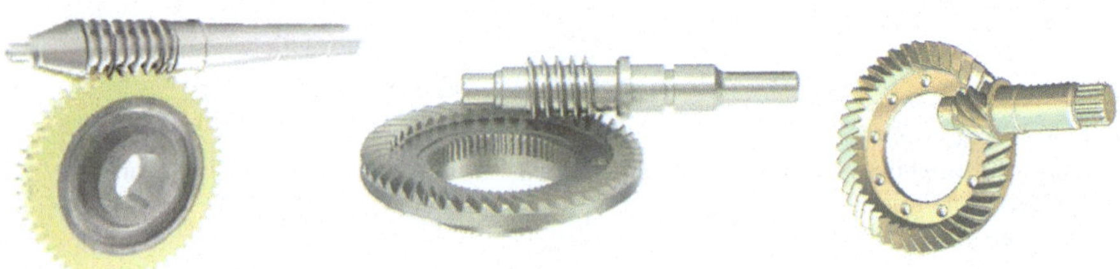

Fig. 10.1 Spiroid gear (in the center) and its analogues—worm (left) and hypoid (right) gears

- possibility of using heat-strengthened steel for the manufacturing of the gearwheel rim instead of antifriction materials;
- possibility of easy adjustment of the backlash up to its full elimination, moreover, directly during the gear operation—by axial displacement of the bevel worm or (in most cases, slightly disturbing the conjugate feature of meshing) of the gearwheel;
- possibility to use widespread and relatively easy to set up and operate gear-cutting equipment.

There is a large number of varieties of spiroid gears in which these advantages are implemented in different drives and production conditions. All this makes the use of spiroid gears attractive for the consumer and convenient and inexpensive for the manufacturer.

10.3 Special Case–Low–Speed Spiroid Gears

One of the most successful discoveries for the application of our developments were gearboxes for control of pipeline valves (Fig. 10.2) [7], operating at low speeds of shaft rotation (high-speed stages - not more than 100–200 rpm), rare switching on, and short-term peak torques. The noted possibility of using heat-strengthened steel for manufacturing the gearwheel was just in time.

At the moment we have about 100 gears in series production with different types, dimensions (gear diameters from 20 to 570 mm), gear ratios (from 1 to 120 per stage), monthly production program usually amounts to 40–50 varieties of them with total number of gears about 2000. One can imagine that, taking into account the limited resources, it is possible only if each of the gears is optimized to the conditions of production, which ensures a quick and cheap preparation of the latter and the subsequent effective manufacture of products. In this case, for each object of production hundreds of variants are considered, and a large number of assessments is formed.

It would seem that low-speed gears do not have high demands on their design and manufacturing. However, this impression is deceptive. Leading manufacturers,

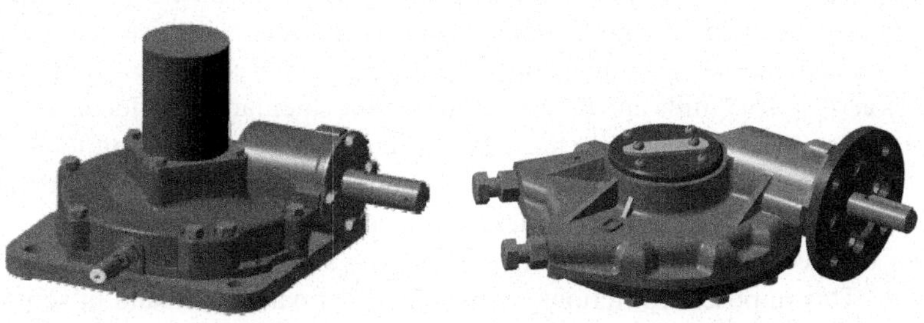

Fig. 10.2 Spiroid gearboxes for pipeline valves—multi-turn (left—for gate valves) and quarter-turn (right—for ball and butterfly valves)

striving to reduce costs and, accordingly, the mass and dimension characteristics of pipeline valve gearboxes have achieved results that look outstanding in comparison with gearboxes from other drive industry segments: the ratio of gearbox mass to the transmitted torque reaches values of 0.005 or even less in the best samples. The requirements to reduce the size and to ensure a high competitiveness of technical solutions in the multi-numbered small batch production gave grounds for the formation of some, in our opinion, important design trends.

10.4 "Unexpected" Design Rules

The best gear is not the best solution for the gearbox.

A reasonable endeavor to minimize gear and gearbox dimensions leads to obvious results: the forces acting in the meshing (consequently acting on bearings and casing parts) are increased and a large number of geometrical constraints arise—intersections of gearbox structural elements. Some of these constraints are dictated by the driven elements, such as the flange connection and the driven shaft. Gearbox components and therefore the gear are forced to be placed around these elements. The second part of the restrictions is related to the risk of mutually overlapping gearbox components. These structural elements are gear links, the gearwheel hub, bearings and parts that hold them together. Figure 10.3 shows fragments of geometric modeling of possible intersections [7].

Because of constraints, the gear is often forced to be made somewhat larger than necessary to develop a given momentum; the methods to improve the gear become limited. Generally speaking, minimizing its size with account of constraints leads to a design solution that didn't seem to be the best.

The best gear is not the best solution for production.

"Costs for gear cutting tools constitute at least 50% of the cost of gear manufacturing," says the Soviet gear manufacturer's handbook [8, p. 105]. One can dispute the accuracy of this estimate, but there is no doubt that the share of these costs is high. One of the efficient and widely tested ways to reduce them is the unification of gear cutting tools, which allows reducing their nomenclature and the cost of production and operation. The great variety of worm gears being produced means that manufacturers must have an equally great variety of gear cutting tools which are complex to produce and operate. We have developed and implemented an alternative approach to unification of spiroid mills and hobs—a method for calculating the localized contact and a new method for gear design [9, 10], where each of the existing tools is used for cutting gearwheels for worms with different numbers of threads and diameters. The effect of its implementation is clearly shown in Fig. 10.4.

With the new approach, the design process is subject to the condition of tool selection. This imposes restrictions on the choice of parameters and in some sense also leads to a design solution that seemed not to be the best, but provides lower production costs.

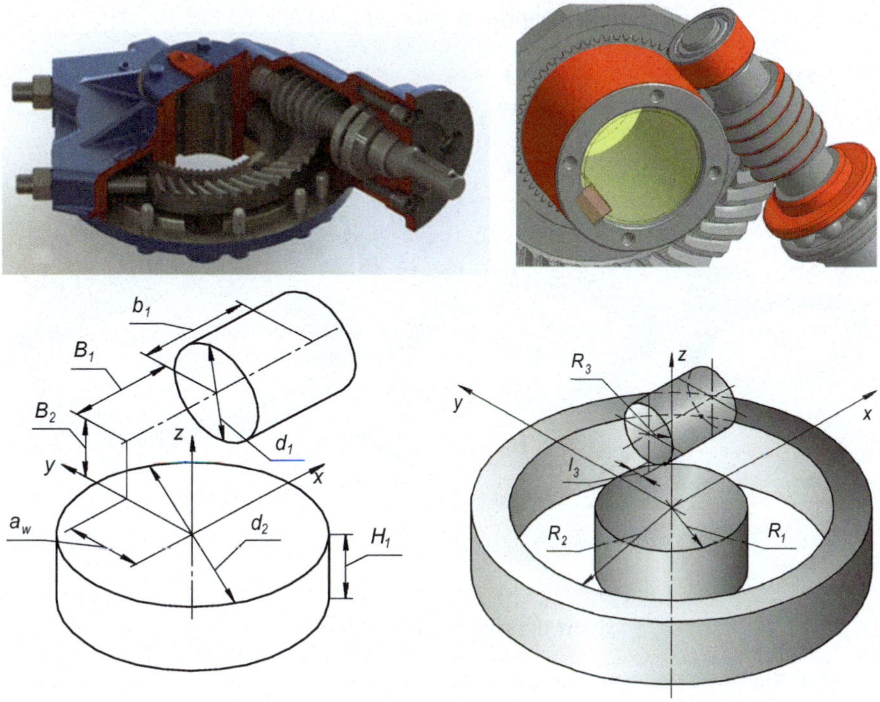

Fig. 10.3 Fragments of modeling of restrictions—intersections of components

Gears which are not the best solution individually are implemented in the best gearbox.

One of the general properties of worm-type gears distinguishing them, for example, from cylindrical and bevel gears, is that the ratio of dimensions of the links can weakly depend on the gear ratio. This, in particular, allows simply to implement different gear ratios in one and the same casing maximizing the use of its space, that is, without laying stock of the latter. Even if the limitations are taken into account, making the best single gear does not mean the best solution for a gearbox with several ratios. Maximum improvement of each implemented gear requires, strictly saying, a unique combination of parameters. The design in these conditions is either a compromise in satisfying the requirement of improving each of the gears, or subject to the requirement of improving the most important of them, the rest being to some extent not optimal.

Of course, when we used the words "the best gear," we meant its most common meaning: the gear with the best performance characteristics. In fact ("Captain Obvious") to such a meaning of "the best" the phrase should be added: "…and convenient for production, cheap in production, manufacturable". Complex consideration can be expressed in the optimization algorithm [10], where design and technological requirements are stated in the form of constraints-equalities and constraints-inequalities. The

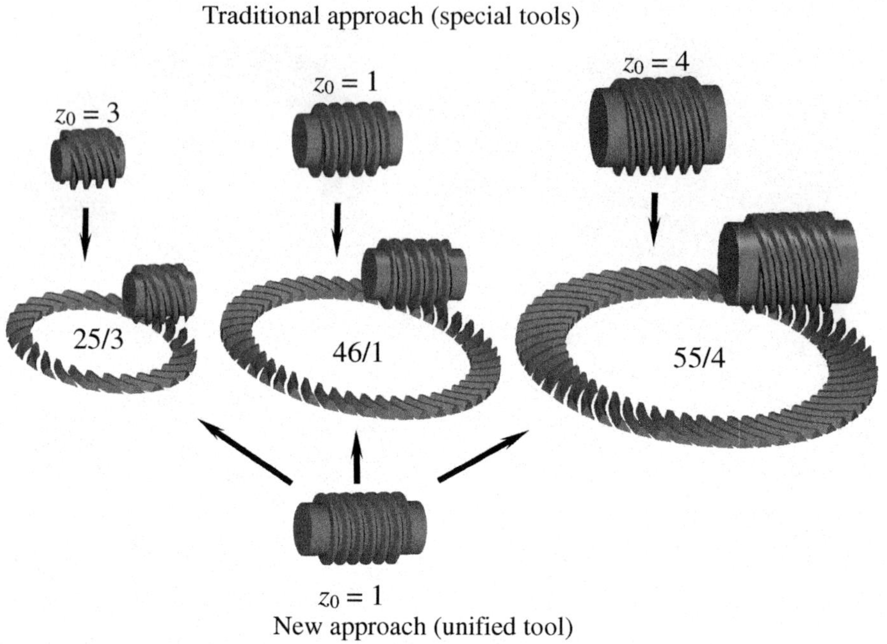

Fig. 10.4 Traditional and new implemented approaches to design and production of worm-type gears

conditions of compliance with the requirements of gear operating characteristics and the degree of contact localization depending on the main design goal are stated either as constraints-inequalities, or as a target function to be minimized or maximized.

10.5 Another Special Case—Spiroid Gears with Small Gear Ratios

It is traditionally assumed that worm gears are convenient where a large reduction in one stage is required. Where the reduction is small ($i \leq 8$), problems arise in a worm gear:

- forces acting in the meshing are less favorably located (Fig. 5a);
- the helix angle of worm threads becomes large (more than 20–25°) which causes difficulties in production, especially in the manufacture and operation of multi-turn hobs.

It became possible to overcome the first difficulty due to the successful distinctive feature of the spiroid gear again—a more favorable direction of the resultant force in the meshing, in particular, exactly at a small gear ratio (Fig. 5b, see also [11]). In this

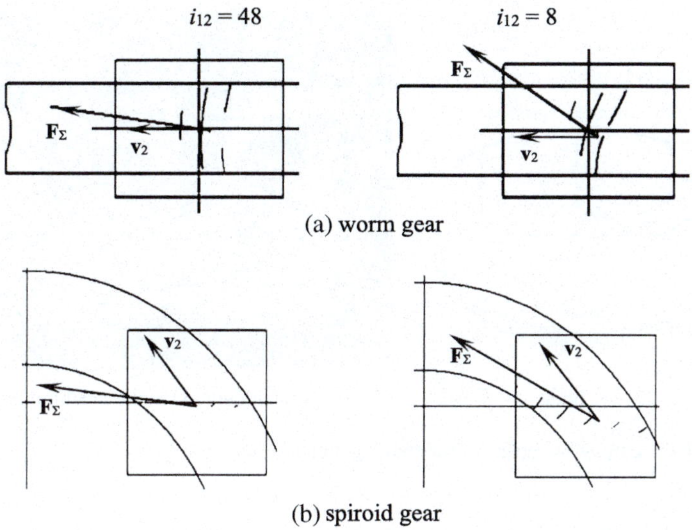

Fig. 10.5 The resultant force in the meshing and vector of gearwheel circumferential speed

respect, the spiroid gear approaches the competitive and more traditional solutions—bevel and hypoid gears. Calculations show that it is possible to reduce contact stresses in spiroid gears [9] and, consequently, either reduce the dimensions or increase the reliability of the solution. But are we basing not only on the calculations, in Fig. 10.6 we can estimate the dimensional effect of implementing spiroid gears instead of bevel and hypoid ones in gearboxes for wedge gate valves.

The main practical motive for development of the discussed alternative is the greater simplicity, flexibility and lower cost of spiroid gear production in comparison with bevel and hypoid gears. This became especially evident with the appearance and implementation of a new technique of spiroid, worm and bevel tooth machining—gear cutting with the help of running cutter heads equipped with carbide inserts [12, 15]—Fig. 10.7. An important side effect of this implementation was the possibility

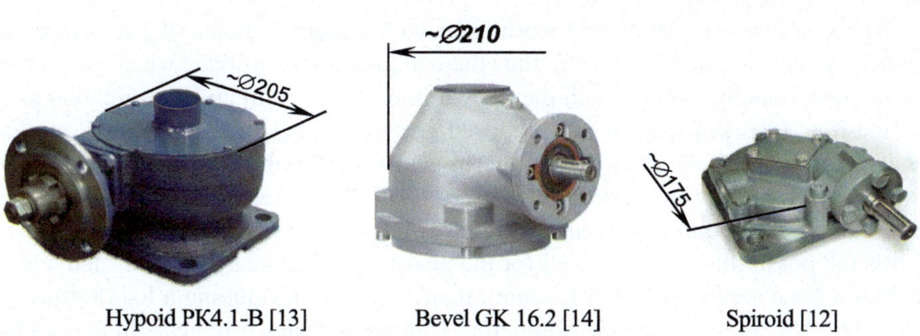

Fig. 10.6 Alternative gearboxes for wedge gate valves

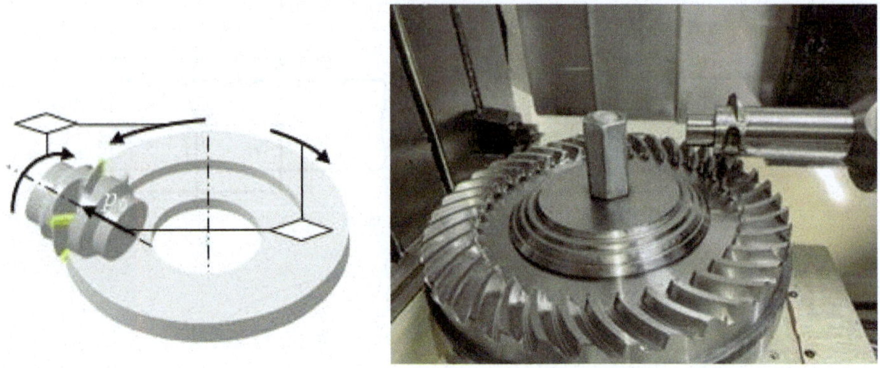

Scheme Implementation at the machining center

Fig. 10.7 Gear cutting by means of the running cutter head

to effectively control the localized contact [9] to provide the desired gear quality and to reduce its noise.

10.6 Steel-Steel Worm Gears

The use of steel for worm (not spiroid) gearwheels occasionally attracts the attention of designers and researchers because steel is stronger and cheaper than traditional bronze and high-strength cast iron. With the exception of low-loaded gears we are almost unaware of any successful practical solutions to this problem. The main reason for this is sharply deteriorated contact conditions in the area of the interaxial line of the gear, the sliding process that takes place virtually along the lines of the conjugated contact (Fig. 8a) which leads to scuffing and, in the case of a steel wheel, to a sharp squeal and a drop in efficiency with rapid contact destruction. We have proposed and put into series production two solutions to this problem—at least for the case of low speeds and heavy loads.

The first is a non-orthogonal worm gear with a special choice of parameters, in which there is no pitch point [16], thus there are no points in space where the vector of relative velocity would touch the helical line of the worm and no so-called axes of meshing (this property of our proposed gear contradicts the established opinion that "axes of meshing exist… in… worm-gear drives with cylindrical and conical worms when the worm is a helicoid" [17, p. 130]). In this case, as Fig. 8b shows, the gear has the desired favorable arrangement of lines of the conjugated contact, which gives the possibility of using steel for the gearwheel. The gearbox with such a gear [18] has been series produced for more than 10 years, developing a load torque up to 72 000 Nm and successfully competing with traditional worm gears in the market of pipeline valve drives.

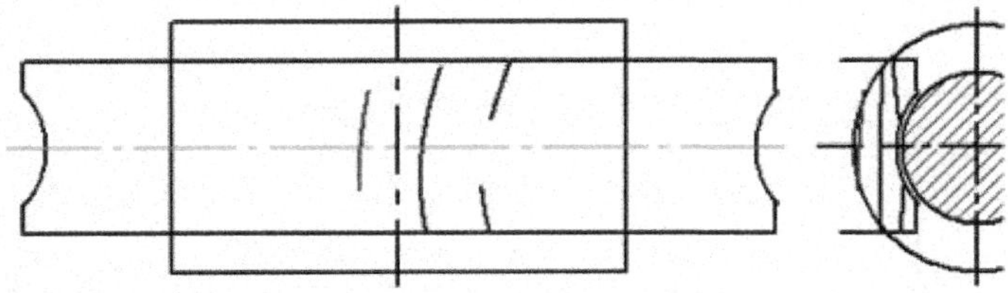

(a) traditional worm gear

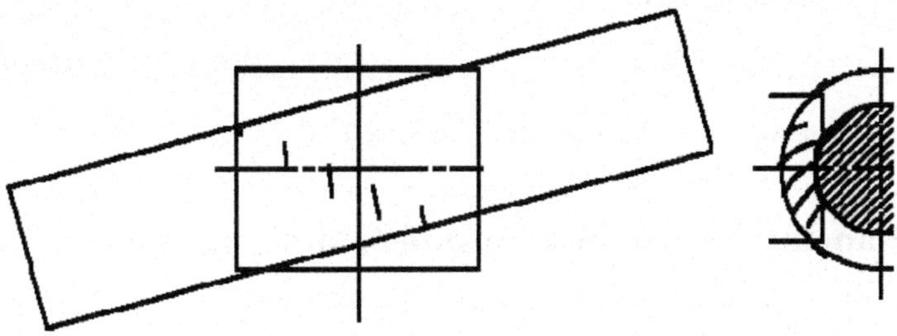

(b) non-orthogonal gear without axes of meshing

Fig. 10.8 Lines of conjugated contact in worm gears

The second solution is the gear which we called QN-gear because of its external resemblance to the "Quarter Note" notation (Fig. 10.9) [19]. The technique itself—to use a part of a traditional worm gearwheel rim and to select parameters in a special way—was proposed by Russian specialists I. M. Egorov and B. Sh. Iofik in [20], but in practice it has not been implemented, apparently, because in case of a bronze rim the sharp reduction of the gearwheel facewidth in comparison with a traditional worm gear could not be compensated by improvement of meshing conditions. Introducing the necessary contact localization created the conditions for using hardened steel and increasing the load capacity of the gear. Thus, when testing the gearwheel of diameter 195 mm we obtained a torque of up to 7000 Nm in quarter-turn mode and motor control, without conceding to the best industrial samples with similar gearwheel diameters [21] made of high-strength cast iron. An additional advantage of worm gears based on QN-gears as compared to traditional ones is a smaller facewidth, which reduces the cost of the gearbox.

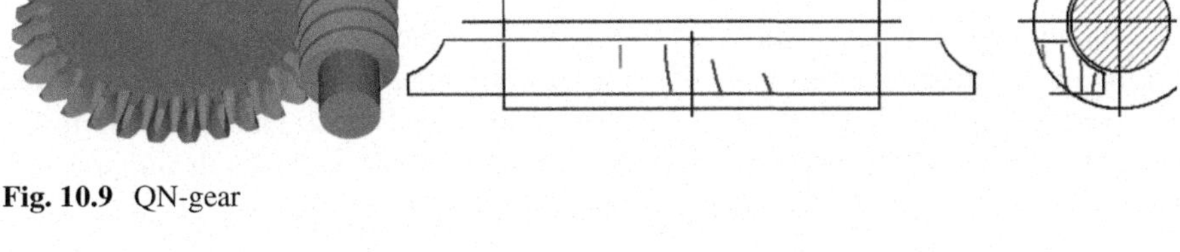

Fig. 10.9 QN-gear

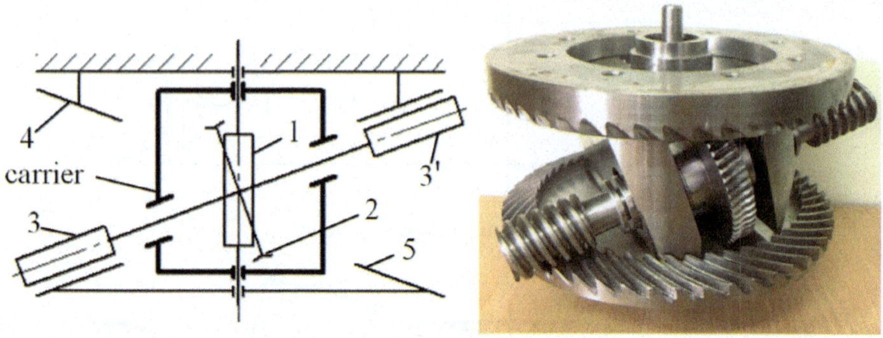

Fig. 10.10 Scheme and prototype of planetary worm and spiroid gear

10.7 Planetary Worm and Spiroid Gear

Planetary gears themselves are often used for large reduction of rotation. The combination of worm type gears and planetary motion in one mechanism (Fig. 10.10) made it possible to achieve extremely high values of the gear ratio with a minimum number of gearwheels [22, 23]. As can be seen in Fig. 10.10, the stages of the mechanism are formed by non-orthogonal worm (input) and spiroid (output) gears. The input link (the sun pinion) here is the worm 1, the output link (the movable crown gearwheel) is the spiroid gearwheel 5 which has a small difference in the number of teeth with the fixed spiroid gearwheel 4. The total gear ratio of the mechanism can be calculated by the formula:

$$i_{\Sigma} = z_{(5)}\left(1 \pm \frac{z_{(2)}z_{(4)}}{z_{(1)}z_{(3')}}\right) \Big/ \left(z_{(4)}\frac{z_{(3)}}{z_{(3')}} - z_{(5)}\right)$$

It ranges from a thousand to several hundred thousand and even more. Such mechanisms should be classified as exotic, used in those rare cases when compactness, extremely low speeds, smoothness and accuracy of rotation are needed. The disadvantage of the mechanism is a relatively low efficiency (0.1…0.6 depending on the gear ratio), that, however, is a general property of mechanisms with a big reduction, which is implemented at a small number of stages.

10.8 Conclusion

The manuscript contains almost no formulas and other mathematical justification (in the design calculations of the considered solutions, of course, they are present and play a crucial role); it gives a number of technical tricks and solutions in the field which is largely considered to be disclosed to an exhaustive depth and practical usefulness both in terms of design and in terms of production. However, the examples given show that the exhaustiveness of the scientific and engineering search often turns out to be apparent. Therefore, the authors hope that these examples will prove to be useful to the engineering community.

References

1. Jáuregui-Correa, J.C., López Cajún, C.S., Sen, M.: Analysis of experimental data from complex multibody system. In: Ceccarelli, M., Hernández Martinez, E. (eds.), Multibody mechatronic systems. Mechanisms and machine science, vol. 25. Springer, Cham (2015)
2. Angeles, J., López-Cajún, C.S.: Kinematic analysis of cam mechanisms. In: Optimization of cam mechanisms. solid mechanics and its applications, vol. 9. Springer, Dordrecht (1991)
3. Angeles, J., López-Cajún, C.S.: The computer-aided drafting and manufacture of cams. In: Optimization of cam mechanisms. Solid mechanics and its applications, vol. 9. Springer, Dordrecht (1991)
4. Goldfarb, V.: Development of the theory and practice of spiroid gears. In: Goldfarb, V., Barmina, N. (eds.) Theory and Practice of Gearing and Transmissions, vol. 34, pp. 55–66. Springer, Cham (2016)
5. Goldfarb, V.I.: What we know about spiroid gearing. In: Proceedings of the international conference on mechanical transmissions, vol. 1, pp. 19–26. Science Press, China (2006)
6. Boloş, V.: Achievements of study concerning worm face gear. Made in Romania. In: Scientific Bulletin of the "Petru Maior" University of Tîrgu Mureş, vol 7 (XXIV), no. 2. (2010)
7. Goldfarb, V.I., et al.: Spiroid gearboxes for pipeline valves. Veche, Moscow (2011)
8. Taits, B.A. (ed.).: Production of gears. Reference book. Mashinostroyeniye, Moscow (1990)
9. Goldfarb, V.I., Trubchev, E.S.: Unification of spiroid hobs. In: Proceedings on the International Conference on Gears, pp. 1755–1759. Munich, Germany (2005)
10. Trubachev, E.S., Savelyevan, T.V., Pushkareva, T.A.: Practice of design and production of worm gears with localized contact. In: Advanced gear engineering, mechanisms and machine science 51, pp. 327–344. Springer, Cham (2018)
11. Goldfarb, V.I., Trubachev, E.S., Pushkareva, T.A., Savelyeva, T.V.: Comparative investigation of worm and spiroid gears with cylindrical worms. In: Uhl, T. (ed.) Advances in Mechanism and Machine Science, IFToMM WC 2019, Mechanisms and Machine Science, vol. 73, pp. 925–935. Springer, Cham (2019)
12. Trubachev, E.S., Bogdanov, K.V., Pushkareva, T.A.: Advanced method of cutting spiroid, worm and bevel gearwheel teeth by running-in cutter heads. J Forsch Ingenieurwes **86**, 709–719 (2022)
13. http://kvark45.com/wp-content/uploads/2018/12/RK4.1-V.pdf. Last accessed 29 April 2023
14. https://www.auma.com/en/products/multi-turn-gearboxes/bevel-gearboxes-gk/. Last accessed 29 April 2023
15. Trubachev, E.S., Bogdanov, K.V., Pushkareva, T.A.: Running-in head with cutting inserts for machining of worm and spiroid gearwheel teeth. Patent RU №2787187. Dated 29.12.2023
16. Trubachev, E.S., Puzanov, V.Y.: New properties of non-orthogonal worm gears. In: Theory and practice of gears and gear-building. Proceedings of scientific and technical conference, pp. 240–244. ISTU Publ., Izhevsk (2008)
17. Litvin, F.L., Fuentes, A.: Gear geometry and applied theory (second edition). Cambridge University Press (2004)
18. Trubachev, E.S., Kuznetsov, A.S., Puzanov V.Y.: New type of worm gearbox. Intellect. Syst. Prod. (2), 79–82 (2014)
19. Trubachev, E.S.: Cylindrical worm gear. Patent RU №2759961. dated 19.11.2021
20. Egorov, I.M., Iofik B.S.H.: Worm cylindrical gear. Patent RU 2 136 987 РФ, МПК F16H 1/16. №99103702/28; claim 03.03.99; publ. 10.09.99

21. https://www.auma.com/en/products/part-turn-gearboxes/gearboxes-gs/#documentsTab. Last accessed 29 April 2023
22. Trubachev, E.S., Mogilnikov, A.V., Barmina, N.A.: Planetary worm and spiroid gearbox. Patent RU №2701637 dated 30.09.2019
23. Trubachev, E.S., Mogilnikov, A.V.: Planetary mechanisms based on worm and spiroid gears. In: Uhl, T. (ed.) Advances in mechanism and machine science. IFToMM WC 2019. Mechanisms and machine science, vol, 73. Springer, Cham (2019)

Chapter 11
Design of Gear Pump of Electro-Hydrostatic Actuator for Robots

Mitsuo Komagata, Ko Yamamoto, and Yoshihiko Nakamura

Abstract An electro-hydrostatic actuator (EHA) is suitable to force controlled robots that interact with surrounding objects because it has high robustness and backdrivability. The major issues of an EHA for robots are internal leakage and its size and weight. Especially, in the design of an EHA, it is important to pay attention to the internal clearances of a hydraulic pump, which is a major source of internal leakage. Since there is a trade-off between internal leakage and friction regarding internal clearances, it is necessary to adjust the clearance properly. In addition, since a pump of an EHA is bidirectional and variable pressure, it is required to consider the change of the state of the internal clearances. This paper summarizes the key points in the design of an external gear pump for an EHA focusing on the internal clearances from the previous researches.

Keywords Hydraulic actuator · Internal leakage · Friction · Backdrivability

11.1 Introduction

Professor Carlos Lopez-Cajun contributed and made an impact to robot kinematics from a theoretical and design point of view. The third author had in depth conversations, when Professor Lopez-Cajun visited his lab in 2012 and when they met

M. Komagata · K. Yamamoto (✉)
Department of Mechano-Informatics, Graduate School of Information Science and Technology, The University of Tokyo, 7-3-1 Hongo, Bunkyo-Ku, Tokyo 113-8656, Japan
e-mail: yamamoto.ko@ynl.t.u-tokyo.ac.jp

M. Komagata
e-mail: komagata@ynl.t.u-tokyo.ac.jp

Y. Nakamura
Professor Emeritus, The University of Tokyo, 7-3-1 Hongo, Bunkyo-Ku, Tokyo 113-8656, Japan
e-mail: 3961323879@g.ecc.u-tokyo.ac.jp

M. Ceccarelli and J. C. Jauregui-Correa (eds.), *State-of-the-Art and Innovations in Mechanism and Machine Science*, Mechanisms and Machine Science 150,
https://doi.org/10.1007/978-3-031-47040-0_11

at conferences and meetings, about anthropomorphic robots. The authors forked a design problem of hydraulic actuators from the study of anthropomorphic humanoid robots, which is the main subject of this paper and clearly benefited from the conversations with Professor Carlos Lopez-Cajun.

Since working area of the conventional robots is usually in a restricted area of industrial facilities and they require fast and accurate performances, they use rigid actuators using gear reducers with high reduction ratio. On the other hand, in the case of robots that interact with various environment, force controllability and durability to impact load are important. Backdrivability is one of the key properties that contributes to these requirements. Conventional highly geared actuators tend to lack backdrivability because of large friction caused by high reduction ratio. Bilateral drive gear by Matsuki et al. [1] achieved high backdrivability at a reduction ratio as high as approximately 100, though such gear transmissions tend to be vulnerable to an impact load because of small contact area of gear teeth.

Series elastic actuators [2] obtain backdrivability by incorporating elastic element before its output end. A drawback is the low operating bandwidth due to the low mechanical stiffness. Variable stiffness actuators [3] can adjust stiffness and obtain backdrivability on demand, while a drawback is that it requires for each degrees of freedom an additional actuator for stiffness adjustment. An electro-hydrostatic actuator (EHA) [4, 5] has a closed hydraulic circuit composed of a hydraulic actuator and a hydraulic pump controlled by a servo motor. It is suitable for backdrivability because its mechanical simplicity without a servo valve allows backflow of hydraulic oil. Since the hydraulic oil is incompressible, transmission stiffness and fast response are achieved. Ko et al. [6] experimentally reported the response frequency of the pressure control a prototyped EHA as high as 100 Hz. As the applications of an EHA to robots, the humanoid Hydra [7] achieved robustness to endure disturbance, and the robot arm Hydracer [8] showed robustness to achieve impulsive hammering motion. The design challenges of EHA actuators have been internal leakage and its size and weight as compared to the alternative actuators.

Internal leakage exists between high and low pressure volumes in a closed hydraulic circuit, which is in one sense necessary to keep the low friction, but minimized since the internal leakage generates heat in the hydraulic circuits and lower the viscosity of hydraulic oil. The requirements of size and weight make the mechanical design for low internal leakage more challenging. This paper introduces the design method of a gear pump focusing on the internal clearances that is the major source of internal leakage and friction by summarizing the previous designs that relate to the internal clearances.

11.2 Difference Between EHA and Conventional Geared Motor

11.2.1 Backdrivability

An EHA has two backdrive states [9]. One is the state that only output axis backdrives, and the other is the state that both input and output axis backdrive. The former is called output backdrive and the latter is called total backdrive. The reason why an EHA has two backdrive states is that an EHA permits backflow of oil without the backward rotation of input axis (pump) by internal leakage, backflow at internal clearances such as between gear and casing. On the other hand, in the case of geared motor, there is no state of output backdrive, and only total backdrive occurs. In short, input axis backdrives in association with output axis.

Output backdrivability depends on the friction of output axis, and total backdrivability depends on the friction of both input and output axis. Since sliding parts of an EHA are immersed in oil, it tends to have smaller friction compared with gear reducers. Therefore, it tends to have higher backdrivability compared with gear reducers. Moreover, an EHA can backdrive at small torque corresponding to output backdrivability that only depends on the friction of output axis.

On the other hand, total backdrivability is also important in terms of safety. When only output axis backdrives, impact power applied from external world must be consumed inside the transmission, which may cause breakage in case of excessive load. To improve output backdrivability, it is necessary to reduce friction at the output axis. To improve total backdrivability, it is necessary to reduce friction of a pump.

11.2.2 Size and Weight

Among the actuators that use servo motor and gear reducers, Harmonic drive and planetary gear reducers have advantage in miniaturization. However, an EHA tends to be heavy and large because it includes many components such as a servo motor, hydraulic pump, hydraulic actuator, and other hydraulic accessories.

On the other hand, an EHA can be compact compared with servo valve system that uses one pump and multiple servo valves. This is because an EHA can arrange all components in a compact unit. For example, in the case of linear type EHA, there are integrated designs [10, 11] that a pump and a motor are arranged right beside the cylinder. In the case of rotary type EHA, there is an integrated design [12] using flat and hollow frameless motor as shown in Fig. 11.1. In this figure, red-colored pump is arranged inside the hollow space of green-colored motor. Therefore, miniaturization of a pump is necessary.

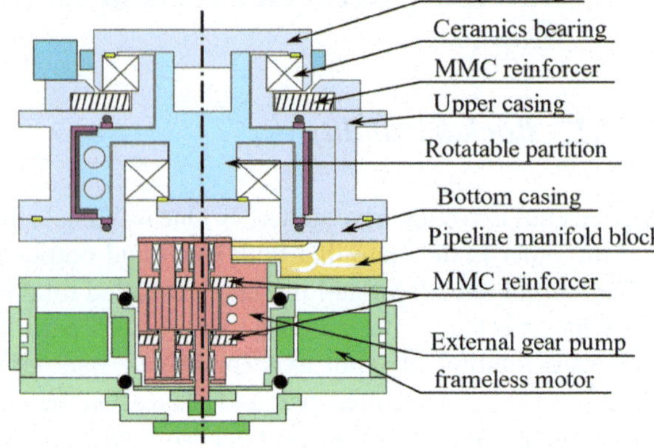

Fig. 11.1 Module design of rotary EHA using flat and hollow frameless motor

11.2.3 Internal Leakage

Internal leakage is one of the major differences between an EHA and geared motor. It is a phenomena specific to fluid system. Internal leakage indicates the backflow from high to low pressure area inside hydraulic system. Increase of internal leakage reduces the volumetric efficiency and reduces the actuator speed. The reduction ratio of an EHA is given as

$$i_{\omega} = k_v / k_p + q_{leak} / k_p \omega_{out} \tag{11.1}$$

where k_v is theoretical flow rate per 1 rad of output side (such as cylinder and vane motor), k_p is theoretical flow rate per 1 rad of input side (pump), q_{leak} is internal leakage flow rate, ω_{out} is an output velocity. Since second term on the right side is the term that depends on internal leakage, it increases as internal leakage increases. Therefore, the output speed of an EHA tends to be inferior to that of geared motor because large internal leakage decreases output speed. In addition, the efficiency of an EHA tends to be inferior to that of geared motor because internal leakage is a loss. Therefore, it is critically important to reduce internal leakage in the design of an EHA.

11.3 Development of a Pump for Robotic EHA

11.3.1 Types of a Pump

Since a hydraulic pump plays an important role in an EHA as the source of pressure, it is important to pay attention to its design. For an EHA to drive robots, hydraulic pump requires high discharge pressure for high-torque output and large discharge amount for high-speed operation. Especially, small friction at a pump is required for high backdrivability, and small and lightweight system is required for robots working standalone.

There are several types of hydraulic pumps such as a piston pump, vane pump, and gear pump. Piston pump has small internal leakage and thus can obtain large discharge pressure. The drawback is the relatively large friction caused by the sliding parts including a piston. Vane pump can obtain large discharge amount. The drawback is the relatively large friction caused by sliding vane on cam ring and low discharge pressure. Gear pump can reduce sliding parts by appropriate design and has advantage in miniaturization because of its simple structure. For highly backdrivable and small hydraulic system, gear pump is expected to be a proper choice.

Hereinafter, we select external gear pump and deal with the design of internal clearances that are important portion relating backdrivability (friction) and internal leakage.

11.3.2 Design of Internal Clearances of a Gear Pump

Inside the gear pump, the clearance between rotating gears and casing is necessary. In this clearance, backflow (internal leakage) is generated from high to low pressure area. On the other hand, in the case of very small clearance, there is a risk of large viscous friction or sliding friction. It indicates the trade-off between internal leakage and backdrivability. Therefore, in the design of internal clearance, it is necessary to reduce internal leakage without losing backdrivability. This paper classifies the internal clearances of a gear pump into the following two portions as shown in Fig. 11.2:

1. Gear side clearance
2. Gear tip clearance

The feature of the former clearance is that its size changes according to the change of internal pressure, and that of the latter clearance is that its size changes according to the change of differential pressure. Therefore, it is necessary to pay attention to

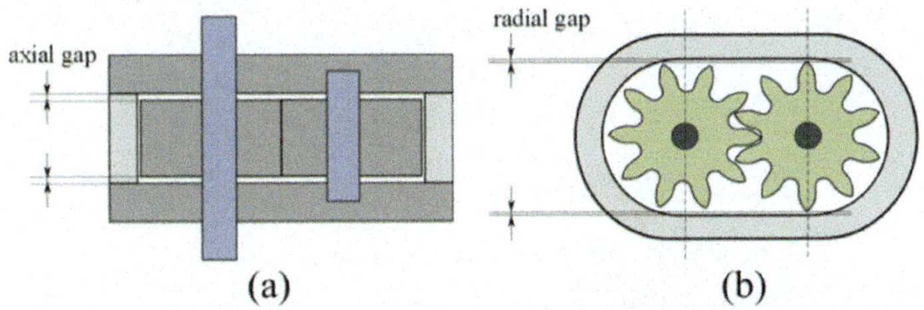

Fig. 11.2 Types of internal clearances of external gear pump. **a** Gear side clearance, **b** gear tip clearance

each feature. In the following sections, we introduce the design relating internal clearance of an external gear pump.

11.3.3 Design of Gear Side Clearance

The representative method to reduce gear side clearance is floating wear plate [13]. It is a method to eliminate the clearance by pushing the floating side-plate toward gears by pressure allocation. The drawback is the friction between rotating gears and the side-plate. It deteriorates the backdrivability.

To solve this problem, we proposed frameless thrust bearing [14] as the method to avoid sliding between gears and sidewalls. This structure is designed as a thin layer composed of small bearing balls (diameter of 1 mm) and accurate thin plate (thickness of 1 mm) as shown in Fig. 11.3, which is suitable to miniaturized design. Since highly accurate bearing balls are commercially available, the clearance can be determined accurately in microns order by manufacturing the thin plate to have accurate thickness. The clearance is obtained by the difference of the thickness of the accurate thin plate and the diameter of the bearing balls.

Regarding the clearance, internal leakage and friction has a trade-off relation. In the case of small clearance, internal leakage decreases and viscous friction increases. This relation is seen in gear side clearance where viscous friction is large because the area of gear-side surface adjacent to the casing is wide. Toward this problem, we proposed the method to obtain the optimal clearance by assuming the simple model of gear side clearance [11].

Concretely, we formulate the internal leakage loss and viscous friction loss generated at gear side clearance. In the following discussion, the size of gear side clearance is assumed to be constant regardless of the position for simplification. First, we formulate the internal leakage loss. Since internal leakage generated at the clearance between the gear side surface and the casing is assumed to be a flow between two parallel plates, it can be regarded as Poiseuille flow assuming laminar flow. In this case, internal leakage loss is proportional to the cube of the clearance.

Here, we assume the internal leakage flow of an external gear pump as shown in Fig. 11.4. The gear profile model is composed of a circle whose diameter is equal to

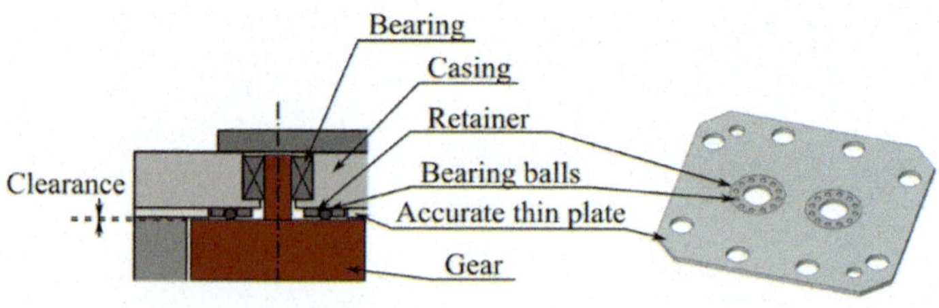

Fig. 11.3 Layer of frameless thrust bearing inside external gear pump for accurate adjustment of internal gear side clearance

gear root diameter and a rectangle whose width is equal to gear thickness. We assume that the internal leakage at gear side clearance is the sum of (1) internal leakage on the circle, (2) internal leakage on the rectangle that is adjacent to the casing, and (3) internal leakage on the rectangle between gears (gear meshing portion is assumed to be a rectangle). When the differential pressure is Δp, internal leakage flow rate is given as

$$q_L(x) = \left[\frac{\{r_{root}^2(2\alpha + \sin 2\alpha) + \pi\alpha^2\}\Delta p}{12\mu(r_{root}^2 - r_{in}^2)} + \frac{(\frac{\pi}{2} - \alpha)\Delta p}{3\mu} + \left(1 + \frac{2}{n}\right)\frac{s\Delta p}{6\mu m_e} \right] x^3 \equiv K x^3 \tag{11.2}$$

where r_{root} is the radius of gear root circle, r_{in} is the radius of the through hole of the casing, μ is the viscosity of oil, n is the number of gear tips that is adjacent to the casing, m_e is the thickness of the gear, x is the gear side clearance, and α is defined as $\alpha = \arcsin(r_{in}/r_{root})$. Internal leakage loss at gear side clearance is obtained from

$$P_q = \Delta p q_L = \Delta p K x^3 \propto x^3 \tag{11.3}$$

On the other hand, when the viscous friction between the gears and the casing is modeled as Couette flow, viscous friction loss is inversely proportional to the clearance x.

As shown in Fig. 11.5, we assumed the gear profile to be the circle whose diameter is equal to pitch circle diameter of the gear. When the gear rotates at the speed of $\dot{\theta}$, two-dimensional shear stress is given as

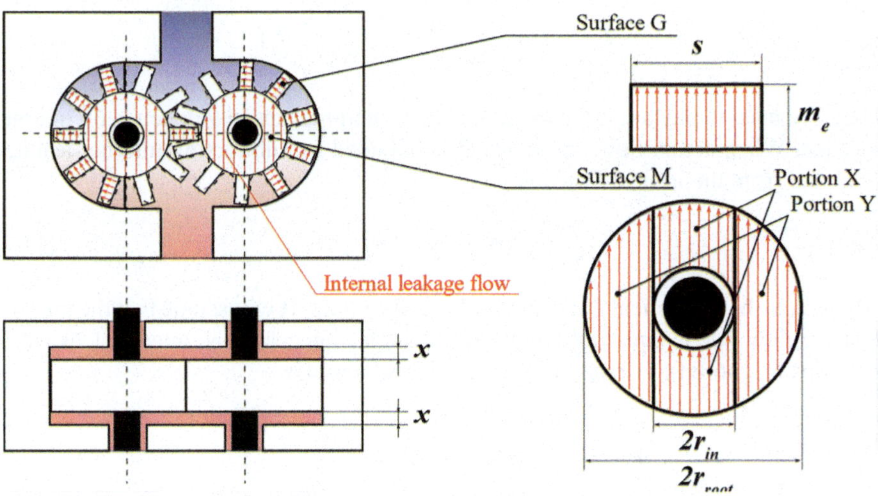

Fig. 11.4 Simplified model of internal leakage flow paths at gear side clearance

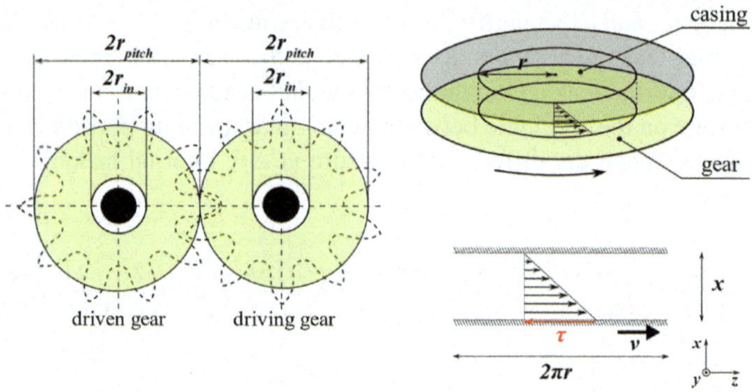

Fig. 11.5 Simplified model to estimate viscous friction at gear side gap. The profiles of drive and driven gears are assumed to be a circle whose diameter is equal to pitch circle diameter of the gears

$$\tau = \mu \frac{\partial v}{\partial x} \tag{11.4}$$

Viscous friction torque applied on the rotating circle is obtained by integrating (11.4) on the circle, which is given as

$$\tau_v(x) = \frac{2\pi\mu\dot{\theta}}{x}\left(r_{pitch}^4 - r_{in}^4\right) \tag{11.5}$$

where r_{pitch} is the radius of pitch circle. Viscous friction loss is obtained from

$$P_v = \tau_v\dot{\theta} = \frac{2\pi\mu\dot{\theta}^2}{x}\left(r_{pitch}^4 - r_{in}^4\right) \tag{11.6}$$

Here, we need to pay attention to the fact that rotational speed $\dot{\theta}$ depends on internal leakage. When actual flow rate is Q, Q is obtained by subtracting internal leakage flow rate from theoretical flow rate.

$$Q = k_d\dot{\theta} - q_L \tag{11.7}$$

where k_d is the theoretical flow rate per 1 rad of a pump. If we assume that the internal leakage flow rate is equal to that at gear side clearance (that is, internal leakage at gear tips is ignored), viscous friction loss is expressed as

$$P_v = \frac{2\pi\mu}{x}\left(r_{pitch}^4 - r_{in}^4\right)\left(\frac{Q + Kx^3}{k_d}\right)^2 \tag{11.8}$$

If $Q \gg q_L$, viscous friction loss is approximated by the following equation, which shows that it is inversely proportional to the clearance.

$$P_v = \frac{2\pi\mu}{x}\left(r_{pitch}^4 - r_{in}^4\right)\left(\frac{Q}{k_d}\right)^2 \propto x^{-1} \tag{11.9}$$

The theoretical optimal clearance to obtain the maximum efficiency (the least loss) is obtained by minimizing the sum of internal leakage loss $P_l(w)$ and viscous friction loss $P_v(w)$.

$$\min_x\left(P_q(x) + P_v(x)\right) \tag{11.10}$$

We obtained the relation between clearance and the loss when the differential pressure is 1 or 14 MPa, and flow rate is 3.63, 7.26, 10.9, 14.5 cm^3/s, and the temperature is 40 °C. Here, we use the loss ratio ξ that is defined as the value dividing the loss $\left(P_q(x) + P_v(x)\right)$ by the sum of output and the loss $\Delta pQ + P_q(x) + P_v(x)$ for normalization.

$$\xi = \frac{\left(P_q(x) + P_v(x)\right)}{\Delta pQ + P_q(x) + P_v(x)} \tag{11.11}$$

Figure 11.6 shows this relation. When the differential pressure is high (Δp = 14 MPa), optimal clearance is narrower than that of low-pressure case (Δp=1 MPa). As the flow rate becomes smaller, optimal clearance gets narrower.

From this result, optimal clearance differs according to differential pressure and flow rate. Since the differential pressure and flow rate of an EHA changes according to the working condition, it is desirable to determine the gear side clearance based on the most frequently used condition.

On the other hand, the pump using fixed side plate that uses no floating side-plate, has a problem of increasing internal clearance due to internal pressure. Internal pressure swells the casing and then increases the clearance. Especially, the inner surface of the casing that is adjacent to the gear side surface has wide area to receive

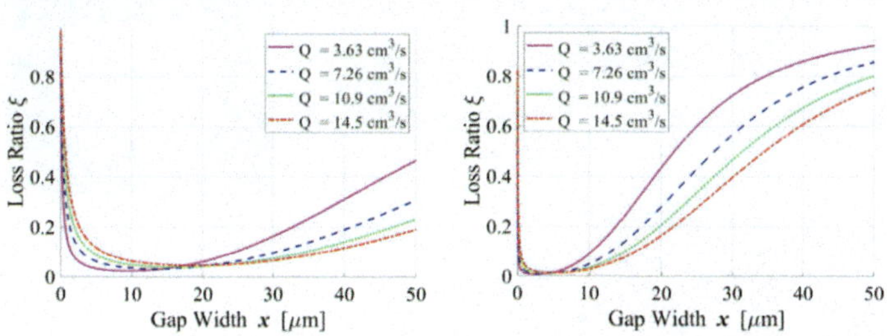

(a) Δp=1 MPa (b) Δp=14 MPa

Fig. 11.6 Simulated relation between gear side clearance and loss ratio. **a** Differential pressure is 1 MPa, **b** differential pressure is 14 MPa pressure by oil, and it is easily deformable by large force. Therefore, it is necessary to suppress deformation by increasing stiffness of the casing. To enhance stiffness without increasing size and weight, we applied fine ceramics that is rigid and lightweight material as shown in Table 11.1 to enhance the casings [14]. Figure 11.7 is the pump structure using ceramics reinforcers, which also includes frameless thrust bearing of Fig. 11.3. From the result of FEM using the designed casings as shown in Fig. 11.8, the upper and bottom casings of the pump with ceramics reinforcers can reduce deformation by 46 or 57% in the condition of 14 MPa internal pressure.

On the other hand, typical ceramics such as alumina has an issue of fragility, they easily broke by an impact. As the countermeasure, MMC (Metal Matrix Composite), the material that combined ceramics and metal, is a better candidate [16]. The drawback of these ceramics-based material is high cost because they are difficult-machining material compared to typical metals.

Table 11.1 Comparison of the mechanical properties of steel, fine ceramics (Alumina), and MMC [15]

	Steel	Alumina (Al_2O_3)	MMC (SA701)
Young's modulus [GPa]	200	370	260
Density [kg/m^3]	7.8×10^3	3.9×10^3	3.0×10^3
Fracture toughness [MPa $\cdot \sqrt{m}$]	–	3	8

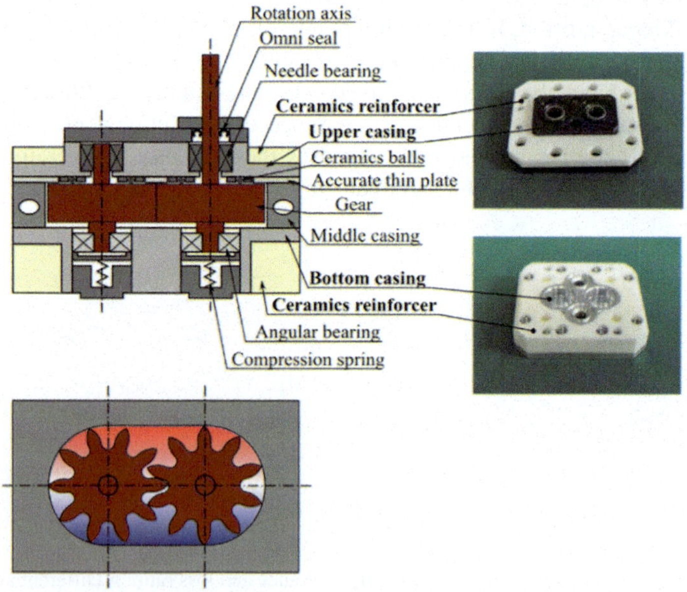

Fig. 11.7 Involute gear pump with ceramics enhanced casings

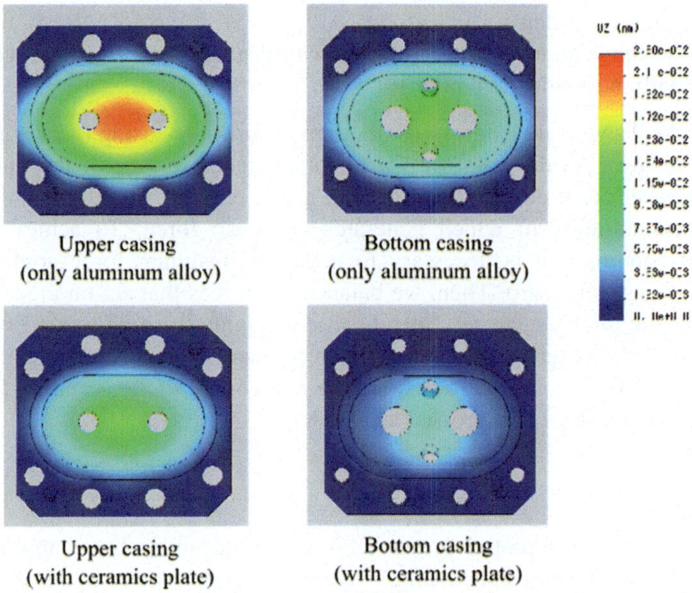

Fig. 11.8 FEM results of the deformation at 14 MPa internal pressure. Right shows the deformation of the upper casing with and without ceramics plate, and left shows that of the bottom casing

11.3.4 Design of Gear Tip Clearance

Differential pressure in an external gear pump generates eccentric force on gears, which changes gear tip clearance. If the gear tips contact with the inner surface of the casing, large force that is proportional to differential pressure acts between the gear and the casing, which generates large friction. As the method to prevent this friction, preliminary running-in operation ensures gear tip clearance [17]. This method scrapes the inner surface of the casing by preliminary pump-rotation to avoid the contact at the gear tips. The condition of working pump that its method is effective is unidirectional and constant pressure. In this condition, gear tip clearance is ideally zero because it doesn't change.

On the other hand, this method is not suitable to a hydraulic pump of an EHA. This is because a pump of an EHA is bidirectional and the differential pressure changes according to working condition. To avoid contact between the gear and the casing of a pump of an EHA, bidirectional running-in operation at the maximum pressure is needed, which increases internal clearances and thus increases internal leakage. Therefore, gear pump of an EHA needs new method to reduce both internal clearance and friction.

Then, we proposed the external gear pump with movable casing inside it [18]. The features of this structure are the following two points:

(A) Two-layered casing
(B) Inner casing can move in accordance with gear eccentricity without resistance (virtually no-loaded condition)

Figure 11.9 shows the conceptual diagram of movable casing (hereinafter called movable casing). As the point to pay attention, inner surface of the inner casing receives non-uniform pressure. By rotating gears, pressure distribution at the inner surface gets un-uniform which generates eccentric force. To achieve virtually no-loaded condition, it is necessary to make the resultant force applied on the movable casing to be zero. Then, we balance the forces that act on inner and outer surfaces of the movable casing. Specifically, we guide the high/low pressure oil to the groove in the outer surface, whose bottom area is S_p, as shown in Fig. 11.10. The forces applied on these grooves are the resist force to the eccentric force. The relation of these forces is expressed as follows:

$$S_p \Delta p = F \tag{11.12}$$

where Δp is the differential pressure, F is the eccentric force applied on the inner surface of the movable casing. Therefore, to determine S_p, it is necessary to estimate F.

In reference [18], F is estimated assuming that the pressure at the curved surface inside the movable casing changes with constant slope. In this case, pressure distribution at the curved surface which is adjacent to gear tips is expressed as

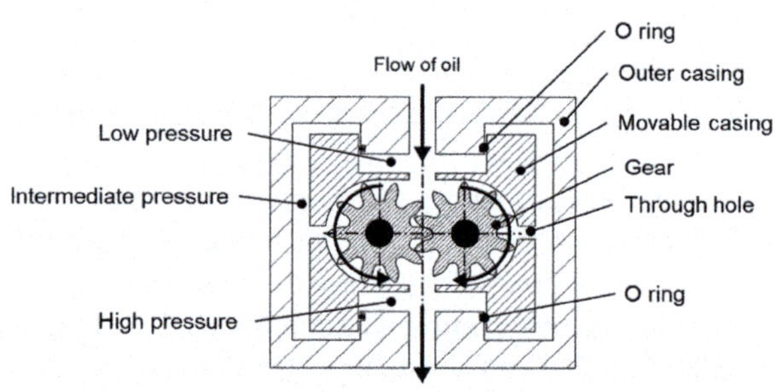

Fig. 11.9 Conceptual diagram of movable casing in an external gear pump

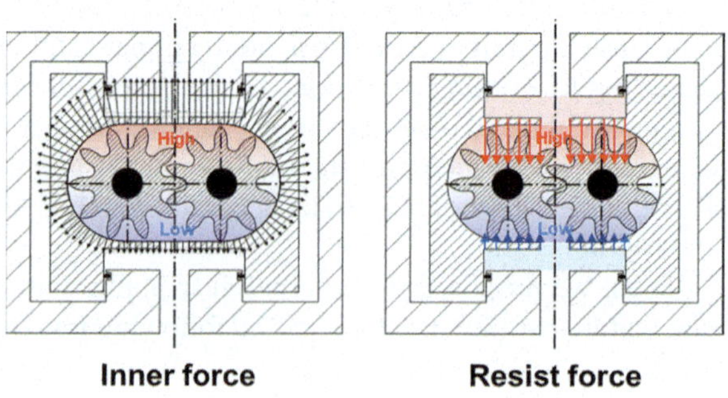

Fig. 11.10 Pressure allocation to achieve virtually no-loaded condition of movable casing

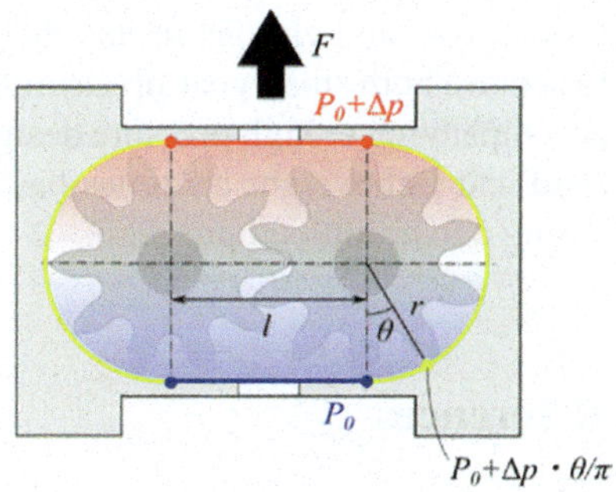

Fig. 11.11 Estimation of the pressure at inner surface of the movable casing

$$P_r(\theta) = P_0 + \Delta p \cdot \theta/\pi \tag{11.13}$$

where P_0 is the pressure of low-pressure side. The eccentric force applied on the inner surface of the casing is given as (Fig. 11.11)

$$F = ld\Delta p - 2d\int_0^{\pi} P_r(\theta)cos\theta \cdot rd\theta = \left(ld + \frac{4rd}{\pi}\right)\Delta p \tag{11.14}$$

Therefore, pressure receiving area is obtained from

$$S_p = ld + \frac{4rd}{\pi} \tag{11.15}$$

By setting the pressure receiving area to be the value calculated from (11.15), reduction of the friction of gear tips is expected. By using the manufactured prototype, friction torque was reduced by 63% compared with the pump whose movable casing was fixed [18].

11.4 Conclusion

Compared with the conventional geared motor, the problem of an EHA for robotic applications is internal leakage and its size and weight. Especially, in the design of an EHA, it is important to pay special attention to internal clearance of a gear pump which is a major source of internal leakage. In terms of its clearance, there is a trade-off relation between internal leakage and friction. We introduced the method to reduce internal leakage without losing backdrivability from two portions of gear side clearance and gear tip clearance. In concrete, as the method to solve the trade-off problem between internal leakage and viscous friction, we introduced the thin frameless thrust bearing which also plays a role of accurate adjustment of gear side clearance, the method to obtain optimal gear side clearance by formulating internal leakage loss and viscous friction loss, and the pump structure using movable casing to achieve both small gear tip clearance and small friction. Moreover, as the problem accompanying to high pressure design, we introduced the reinforcer of a pump using rigid and lightweight ceramics-based material to suppress the deformation of the casing due to internal pressure.

References

1. Matsuki, H., Nagano, K., Fujimoto, Y.: Bilateral drive gear—a highly backdrivable reduction gearbox for robotic actuators. IEEE/ASME Trans. Mechatron. **24**(6), 2661–2673 (2019)
2. Pratt, G.A., Williamson, M.M.: Series elastic actuators. In: Proc. of IEEE/RSJ Int'l Conf. on Intelligent Robots and Systems, pp. 339–406 (1995)
3. Jafari, A., Tsagarakis, N.G., Vanderborght, B., Caldwell, D.G.: A novel actuator with adjustable stiffness (AwAS). In: Proc. of IEEE/RSJ Int'l Conf. on Intelligent Robots and Systems, pp. 4201–4206 (2010)
4. Bobrow, J.E., Desai, J.: A high torque to weight ratio robot actuator. Robotica **13**, 201–208 (1995)
5. Habibi, S., Goldenberg, A.: Design of a new high-performance electrohydraulic actuator. IEEE/ASME Trans. Mechatron. **5**, 158–164 (2000)
6. Ko, T., Kaminaga, H., Nakamura, Y.: Current-pressure position triple-loop feedback control of electro hydrostatic actuators for humanoid robots. Adv. Robot. **32**(24), 1269–1284 (2018)
7. Ko, T., Murotani, K., Yamamoto, K., Nakamura, Y.: Whole-body compliant motion by sensor integration of an EHA-driven humanoid hydra. Int. J. Humanoid Rob. **18**(01), 2150002 (2021)
8. Komagata, M., Imashiro, Y., Suzuki, R., Oishi, K., Yamamoto, K., Nakamura, Y.: Experimental study on impact resistance of multi-DOF electro-hydrostatic robot systems using Hydracer, a 6DOF arm. In: Proc. of 2022 IEEE/RSJ Int'l Conf. on Intelligent Robots and Systems, pp. 7352–7358 (2022)
9. Kaminaga, H., Ono, J., Nakashima, Y., Nakamura, Y.: Development of backdrivable hydraulic joint mechanism for knee joint of humanoid robots. In: Proc. of Int'l Conf. of Robotics and Automations, pp. 1577–1582 (2009)
10. Altare, G., Vacca, A.: A design solution for efficient and compact electro-hydraulic actuators. Procedia Eng. **106**, 8–16 (2015)
11. Komagata, M., Yamamoto, K., Nakamura, Y.: Compact, backdrivable, and efficient design of linear electro-hydrostatic actuator module. Adv. Robot. **36**(19), 1030–1047 (2022)
12. Nakanishi, T., Komagata, M., Yamamoto, K., Nakamura, Y.: Toward high power-to-weight ratio electro-hydrostatic actuators for robots. In: International Symposium on Experimental Robotics (2020)
13. Koc, E., Hooke, C.J.: An experimental investigation into the design and performance of hydrostatically loaded floating wear plates in gear pumps. Wear **209**, 184–192 (1997)
14. Komagata, M., Ko, T., Nakamura, Y.: Design and development of compact ceramics reinforced pump with low internal leakage for electro-hydrostatic actuated robots. In: Proc. of the 15th IFToMM World Congress on Mechanism and Machine Science (2019)
15. Japan Fine Ceramics Co., Ltd.: SA701 details. https://www.japan-fc.co.jp/en/products/cate01/cate0104/sicasa-1.html
16. Komagata, M., Imashiro, Y., Yamamoto, K., Nakamura, Y.: Preferred oil and ceramics options for EHA drive systems and computed torque control of an EHA-driven robot manipulator. In: Proc. of 30th IEEE International Conference on Robot and Human Interactive Communication, pp. 540–545 (2021)

17. Manco', S., Nervegna, N.: Simulation of an external gear pump and experimental verification. Proc. JFPS Int. Symp. Fluid Power **1989**(1), 147–160 (1989)
18. Komagata, M., Yamamoto, K., Nakamura, Y.: Design of movable casing for friction reduction of external gear pump for hydraulic actuators. In: Proc. of the 16th IFToMM World Congress (2023)

Chapter 12
Overview of Special Wire Mechanisms Used for Self-balancing Mechanisms

Lovasz Erwin-Christian, Moldovan Cristian Emil, and Mateaş Marius Corneliu

Abstract Wire mechanisms are a special type of centroidal mechanism using a wire, which is enveloped or developed on a noncircular wheel, when one side of the wire is fixed on the noncircular wheel. The variable instantaneous length of the links: radius of a noncircular wheel and free length of the wire allows the generation of a variable moment or transmission function. Self-balancing methods using special type of belt mechanisms can be used in many technical applications in the field of mechanical devices, measurement tools, robotics, etc. The chapter shown the self-balanced measurement head of a coordinate measuring device, the Archimedean force compensatory mechanism in a constant pressure chamber, the force equilibrium mechanism for a loading platform and the self-balancing Conco-Balancer manipulator.

Keywords Wire mechanism · Noncircular wheel · Self-balancing · Measurement head · Archimedean force compensator · Loading platform · Conco-Balancer

12.1 Introduction: State-of-Art

Professor Carlos Lopez-Cajun was a distinguished personality in Mechanism and Machine Science and a great IFToMMist in the worldwide IFToMM community. He served IFToMM as Chair of Mexico IFToMM member organization, member of the Permanent Commission on History of MMS, member of the Executive Council and also as Secretary General in 2008–2011. His work covered aspects of mechanisms design, cam transmissions, design of service robots, machine diagnostics,

L. Erwin-Christian (✉) · M. C. Emil · M. M. Corneliu
Department of Mechatronics, Politehnica University of Timişoara, Timişoara, Romania
e-mail: erwin.lovasz@upt.ro

M. C. Emil
e-mail: cristian.moldovan@upt.ro

M. M. Corneliu
e-mail: marius.mateas@upt.ro

M. Ceccarelli and J. C. Jauregui-Correa (eds.), *State-of-the-Art and Innovations in Mechanism and Machine Science*, Mechanisms and Machine Science 150,
https://doi.org/10.1007/978-3-031-47040-0_12

vehicle mechanics and history of mechanisms and machines. This paper wants to be a contribution to the tribute of his activity in promoting and enhancing MMS.

The technical literature shows many solutions for self-balancing of the manipulators and mechanisms. These solutions were developed for statically or dynamically conditions using counterweight, springs, wire mechanisms and actuators. Baradat et al. in [1] and Arakelian in [2] systematized the balancing solutions for robotic systems by means of balancing using counterweights, springs, pneumatic or hydraulic cylinders, electromagnetic device, etc.

In [3] was given for reducing the vibrations a solution for dynamic balancing through inertial forces and moments optimisation. The dynamic balancing was formulated as an optimisation problem considering the balancing of the shaking forces accomplished through analytically balancing constraints. The optimization function based on the sensitivity analysis of shaking moment in respect to the kinematic parameters of the links was used in order to minimise the shaking moment. The dynamic balancing of four-, five-, and six-bar linkages with prismatic pairs was presented by Feng in [4] using a combination of mass redistribution and the addition of two inertial counterweights. The inertial counterweights were ordinary geared trains or planetary-gear trains. A complete balancing of planar linkages was developed by Ye and Smith in [5] as an equivalence method, which considered the effects of inertial moments and forces of a link by an equivalent simple links. The complete shaking forces and moments balancing of spatial mechanisms was presented in [6] using the methods of multibody dynamics.

Kazerooni in [7] treated a statically balanced method without using the gravity forces (without any counterweights) and choosing of smaller actuators and amplifiers. In [8] Wang and Gosselin studied static balanced manipulators with revolute actuators by using counterweights or springs. A hybrid methodology for balancing of spatial manipulator, that combines balancing using counterweights, springs and auxiliary parallelograms is presented by Agrawal and Fattah in [9]. Wire mechanism used for static force balancing of mechanisms was treated by Hain in [10] and Perju in [11]. In [11] Perju present an application of gear boxes with variable load for testing or running by using special types of belt mechanism with a spring.

Streit and Shin in [12] provided a mathematical algorithm for comparing the complexity of equilibration methodologies. For a few topologies of 3-DOF planar parallel mechanisms has been developed the dynamic analysis using counterweights or springs or a combination of both by Laliberte et al. [13], to illustrate the behaviour of static balancing and to characterise of each of the mechanism topologies presented here. Static balancing can be used successfully in several mechatronic systems, including manual devices, flight simulators, robotic manipulators, and others, reducing substantial the actuator torques and power.

12.2 Special Self-balancing Wire Mechanisms

The state-of-art in the field of the self-balancing mechanisms may conclude that the self-balancing methods using special type of wire mechanisms can be used in many technical applications in the field of mechanical devices, measurement tools, robotics, etc. Based on the experience of the authors in study of the wire (belt) mechanisms some applications of the special type of wire mechanism using a spring were published by the author in the papers [14–17].

In the following are presented some applications of the special wire mechanism using a spring for achieving static force equilibrium of a loading platform and in a constant pressure chamber respectively self-balancing of a measurement tool and a manipulator.

12.2.1 Principle of Force Equilibrium

The special type of wire mechanism illustrated in Fig. 12.1 can be used for achieving the force equilibrium an elastic force by considering a spring (a) or a gravitational force by considering a counterweight (b). The main forces F, which must be equilibrated, are usually of gravitational type.

In both cases the equilibrium equation is of moment type. If a counterweight is used, the equilibrating mechanism must ensure its displacement, in horizontal plane, so that the total moment it vanishes. That means the counterweight moves in horizontal plane opposite and proportional with the projection of the centre of gravity of the equilibrated element/mechanism in the same plane. If the centre of gravity of the equilibrated element it moves only vertical (like at an elevator), then the counterweight will move also vertically only, i.e. at constant arm.

In this case the equilibrium moment's equation it has the form:

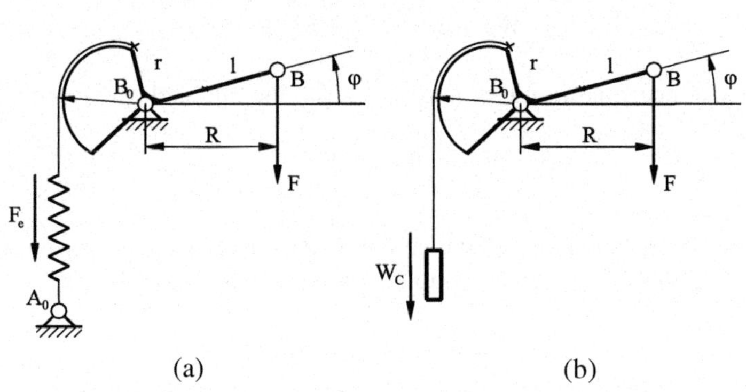

Fig. 12.1 Belt mechanism with a spring **a** or counterweight **b** used for static force equilibrium

$$F \cdot R = Q \cdot r, \tag{12.1}$$

where F is the force or weight to be equilibrated,

Q—the weight of the counterweight W_C,

R and r—the arms of the two forces (variable or constant).

The second solution is to use an elastic force, given by a spring, for equilibration. In this case in the Eq. (12.1) Q means the elastic force of the spring F_e. The elastic force is usually a linear variable force of the form:

$$F_e = F_{e0} + k \cdot s, \tag{12.2}$$

where F_{e0} is the pretension elastic force of the spring,

k—the spring's elasticity constant,

s—the spring's elongation.

The synthesis problem consists in computing the equation of a noncircular wheel $r = r(\varphi)$ for a selected spring so that the equilibrium should be done. This kind of equilibrating mechanisms belongs to the class of, so-called, wire mechanisms or centroidal mechanisms.

12.2.2 *Self-balancing of a Measurement Head*

The simplest equilibrating mechanism is the self-balancing of a body in vertical movement for its entire working range [14, 15], as is the case of vertical displacement of the measuring head of a coordinate measuring device (Fig. 12.2).

In this case the equilibrium Eq. (12.1) becomes:

$$W \cdot R_0 = (F_{e0} + k \cdot s) \cdot r. \tag{12.3}$$

with the clearance condition $r \leq R_0$ ($r_{\max} = R_O$) the initial elastic force will be:

$$F_{e0} = W. \tag{12.4}$$

If the stroke H of the equilibrated element is not very large, the maximum angle of the pulley assembly $\varphi_{\max}$ can be up to 2π radians. In this case:

$$R_0 = H/\varphi_{\max} = H/2\pi \tag{12.5}$$

For a long stroke, but a reasonable clearance, the noncircular wheel of the pulley assembly is made as a conical helix (see Fig. 12.2c) allowing more than one full rotation.

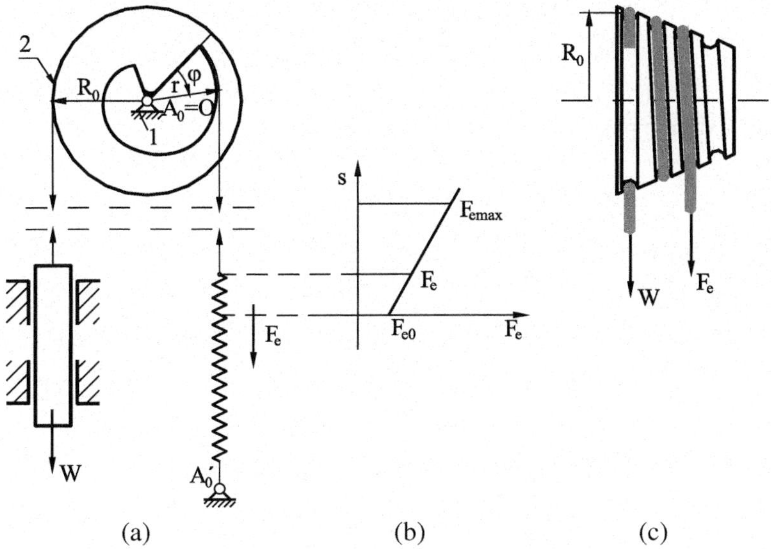

Fig. 12.2 Self-balanced measuring head of a coordinate measuring device

With the constant value of the left side term of Eq. (12.3),

$$W \cdot R_0 = C \tag{12.6}$$

and the wrapping condition:

$$r \cdot d\varphi = ds, \tag{12.7}$$

the Eq. (12.3) becomes:

$$C \cdot d\varphi = (F_{e0} + k \cdot s) \cdot ds. \tag{12.8}$$

By integrating the Eq. (12.8) results:

$$k \cdot s^2 + 2 \cdot F_{e0} \cdot s - 2 \cdot C \cdot \varphi = 0, \tag{12.9}$$

from which, the current arc of the noncircular wheel it results:

$$s = \frac{1}{k}\left(\sqrt{F_{e0}^2 + 2Ck \cdot \varphi} - F_{e0}\right). \tag{12.9'}$$

The equation of the noncircular wheel profile, in polar coordinate, is obtained by derivation of the Eq. (12.9) according to (12.7) i.e.:

$$r = \frac{ds}{d\varphi} = \frac{C}{\sqrt{F_{e0}^2 + 2Ck \cdot \varphi}}. \tag{12.10}$$

In the extreme position the equilibrium equation is:

$$r_{\min} \cdot F_{e\max} = r_{\max} \cdot F_{e\min} = C; \quad F_{e\min} = F_{e0} \tag{12.11}$$

with the boundary conditions: $r_{max} = r|_{\varphi=0} \leq R_0$, $r_{min} = r|_{\varphi=\varphi_{max}}$.

For $\varphi_{\max} = 2\pi$

$$r_{\min} = \frac{C}{\sqrt{F_{e0}^2 + 4\pi Ck}}. \tag{12.12}$$

with these values, from Eq. (12.9) results:

$$k^2 \cdot s_{\max}^2 + 2F_{e0}k \cdot s_{\max} - 4\pi k R_0 F_{e0} = 0. \tag{12.13}$$

where by admitting one of the two variable ($s_{\max}$ or k) the second can be found from Eq. (12.13) or Eq. (12.9).

Numerical example

By considering the characteristic values for self-balancing measurement head given in Table 12.1, the stroke H of the equilibrated element and the maximum elongation of the spring s_{max} can be calculated:

$$H = 2\pi R_0 = 314\,mm \; s_{\max} = \tfrac{1}{k}\sqrt{F_{e0}^2 + 4\pi Ck} - F_{e0} = 134.12\,mm.$$

In Fig. 12.3 the variation of the spring's elongation in respect with current angle φ of pulley assembly are presented, and in the Fig. 12.4 the radius of noncircular profile is represented in polar coordinates, i.e. the actual profile of the noncircular wheel.

Table 12.1 Characteristic values for self-balancing measurement head

Nr.	Characteristic dimension	Notation	Value
1	Weight of measurement head	W	100 N
2	Pretension force of the spring	$F_{e0} = F_{\min}$	100 N
3	Radius of the circular wheel	R_0	50 mm
4	Spring's elasticity constant	k	2 N/mm
5	Maximum rotation angle	$\phi_{\max}$	360°

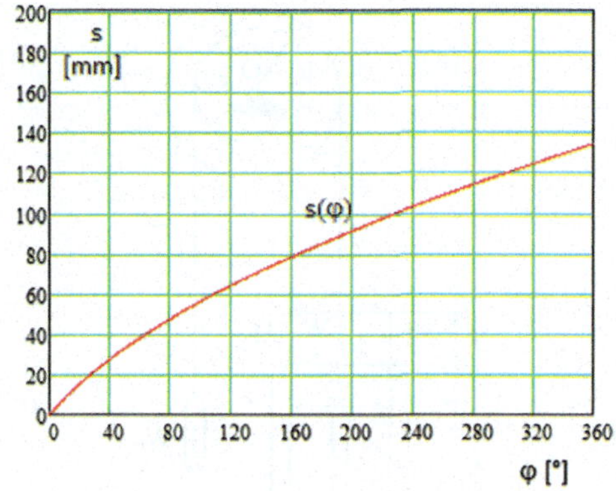

Fig. 12.3 The spring elongation $s = s(\varphi)$

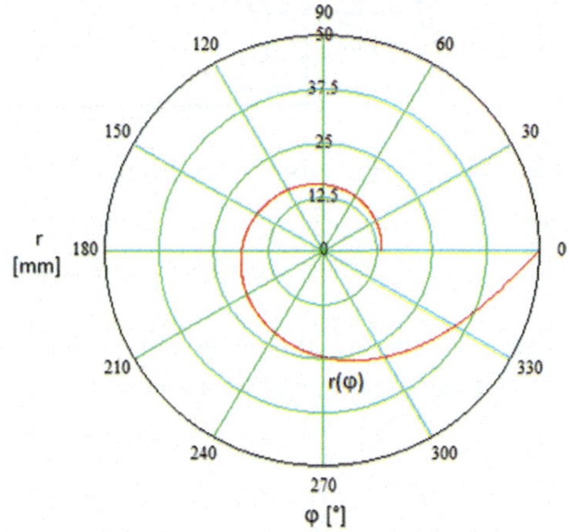

Fig. 12.4 The profile of noncircular wheel $r = r(\varphi)$

12.2.3 Force Equilibrium in a Constant Pressure Chamber

In some technical applications, a constant pressure of a gas is necessary to be maintained in a chamber with variable volume. This is the case with biogas reservoir, or the equipment for gas meters' calibration [14, 16]. Usually the pressure-tight (packing) is made using a bell and a vessel with a liquid. The gas pressure under the bell depends on the weight of it and of the immersion dipping of the bell's wall.

In order to have a constant gas pressure, the apparent bell's weight has to be constant. That means the Archimedean force has to be compensated. The compensatory mechanisms can use a wire mechanism with massless wire and a noncircular wheel or a simple pulley mechanism with circular wheel and a heavy wire. In the following is presented the first type of compensatory mechanism. In Fig. 12.5 a constant pressure chamber, using a bell and a vessel with liquid filling, is presented. The value of pressure inside the bell is:

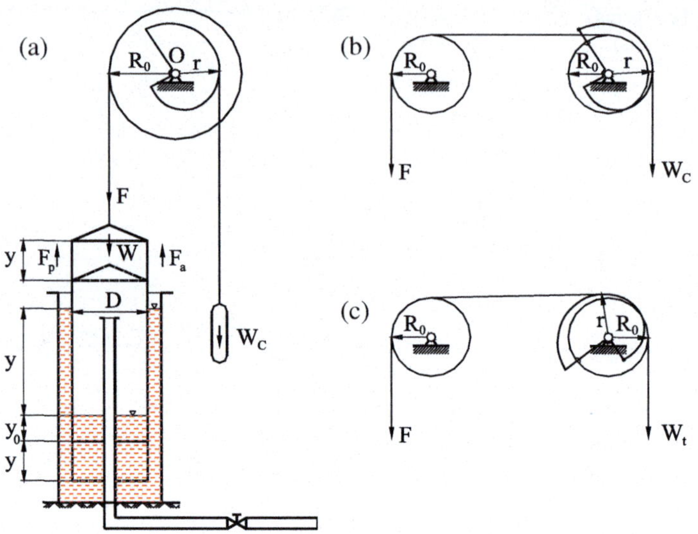

Fig. 12.5 Compensatory wire mechanism of the Archimedean effect

$$p = \rho \cdot g \cdot y, \tag{12.14}$$

where ρ is the liquid density and g—local gravitational acceleration.

The apparent weight of the bell is:

$$F = W - F_p - F_a, \tag{12.15}$$

where W is the bell's weight,

$$F_p = p \cdot \pi \cdot D^2/4 \tag{12.16}$$

is the acting force due to the pressure inside the bell,

$$F_a = A_i \cdot \rho \cdot g \cdot (y_0 + y) \tag{12.17}$$

is the Archimedean (accessional) force due to the bell's immersion into the liquid with $(y_0 + y)$ dipping, y_0 being the initial immersion dipping,

$$A_i = \pi \cdot D_m \cdot b = \pi \cdot (D + b) \cdot b \tag{12.18}$$

is the ring-shaped transversal aria of the bell's wall of thickness b.

The equilibrium condition of the two coupled wheels of the compensatory mechanism with massless wire (Fig. 12.5a), has the form of the relationship (12.1).

The correlation between the vertical displacement of the bell and the rotation angle of the two wheels is given by the obvious relationship:

$$y = R_0 \cdot \varphi. \tag{12.19}$$

With these values, from the equilibrium condition, the equation of the noncircular wheel's profile, in polar coordinates, results, in the form:

$$r = R_0 \cdot \frac{F}{G_c} = R_0 \cdot \frac{W - F_p - A_i \cdot \rho \cdot g(y_0 + R_0\varphi)}{W_c} = R_0(k_1 + k_0 \cdot \varphi), \tag{12.20}$$

i.e. an Archimedean spiral.

Obviously, the Eq. (12.20) has to satisfy the condition $r \geq 0$, even for maximum bell's displacement y_{max} i.e.:

$$W - F_p - A_i \cdot \rho \cdot g \cdot (y_0 + y_{max}). \tag{12.21}$$

Accepting:

$$W - F_p - A_i \cdot \rho \cdot g \cdot y_0 = W_c, \tag{12.22}$$

the condition (12.21) becomes:

$$W_c \geq A_i \cdot \rho \cdot g \cdot yi_{0_{maxmax}}, \tag{12.23}$$

and the Eq. (12.20) will have the form:

$$r = R_0 \cdot \left[1 - \frac{A_i \cdot \rho \cdot g \cdot R_0}{W_c} \cdot \phi\right]. \tag{12.24}$$

For a large diameter D of the bell, a solution like in Fig. 12.5a is not a practical one. In this case, some arrangements like in Fig. 12.5b or c should be used. The solution represented in Fig. 12.5c has the advantage that the counterweight moves on a vertical line. In this case the counterweight may be guided, in a tube for example, in order to prevent its oscillation.

Numerical example

By considering a bell having the characteristic values given in Table 12.2, the bell's stoke y_{max} can be computed from the bell volume relationship as y_{max}.

The profile's Eq. (12.24) of the noncircular wheel can be written simplified as:

$$r = R_0 \cdot (1 - k\varphi). \tag{12.25}$$

Choosing the value for the radius of the circular bell (see Table 12.2), the maximum rotating angle of the two wheels will be:

$$\varphi_{max} = \frac{y_{max}}{R_0} = \frac{1,02}{0,3} = 3,4\ rad = 194,9°$$

Table 12.2 Characteristic values for self-balancing of a constant pressure chamber

Nr	Characteristic dimension	Notation	Value
1	Bell diameter	D	0.5 m
2	Thickness of the bell	b	5 mm
3	Density of the packing liquid	ρ	$0.88\ 10^3\ kg/m^3$
4	Gas volume	V	$0.2\ m^3$
5	Radius of the circular wheel	R_0	0.3 m

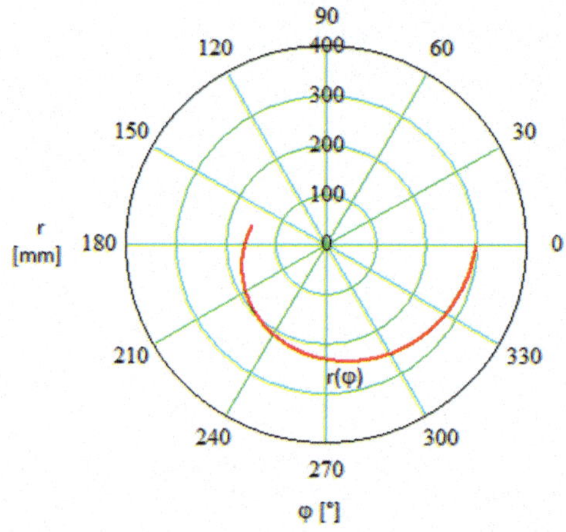

Fig. 12.6 The profiled wheel for compensatory belt mechanism

Admitting the boundary conditions for the noncircular wheel $r0_{\max}$ for $\phi = 0$ and $r_{\min} = 0,5 \cdot R_0$, the Eq. (12.25) becomes:

$$r = 300 \cdot (1 - 0,147 \cdot \varphi)[mm] \tag{12.26}$$

where φ increasing in clockwise direction.

In Fig. 12.6 the profile of the noncircular wheel is shown.

12.2.4 Force Equilibrium of a Bistable Type Mechanism

The synthesis of a bistable type mechanism is more complicated as the previous case. As an example, is taken into account the mechanism of a loading platform. The platform must be in equilibrium in a certain position only, which is the median angular position φ_m between the angular positions $\varphi_{\min}$ and $\varphi_{\max}$. Above this position on the platform it acts an ascension force and under it a symmetrical, equal and opposite force. At the ends of stroke these forces must have prescribed absolute values (F_m).

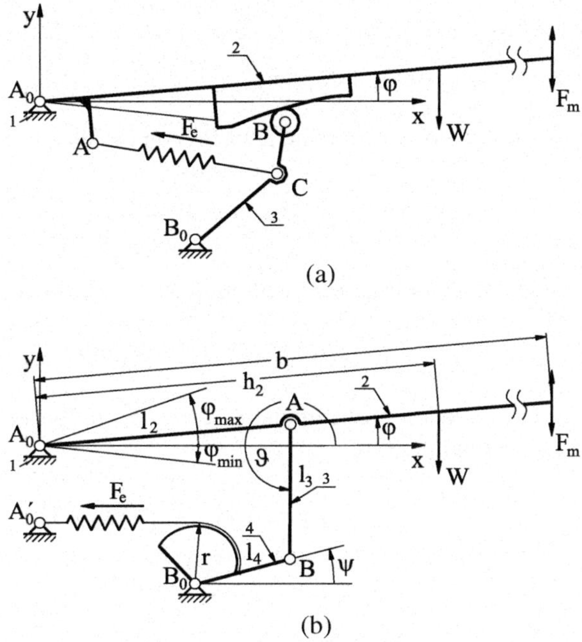

Fig. 12.7 Two solutions for a loading platform mechanism

An existing solution for this kind of mechanism is a cam-follower one, with the follower as the driving element (Fig. 12.7a), with its specific disadvantages.

In [14, 16] was proposed a four-bar linkage, actuated by means of a spring and a noncircular wheel (a wire mechanism—Fig. 12.7b) to accomplish the imposed conditions. In order to compute the kinematic parameters of the four-bar linkage is considered the closed loop equation:

$$l_2 \cdot e^{i \cdot \varphi} + l_3 \cdot e^{i \cdot \vartheta(\varphi)} - l_4 \cdot e^{i \cdot \psi(\varphi)} - (x_{B_0} + i \cdot y_{B_0}) = 0. \tag{12.27}$$

and its complex conjugate, the current position of links 3 and 4 respectively, in function of the independent variable φ, it results in the forms:

$$\psi(\varphi) = 2 \cdot \arctan \frac{B_1(\varphi) + \sqrt{A_1^2(\varphi) + B_1^2(\varphi) - C_1^2(\varphi)}}{A_1(\varphi) - C_1(\varphi)}, \tag{12.28}$$

$$\vartheta(\varphi) = 2\pi + 2 \cdot \arctan \frac{B_2(\varphi) + \sqrt{A_2^2(\varphi) + B_2^2(\varphi) - C_2^2(\varphi)}}{A_2(\varphi) - C_2(\varphi)}, \tag{12.29}$$

where

$$\begin{aligned} &A_1(\varphi) = 2l_4\left(x_{B_0} - l_2 \cdot \cos\varphi\right),\ B_1(\varphi) = 2l_4\left(y_{B_0} - l_2 \sin\varphi\right), \\ &C_1(\varphi) = (x_{B_0}^2 + y_{B_0}^2) + l_2^2 - l_3^2 + l_4^2 - 2l_2(x_{B_0}\cos\varphi + y_{B_0}\sin\varphi), \end{aligned} \tag{12.30}$$

$$\begin{aligned} A_2(\varphi) &= 2l_3\big(-x_{B_0} + l_2 \cdot \cos\varphi\big),\ B_2(\phi) = 2l_3\big(-y_{B_0} + l_2 \sin\varphi\big), \\ C_2(\varphi) &= (x_{B_0}^2 + y_{B_0}^2) + l_2^2 + l_3^2 - l_4^2 - 2l_2(x_{B_0}\cos\varphi + y_{B_0}\sin\varphi). \end{aligned} \tag{12.31}$$

The reacting force in the joint B is on BA direction and has the value:

$$R_B(\varphi) = W\frac{-h_2 \cdot \cos\varphi}{l_2 \cdot \sin(\vartheta(\varphi) - \varphi)}. \tag{12.32}$$

The moment in respect with joint B_0 of this force is:

$$M_1(\varphi) = R_B(\varphi) \cdot l_4 \cdot \sin(\vartheta(\varphi) - \psi(\varphi)). \tag{12.33}$$

The variation of manipulating force F_m in the working span of the platform is accepted in the form:

$$F_m(\varphi) = \begin{cases} -F_m + 2\lambda F_m\Big(1 - \cos\Big(\pi\frac{\varphi - \varphi_{\min}}{2\lambda\Delta\varphi}\Big)\Big) & \varphi \in (\varphi_{\min},\ \varphi_{\min} + \lambda\Delta\varphi) \\ -F_m + 2F_m\Big[\lambda + (1-\lambda)\cos\Big(\pi\Big(\frac{1}{2} + \frac{(\varphi - \varphi_{\min}) - \lambda\Delta\varphi}{2(1-\lambda)\Delta\varphi}\Big)\Big)\Big] & \varphi \in (\varphi_{\min} + \lambda\Delta\varphi,\ \varphi_{\max}) \end{cases} \tag{12.34}$$

for $\lambda = 12/25$, which gives a supplementary moment on driving link:

$$M_2(\varphi) = F_m(\varphi) \cdot \frac{-b \cdot \cos\varphi}{l_2 \cdot \sin(\vartheta(\varphi) - \varphi)} \cdot l_4 \cdot \sin(\vartheta(\varphi) - \psi(\varphi)). \tag{12.35}$$

The total moment acting on this link becomes:

$$M(\varphi) = M_1(\varphi) + M_2(\varphi). \tag{12.36}$$

This moment have to be equilibrated by the moment of the elastic force of the spring i.e.:

$$M(\varphi) = (F_{e0} + k \cdot s) \cdot r. \tag{12.37}$$

With the wrapping condition (12.7) for driving link, this equation can be written in the form:

$$M(\varphi)\frac{d\psi}{d\varphi}d\varphi = (F_{e0} + k \cdot s) \cdot ds. \tag{12.38}$$

where

$$\frac{d\psi}{d\varphi} = \psi'(\varphi) = \frac{l_2 \cdot (-x_{B_0}\sin\varphi + y_{B_0}\cos\varphi + l_4\sin(\psi(\varphi) - \varphi))}{l_4 \cdot (-x_{B_0}\sin\psi(\varphi) + y_{B_0}\cos\psi(\varphi) + l_2\sin(\psi(\varphi) - \varphi))}. \tag{12.39}$$

By integrating Eq. (12.38) it is obtained:

$$s(\varphi) = \frac{1}{k}\left(\sqrt{F_{e0}^2 + 2kM_{int}} - F_{e0}\right). \tag{12.40}$$

The equation of the profiled driving element, in polar coordinates, results as derivate of relationship (12.40):

$$r = \frac{ds}{d\psi} = \frac{1}{\psi'(\varphi)}\frac{ds}{d\varphi} = \frac{M'_{\text{int}}}{\psi'(\varphi)\frac{1}{\sqrt{F_{e0}^2+2kM_{\text{int}}}}}, \tag{12.41}$$

where

$$M'_{\text{int}} = M(\varphi)\cdot\psi'(\varphi). \tag{12.42}$$

Numerical example

For the loading ledge mechanism are considered the characteristic values in Table 12.3.

The geometrical parameters of the four-bar linkage were chosen as in Table 12.3 to be fulfilled the following conditions:

- because the working angular span is relatively small (25°), it is favorable to amplify the rotation angle by means of the four-bar linkage.
- it has been done approximately of two times, with the input and output links parallel to each other in the median position of the platform.

Table 12.3 Characteristic values for the loading platform mechanism

Nr.	Characteristic dimension	Notation	Value
1	Weight of the loading platform	W	5.800 N
2	Manipulating force	F_m	200 N
3	Minimum angular position	ϕ_{min}	−5°
4	Maximum angular position	ϕ_{max}	20°
5	Median angular position	ϕ_m	7,5°
6	Weight position on the loading platform	h_2	1.050 mm
7	Manipulating force position on the loading platform	$l_m = b$	2.100 mm
8	Length of the link 2	l_2	600 mm
9	Length of the link 3	l_3	340 mm
10	Length of the link 4	l_4	300 mm
11	Coordinate of the joint B	x_{B0} y_{B0}	300 mm 300 mm

The reacting force in the joint B and his corresponding moment in respect with joint B_0 are represented in Fig. 12.8. In order to integrate the Eq. (12.38) the left side function is approximated with a polynomial one in the form:

$$M(\varphi)\frac{d\psi}{d\varphi} = 23421643.66 \cdot \varphi^3 - 11721538.75 \cdot \varphi^2 - 1884454.02 \cdot \varphi + 6425394.55. \quad (12.43)$$

The profiled element is presented in Fig. 12.9a, and the computed manipulating force (F_m) with this profile is presented in Fig. 12.9b. As can be seen in Fig. 12.9b, the error due to the function approximation for the manipulating force is less than 2%.

In all the cases the influence of the transmission's angle variation at the noncircular wheel was neglected because this variation is less than ±15°, which implied an error under 3.5%.

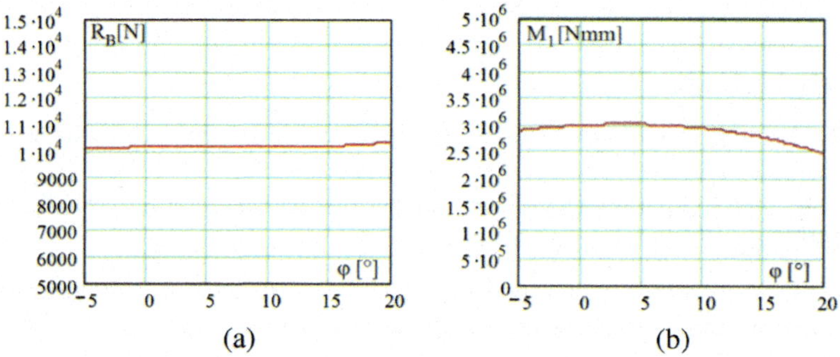

Fig. 12.8 The reacting forces **a** and its moment **b** on driving link

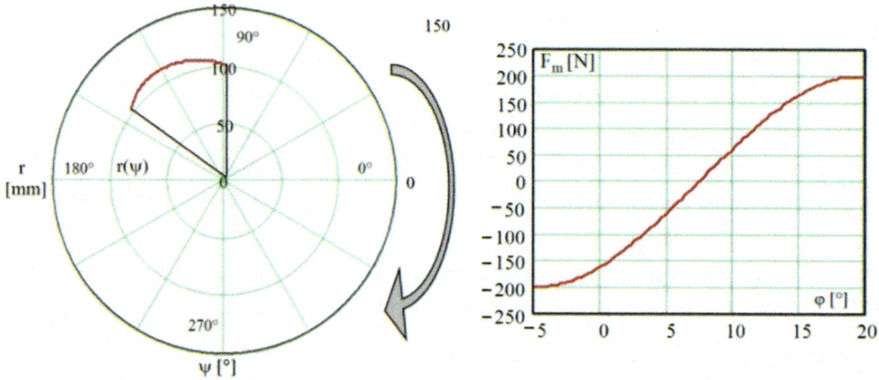

Fig. 12.9 The synthesized noncircular wheel profile **a** and the resulting manipulating force **b**

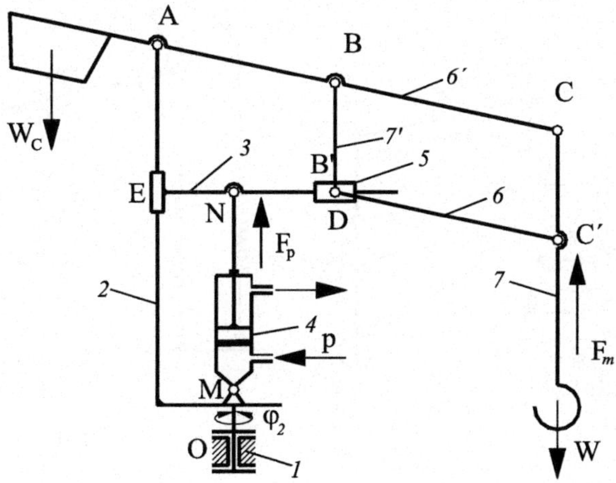

Fig. 12.10 Kinematic schema of Conco-Balancer [14, 17]

12.2.5 *Force Equilibrium by Self-balancing Conco-Balancer Manipulator*

The Conco-Balancer manipulator is used in industrial activities for handling products or to support devices and it is manipulated by an operator. For an easy manipulation by the operator, it must be self-balanced. The gravitational self-balancing with the balancing weight W_C compensates for the handling weight W in the workspace in different positions (Fig. 12.10). Other ways by increasing the pressure in the pneumatic cylinder (4) the handling weight W is compensated for through the pneumatic force F_p.

The self-balancing using the pneumatic force is limited by setting the height only for a certain position. The repositioning of the handling weight W in the workspace is done by the operator acting with his own muscular force.

The technical solution for self-balancing with counterweight, springs and actuators can be substituted through a wire mechanism. The variable arm of the elastic force F_e given by the noncircular profile allows the generation of a balancing force, which reduces the manipulation force of the operator [14, 17].

The self-balancing solution using wire mechanism (Fig. 12.11) contains a noncircular wheel fixed with the element (6′). This alternative solution simplified the construction of the Conco-Balancer manipulator, respectively ensure the equilibrium in all the positions and an easy manipulation in the workspace.

By acting the pneumatic cylinder, the manipulator is balanced in the median position $\varphi_m = (\varphi_p + \varphi_n)/2$ (see Fig. 12.11b). Around the median position on the manipulator act the variable moments generated by the spring force F_e and the prescribed manipulation force F_m.

The moment equilibrium equation of the above-mentioned forces:

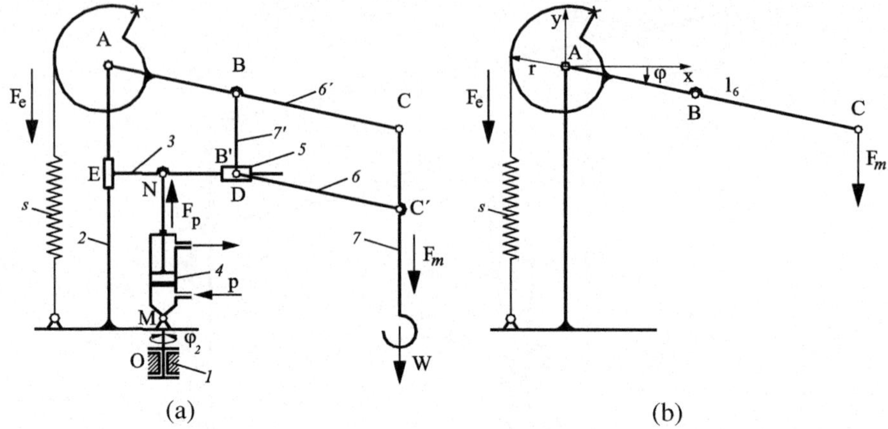

Fig. 12.11 Kinematic schema of the Conco-Balancer with self-balancing wire mechanism [14, 17]

$$F_m \cdot l_6 \cdot \cos\varphi = F_e \cdot r, \tag{12.44}$$

where l_6 is the length of the manipulator arm.

In Eq. (12.44) are neglected the weights of the manipulator arms and the spring has a linear variable force characteristic (12.2).

The synthesis of the self-balancing wire mechanism assume the computation of the noncircular wheel equation $r = r(\phi)$ for a chosen spring characteristic.

The equilibrium Eq. (12.44) with (12.2) becomes:

$$F_m \cdot l_6 \cdot \cos\varphi = (F_{e0} - k \cdot s) \cdot r. \tag{12.45}$$

By using the wrapping condition (12.7) the Eq. (12.45) becomes:

$$F_m \cdot l_6 \cdot \cos\varphi \cdot d\varphi = (F_{e0} - k \cdot s) \cdot ds. \tag{12.46}$$

Through the integration of the differential Eq. (12.46) results:

$$k \cdot s^2 - 2 \cdot F_{e0} \cdot s - 2 \cdot F_m \cdot l_6 \cdot \sin\varphi = 0, \tag{12.47}$$

from which, the current arc of the noncircular wheel is computed:

$$s = \frac{1}{k}\left(\sqrt{F_{e0}^2 + 2F_m \cdot l_6 \cdot k \cdot \sin\varphi} + F_{e0}\right). \tag{12.48}$$

By derivation of the current arc of the noncircular wheel (12.48) in respect to the angle φ results the equation in polar coordinate of the noncircular wheel profile, i.e.:

$$r = \frac{ds}{d\varphi} = \frac{F_m \cdot l_6 \cdot \cos\varphi}{\sqrt{F_{e0}^2 + 2F_m \cdot l_6 \cdot k \cdot \sin\varphi}}. \tag{12.49}$$

The equilibrium condition in the extreme position, symmetrical disposed to the median position on the horizontal direction, i.e. $\varphi_{\min} = -\varphi_{\max}$, is:

$$r_n \cdot F_{e\max} = r_p \cdot F_{e\min} = C, \tag{12.50}$$

with:

$$F_{e\min} = F_{e0}, \tag{12.51}$$

$$C = F_m \cdot l_6 \cdot \cos\ \varphi_{\max}, \tag{12.52}$$

$$r_n = r\big|_{\varphi=\varphi\min},\ r_p = r\big|_{\varphi=\varphi\max} \tag{12.53}$$

One of the two variables from the differential Eq. (12.46): maximum wrapping length s_{max} or spring elastic constant k must to be chosen, the second will be computed for $\varphi \in (-\varphi_{\min}, \varphi_{\max})$ and $s \in (0, s_{\max})$:

$$k \cdot s_{\max}^2 + 2F_{e0} \cdot s_{\max} - 2F_m l_6 \cdot (\sin\varphi_{\max} - \sin\varphi_{\min}) = 0. \tag{12.54}$$

Numerical example

For the Conco-Balancer manipulator with self-balancing wire mechanism are considered the characteristic values given in Table 12.4.

The manipulation force is imposed constant in the angular range $\varphi \in (-\varphi_{\min}, \varphi_{\max})$ (see Fig.12.12a) and the corresponding moment $M(\varphi)$ is represented in Fig. 12.12b.

Table 12.4 Characteristic values for the Conco-Balancer manipulator

Nr.	Characteristic dimension	Notation	Value
1	Weight to be manipulated	W	1.500 N
2	Manipulating force	F_m	50 N
3	Pretension elastic force of the spring	F_{e0}	100 N
4	Spring's elasticity constant	k	20 N/m
5	Minimum angular position	$\varphi_{\min}$	−60°
6	Maximum angular position	$\varphi_{\max}$	60°
7	Median angular position	φ_m	0°
8	Length of the link 6	l_6	1.200 mm

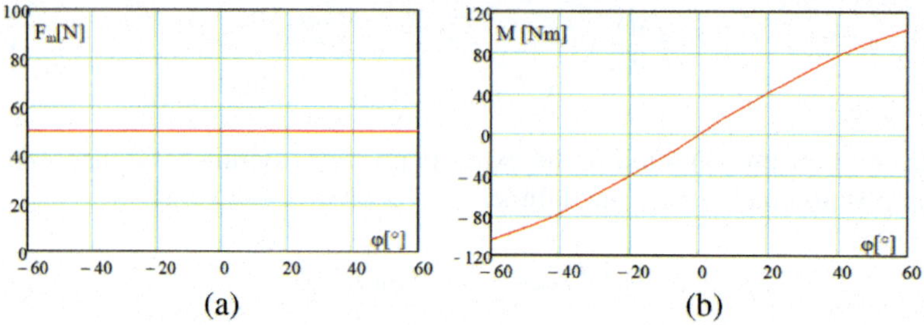

Fig. 12.12 Manipulation force $F_m = F_m(\varphi)$ **a** and his corresponding moment $M = M(\varphi)$ **b**

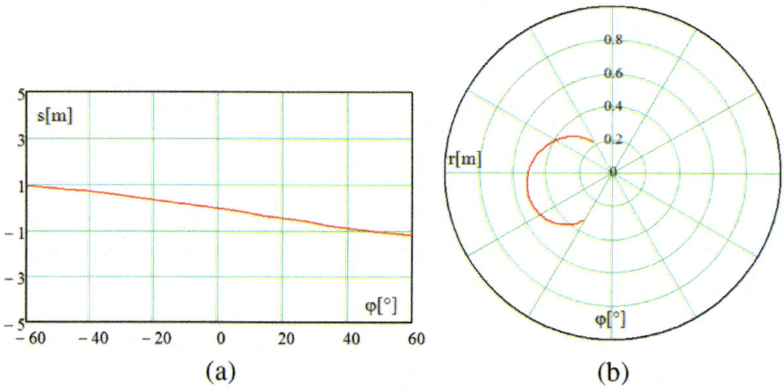

Fig. 12.13 The variation of the spring's elongation **a** and the profile of the noncircular wheel **b**

Figure 12.13a shows the variation of the spring's elongation in respect with current angle φ of the manipulator arm and Fig. 12.13b the noncircular profile represented in polar coordinate $r = r(\varphi)$.

12.3 Scientific Contributions

The research shows some synthesis application of the wire mechanisms also named band mechanisms, which represents the main research topic of the author.

The equilibrating mechanisms based on elastic forces acting with variable arms can be used successfully in order to reduce the total weight, clearance needed and the cost.

In the theoretical approaches the friction forces have been neglected and the solutions represent the first approximation only. Generally, in such cases the friction forces extend the equilibrium's range, so that these forces have a favourable influence in the presented applications.

The examples using wire mechanisms have shown solution for self-balancing of measuring head of a coordinate measuring device, the compensation of the Archimedean effect, static equilibrium of a loading platform and the self-balancing of a Conco-balancer manipulator.

All examples show explicitly the theoretical background for the moment compensation by means of wire mechanisms having a profiled noncircular wheel and are followed with a corresponding numerical example.

Acknowledgements This work was developed at the Politehnica University of Timişoara at Department of Mechatronics and Technische Universität Dresden at Institut für Festkörpermechanik within the framework of two PhD thesis and one Habilitation thesis, respectively by the Mechatronics and Robotics research team.

References

1. Baradat, C., Arakelian, V., Briot, S., Guegan, S.: Design and prototyping of a new balancing mechanism for spatial parallel manipulators. J. Mech. Des. **130**(7), 072305 (2008)
2. Arakelian, V.: Gravity compensation in robotics. Adv. Rob. J. **30**(2), 1–18 (2015)
3. Alici, G., Shirinzadeh, B.: Optimum dynamic balancing of planar parallel manipulators based on sensitivity analysis. Mech. Mach. Theory **41**(12), 1520–1532 (2006)
4. Feng, G.: Complete shaking force and shaking moment balancing of 26 types of four-, five- and six-bar linkages with prismatic pairs. Mech. Mach. Theory **25**(2), 183–192 (1990)
5. Ye, Z., Smith, M.R.: Complete balancing of planar linkages by an equivalence method. Mech. Mach. Theory **29**(5), 701–712 (1994)
6. Nguyen, V.K., Nguyen, P.D.: Balancing conditions of spatial mechanisms. Mech. Mach. Theory **42**(9), 1141–1152 (2007)
7. Kazerooni, H.: Design and analysis of the statically balanced direct-drive robot manipulator. Rob. Comput. Integr. Manuf. **6**(4), 287–293 (1989)
8. Wang, J., Gosselin, C.M.: Static balancing of spatial three-degree-of-freedom parallel mechanisms. Mech. Mach. Theory **34**(3), 437–452 (1999)
9. Agrawal, S.K., Fattah, A.: Gravity-balancing of spatial robotic manipulators. Mech. Mach. Theory **39**(12), 1331–1344 (2004)
10. Hain, K.: Feder-Getriebe und Band-Getriebe für den Kraftausgleich. VDI-Zeitschrift **97**(9), 278 (1955)
11. Perju, D., Nicoara, I., ELovasz, C., Creţu, E.: Mecanism programator cu elemente de lungime variabila. In: Proceedings of the 7-th National Symposium, MTM'96, Timisoara (1996)
12. Streit, D.A., Shin, E.: Equilibrators for planar linkages. Trans. ASME, J. Mech. Des. **115**, 604–611 (1993)
13. Laliberte, T., Gosselin, C.M., Jean, M.: Static balancing of 3-DOF planar parallel mechanisms. IEEE/ASME Trans. Mechatron. **4**(4), 363–377 (1999)
14. Lovasz, E.-C.: Contributions on development of mechanism science with applications in robotics, Mechatronics and Mechanical Engineering, Habilitation Thesis, Politehnica University of Timişoara (2016)
15. Perju, D., Modler, K.-H., Lovasz, E.-C., Mesaros-Anghel, V.: On the mechanisms' synthesis of geometrical and static equilibrium conditions. In: Proc. of the 9-th International Symp. MTM'2004, Acta Technica Napocensis, Series: Applied Mathematics and Mechanics, **48**(1), 155–161 (2004)

16. Perju, D., Modler, K.-H., Mateaş, M., Lovasz, E.-C.: On the compensatory mechanisms of the Archimedean effect. In: Proceedings of the 8-th International Conference COMEFIN-8", Acta Technica Napocensis, Series: Applied Mathematics and Mechanics, **49**(3), 607–610 (2006).
17. Lovasz, E.-C., Perju, D., Dehelean, N., Dehelean, L.M., Maniu, I., Moldovan, C.: Self-balanced conco-balancer manipulator with band mechanism. Solid State Phenom. **166–167**, 259–265 (2010)

Chapter 13
Design of a Five DOF Contactless Robot for Facade Inspection

Ginna Marcela García-Rodríguez, Eduardo Castillo-Castañeda, and Med Amine Laribi

Abstract Service tasks on vertical surfaces such as building facades continue to be performed manually by specialized technicians operating from complex scaffolding. This way of performing these tasks, in addition to being inefficient, is costly and dangerous for personnel who must work under the risk of falls. The façade inspection task, as well as façade cleaning task, can be achieved by following zig-zag motions to cover the whole working surface while keeping the tool perpendicular to the façade surface. The aim of this work is to develop a robotic inspection device of 5 degrees of freedom based on a Cartesian configuration with three translations and a pan-tilt type system with two rotations. The robotic device will be placed in front of the façade of a building, which can be previously characterized by a cloud of points. The kinematics of the robots is also presented as well as the mechanical elements to build a scalable prototype that can be sized according to the building dimensions.

Keywords Service robot · Cartesian configuration · Façade inspection · Inspection trajectories

13.1 Introduction

The construction of increasingly tall buildings, technological advances, and the realization of mandatory periodic inspections of buildings, are factors that promote the development of practical and innovative systems that facilitate the execution of these tasks. The robotization of inspection tasks leads to the optimization of processes in relation to time and cost, reduction of operational risks, and minimization of affections around the target structure of the inspection. The use of climbing robots

G. M. García-Rodríguez · E. Castillo-Castañeda
Instituto Politécnico Nacional, CICATA-Unidad Querétaro, Santiago de Querétaro, Mexico
e-mail: ecastilloca@ipn.mx

M. A. Laribi (✉)
Université de Poitiers, Institut PPRIME—DMSC, Chasseneuil-du-Poitou, France
e-mail: med.amine.laribi@univ-poitiers.fr

M. Ceccarelli and J. C. Jauregui-Correa (eds.), *State-of-the-Art and Innovations in Mechanism and Machine Science*, Mechanisms and Machine Science 150,
https://doi.org/10.1007/978-3-031-47040-0_13

designed for inspection, maintenance, and many other tasks related to the construction sector, has been acquiring great relevance in recent times. Service robotics is an area of great opportunity that, since 2019, has been the one with the greatest development and the one that has generated the greatest increase in sales. Cleaning and inspection tasks are part of service robotics; However, there are very few commercial robots that perform these tasks on vertical surfaces [1].

At first, the use of inspection robots was established for hazardous environments with serious risks to the health and integrity of human operators. Currently, the field of application of this type of robots is booming and not only for dangerous industrial actions, but also for other areas of inspection and maintenance such as: construction, the electrical sector, forestry, among others. The objective is none other than to facilitate the work of workers by minimizing risks, execution time, and costs [2]. The construction sector has produced in recent decades a remarkable increase in the development of very tall buildings and structural ensembles with peculiar architecture, which, under traditional techniques, make access to the operational surface of inspection and maintenance difficult, expensive, and dangerous [3].

The integration of autonomous robotic systems in these tasks on vertical surfaces is presented as a technological challenge itself, in relation to the support of the system on the surface, to its displacements along it, and even more important, to have an obstacle detection system, necessary for the definition of the trajectory.

Among the main robotic developments focused on inspection and cleaning, there are some works such as: robot with sliding guide integrated to the glass facades with suction cup adhesion driven by pneumatic actuators [4]; prototype with passive suction cup adhesion [5] and device that uses negative pressure between the robot and the wall through vacuum pumps to adhere to the surface [6]. There are also robots with magnetic grip for metallic tanks [7]. Rope or rail grip is another method where the robot is held from the wall or ceiling via a rope or cable, allowing mobility and safe execution of tasks [8]. Robots with bioinspired adhesion have also been proposed, with the ability to scale vertical walls even on uneven surfaces such as concrete [9]. However, the force of gravity and the irregularity of the surfaces (for materials other than glass) of the facades, have become a challenge to achieve the high levels of adaptability and flexibility required by climbing robots in the execution of vertical tasks. Most traditional inspection systems are limited to structures such as scaffolding, hanging platforms through cables, articulated arms, and lifting platforms.

Regarding autonomous systems, the drones that have acquired great importance and popularity in recent years in multisectoral tasks [10], mainly in inspection. However, limitations in legislation, payload and stability against external environmental agents, due to their operational exposure to the elements, are some of the challenges they have nowadays. Recognizing the interest of prof Carlos Lopez-Cajun in investigating on designs of Robots, but with special attention to new solutions, this contribution refers to his interest in parallel manipulators and cable driven systems.

The aim of this work is to develop a robotic inspection device of 5 degrees of freedom based on a Cartesian configuration with three translations and a pan-tilt type system with two rotations. The robotic device will be placed in front of the façade of a building, which can be previously characterized by a cloud of points.

Fig. 13.1 Inspection of a façade: **a** traditional way, **b** ideal trajectory

13.2 Inspection Task for Building Facades

Traditionally, building inspection tasks are performed manually following zig-zag trajectories, the technician moves horizontally until he covers the total length of the building, then he performs a downward displacement and covers again the total length of the building [11], see Fig. 13.1a. This trajectory is made from top to bottom until it covers the entire height of the façade, see Fig. 13.1b.

From these basic inspection task motions, the fundamental requirements were obtained to select the mobility of the proposal presented in this paper. The first refers to its locomotion, which must have 3 GDL to efficiently cover the area to be inspected, that is, to provide adaptability to the surface (concavities and convexities) and maneuverability both horizontally and vertically. The second requirement is the positioning capacity of its end effector (for the present case, the inspection sensor), so that it should be perpendicular to any point of the vertical surface. For the first requirement, the simplest option to obtain three translations is the cartesian configuration. For the second, a pan-tilt system was selected for end effector orientation, where the inspection sensor, a pachometer, will be mounted to evaluate the condition of a façade.

13.3 Conceptual Design of the Proposal

Compared to other proposals of service robotics, where the robot climbs to the surface of the building, our concept consists of a robot located in front of the façade, without making direct contact with it, see Fig. 13.2.

To define the CAD model of the proposal, aspects such as: locomotion, adaptability to the surface, trajectories of the end effector, horizontal and vertical maneuverability, simplicity of assembly, weight and type of control were considered. The needs that the robot must cover are focused: on the reach to all points of the façade

Fig. 13.2 Full view of the conceptual design proposal

surface considering the different profiles (concavities and convexities); the monitoring of the determined inspection path; the range of orientations, positions, and velocities; the efficiency and effectiveness of the inspection task.

13.4 Forward and Inverse Kinematics

The kinematic analysis of the device is simplified considering that it is composed of two parts: a Cartesian configuration structure and a pan-tilt system. Figure 13.3 shows the diagram of the robot of 5 DOF: 3 translations and 2 rotations. The pose of the end effector is defined by *Px*, *Py*, *Pz*, ϕ, θ.

The corresponding Denavit-Hartenberg parameters are shown in Table 13.1.

The dimensions of the robot are defined by the constants a_1, a_2, a_3, a_4, a_5, a_6, Joint variables by translational distances d_1, d_2, d_3 and by angles θ_1, θ_2. The first three homogeneous transformation matrices corresponding to the three translations of the robot are shown in Fig. 13.4.

The last two homogeneous transformation matrices corresponding to the two rotations of the robot are shown in Fig. 13.5.

By multiplying the above transformation matrices, one can establish the closed-loop equation of the mechanism:

$$^{0}T_N = {}^{0}T_1 {}^{1}T_2 {}^{2}T_{N-1} \dots {}^{N-1}T_N \tag{13.1}$$

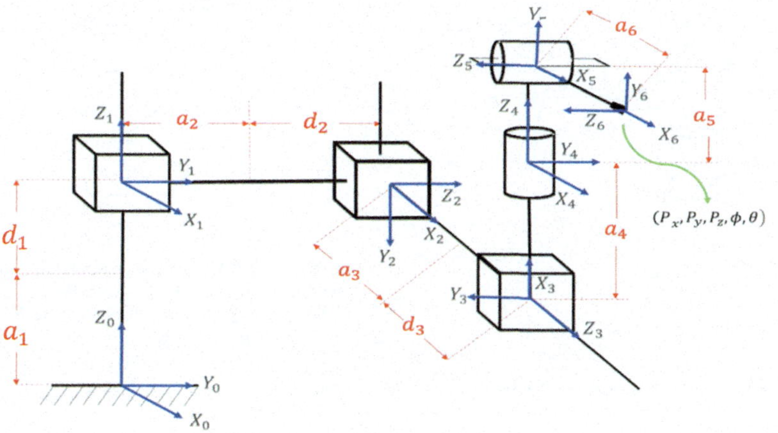

Fig. 13.3 Diagram of the 5 DOF robot for inspection tasks

Table 13.1 Denavit-Hartenberg parameters of the robot

Joint i	θ_i	d_i	a_i	α_i
1	0	d_1	a_1	0
2	0	d_2	a_2	0
3	0	d_3	a_3	0
4	θ_1	0	a_4	0
5	θ_2	0	a_5	0

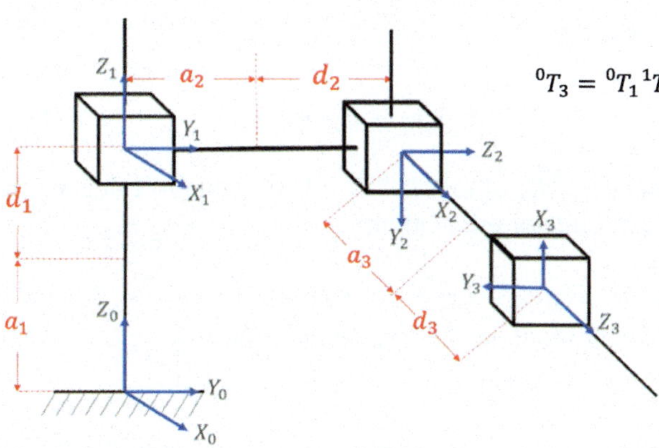

$$
{}^{0}T_3 = {}^{0}T_1\,{}^{1}T_2\,{}^{2}T_3 = \begin{bmatrix} 0 & 0 & 1 & a_3 + d_3 \\ 0 & -1 & 0 & a_2 + d_2 \\ 1 & 0 & 0 & a_1 + d_1 \\ 0 & 0 & 0 & 1 \end{bmatrix}
$$

Fig. 13.4 Transformation matrix from frame 0 to frame 3

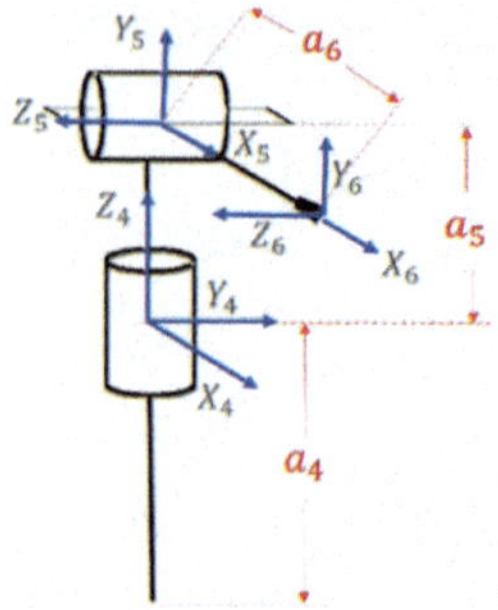

$$ {}^4T_6 = {}^3T_4\,{}^4T_5\,{}^5T_6 = \begin{bmatrix} c\theta_1 c\theta_2 & -c\theta_1 s\theta_2 & s\theta_1 & a_6 c\theta_1 c\theta_2 \\ s\theta_1 c\theta_2 & -s\theta_1 s\theta_2 & -c\theta_1 & a_6 s\theta_1 c\theta_2 \\ s\theta_2 & c\theta_2 & 0 & a_4 + a_5 + a_6 s\theta_2 \\ 0 & 0 & 0 & 1 \end{bmatrix} $$

Fig. 13.5 Transformation matrix from frame 4 to frame 6

$$ {}^0T_6 = \begin{bmatrix} c\theta_1 c\theta_2 & -c\theta_1 s\theta_2 & s\theta_1 & a_3 + d_3 + a_6 c\theta_1 c\theta_2 \\ s\theta_1 c\theta_2 & -s\theta_1 s\theta_2 & -c\theta_1 & a_2 + d_2 + a_6 s\theta_1 c\theta_2 \\ s\theta_2 & c\theta_2 & 0 & a_1 + d_1 + a_4 + a_5 + a_6 s\theta_2 \\ 0\,0\,0 & & & 1 \end{bmatrix} \tag{13.2} $$

In this way, the position P_x, P_y, P_z of the end effector will be given by the last column of the previous matrix:

$$ P_x = a_3 + d_3 + a_6 c\theta_1 c\theta_2 \tag{13.3} $$

$$ P_y = a_2 + d_2 + a_6 s\theta_1 c\theta_2 \tag{13.4} $$

$$ P_z = a_1 + d_1 + a_4 + a_5 + a_6 s\theta_2 \tag{13.5} $$

Finally, with the rotational part ${}^0R_6$ of matrix ${}^0T_6$ it is possible to calculate the Euler angles that define the end effector orientation:

$$ R(\psi, \varphi, \theta) = \begin{bmatrix} c\varphi c\theta & s\varphi s\psi c\theta - c\psi s\theta & s\varphi c\psi c\theta + s\psi s\theta \\ c\varphi s\theta & s\varphi s\psi s\theta + c\psi c\theta & s\varphi c\psi s\theta - s\psi c\theta \\ -s\varphi & c\varphi s\psi & c\varphi c\psi \end{bmatrix} \tag{13.6} $$

For the angle φ, matching the element (3, 1) of matrix ${}^0T_6$ with the element (3, 1) of matrix $R(\psi, \varphi, \theta)$:

$$ s\theta_2 = -s\varphi \tag{13.7} $$

$$ \varphi = -\theta_2 \tag{13.8} $$

For the angle θ, matching the element (2, 1) of matrix ${}^0T_6$ with the element (2, 1) of matrix $R(\psi, \varphi, \theta)$:

$$s\theta_1 c\theta_2 = c\varphi s\theta \tag{13.9}$$

$$s\theta = \frac{s\theta_1 c\theta_2}{c\varphi} \tag{13.10}$$

Replacing the previously calculated value for the angle φ:

$$s\theta = \frac{s\theta_1 c\theta_2}{c(-\theta_2)} = \frac{s\theta_1 c\theta_2}{c(\theta_2)} = s\theta_1 \tag{13.11}$$

$$\theta = \theta_1 \tag{13.12}$$

From the equations obtained for direct kinematics, the joint variables can be obtained:

$$\theta_1 = \theta \tag{13.13}$$

$$\theta_2 = -\varphi \tag{13.14}$$

$$d_1 = P_z - a_1 - a_4 - a_5 - a_6 s\theta_2 \tag{13.15}$$

$$d_2 = P_y - a_2 - a_6 s\theta_1 c\theta_2 \tag{13.16}$$

$$d_3 = P_x - a_3 - a_6 c\theta_1 c\theta_2 \tag{13.17}$$

13.5 Detailed Robot Design

Based on the design requirements, the detailed CAD design of the robot was performed using AUTOCAD and SKETCHUP software. The design includes: a mobile structure, a mobile platform, a pan-tilt system, and a final effector, Fig. 13.6 shows two views of the final detailed design of the robot that is fully scalable to fit the dimensions of a building's façade.

The mobile structure has wheels, with a braking system, in its four bases for its manual displacement on the ground. Inside, it supports the mobile platform that moves vertically driven by a triple pulleys system mounted at the top of the mobile structure and slides through two vertical rails. The mobile platform moves horizontally through a rack and pinion system, and it moves inward and outward thanks to a

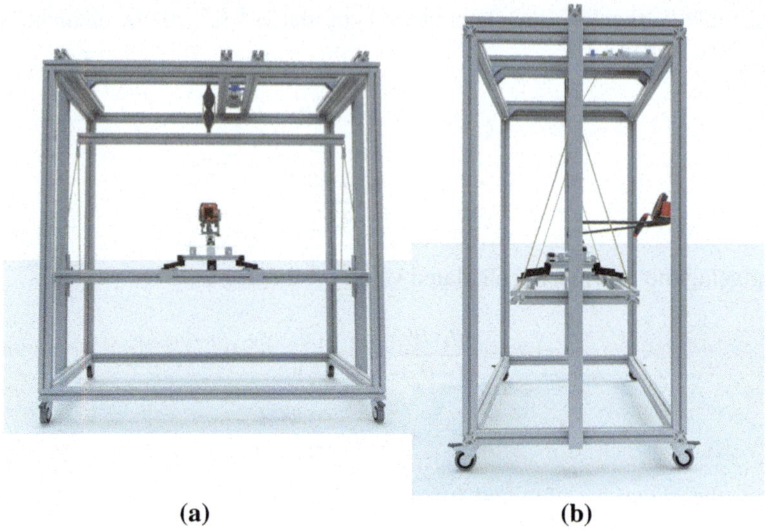

Fig. 13.6 Views **a** frontal and **b** lateral of a scalable prototype proposal

screw-nut transmission. This las motion is very important to adapt to façade profile. Also, the mobile platform supports the pan-tilt system as well as the end effector that holds the pachometer. The Fig. 13.7 shows the main mechanical elements of our proposal to generate the mobility required for an inspection task.

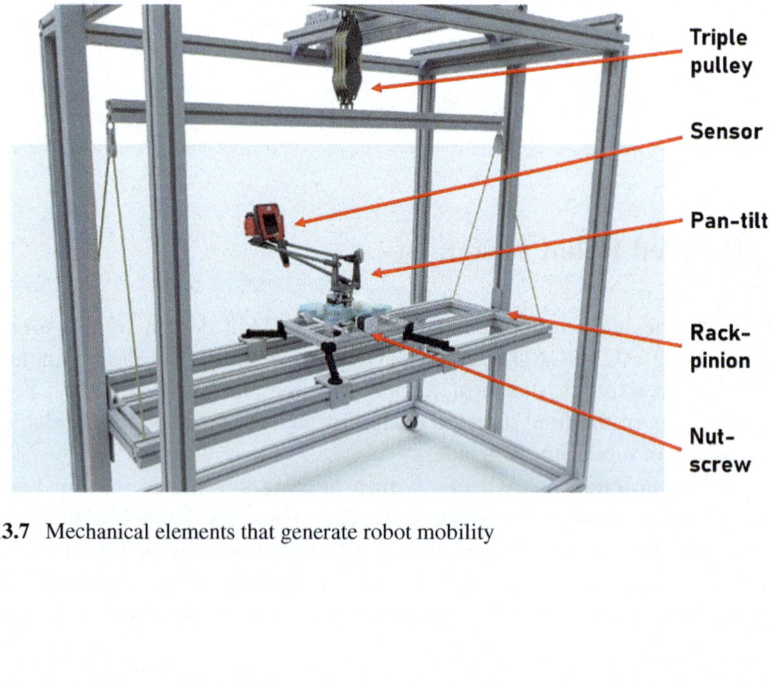

Fig. 13.7 Mechanical elements that generate robot mobility

13.6 Construction of a Laboratory Prototype

To validate the design concept, a laboratory prototype was manufactured. During the process, the following manufacturing requirements were considered: Means to move on the ground, adaptability to the façade profile, zigzag of the end effector, horizontal and vertical maneuverability, simplicity of assembly, weight, and actuators.

The mobile structure of the robot is built in aluminum structural profile with 40 × 40 mm T-groove. At each end of the base of the mobile structure, four 3" rubber wheels are attached, each mounted on a rotating plate with brake. These wheels have thermoplastic properties, polypropylene rims, reinforced bushings, and a load capacity of up to 75 kg each. The mobile platform slides vertically through two iron U-rails and a couple of four-wheel trolley, it is driven by a triple pully with 120 kg capacity and a DC motor. The mobile platform slides horizontally through four 6063T5 aluminum profiles and UHMW antifriction pads, it is driven by a rack and pinion system and a DC motor fixed to the pinion. The rack is made of low carbon steel, type TR with pressure angles of 20°, with an overall length of 1120 mm and a width of 12 mm.

The pan-tilt system, driven by two steppers motors, is mounted on a square acrylic plate, which was mounted to the nut of the screw-nut transmission mechanism which is driven by a DC motor. The square acrylic plate has four legs that were designed and printed in 3D (black color), with a density of 40%. The screw-nut system consists of two 400 mm linear motion rod shaft guides, one 8 mm feed screw, four SK8 shaft supports, four SCS8UU bearings, two KP08 flexible shaft couplings and 2 flexible shaft couplings (Fig. 13.8) .

13.7 Conclusions

This research proposes a 5-DOF for inspection tasks on vertical surfaces. The kinematics are simple since the translations and rotations are all decoupled. The manufactured prototype is light but robust, easy for assembling and enough scalable to adapt to building sizes. Some mobility tests were performed to validate the mechanical assembling. The prototype can perform inspection and some other tasks such as cleaning by changing the sensor with a high-pressure washer. A control system will ensure that the robot follows the zigzag trajectory during an inspection task, a human operator will be able to monitor the task from the ground through a fixed camera mounted on the structure.

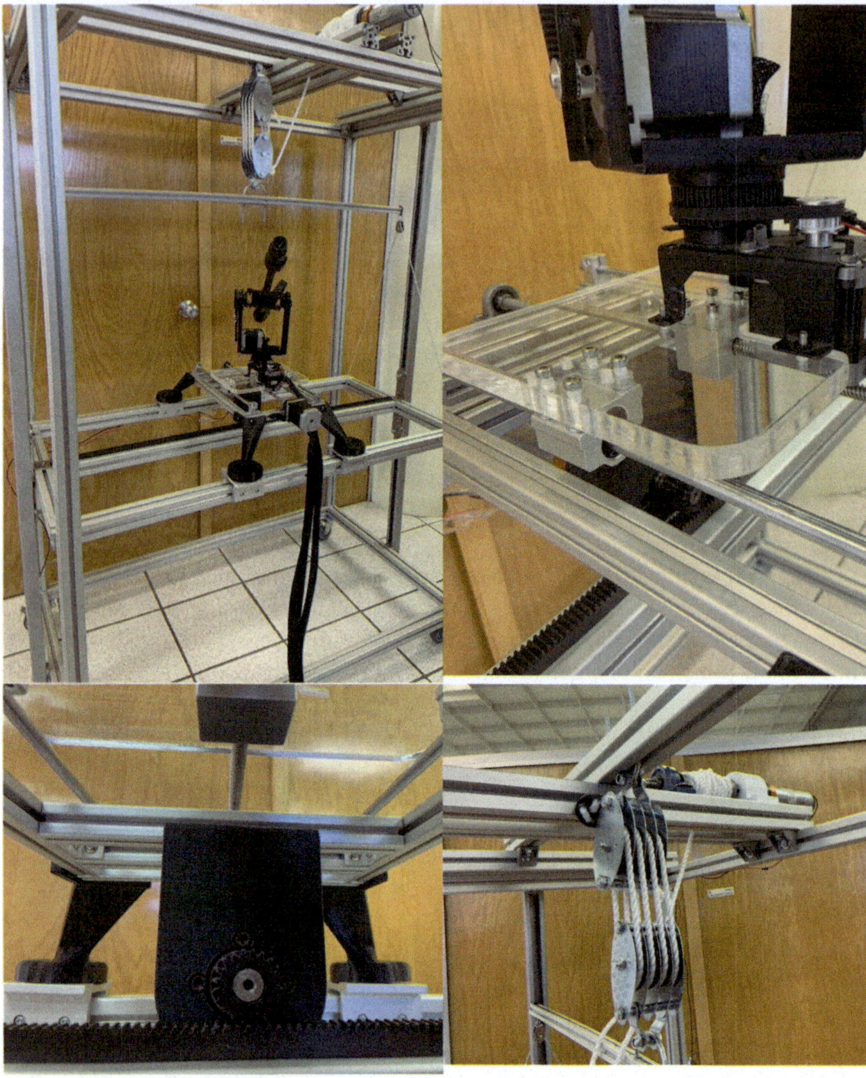

Fig. 13.8 Views of the manufactured prototype

References

1. IFR Press Conference. Frankfurt (2020). https://ifr.org/DOWNLOADS/PRESS2018/PRESENTATION_WR_2020.PDF
2. Riobo, J. et al.: New robotic systems of inspection and intervention in façade rehabilitation (2018). https://www.researchgate.net/publication/327560208_Nuevos_sistemas_roboticos_de_inspeccion_e_intervencion_en_rehabilitacion_de_fachadas_New_robotic_systems_of_inspection_and_intervention_in_facade_rehabilitation

3. Nansai, S., Mohan, R.: A survey of wall climbing robots: recent advances and challenges. Robotics **5**(3). https://doi.org/10.3390/robotics5030014
4. Minmoon, S., Jaemyunghuh, S.D., Soohan, Ch.: Vertical motion control of building façade maintenance robot with built-in guide rail (2015)
5. Kawasaki, S., Kikuchi, K.: Development of a small legged wall climbing robot with passive suction cups. In: Proceedings of the 3rd International Conference on Design Engineering and Science—ICDES, pp. 112–116. Pilsen, Czech Republic (2014)
6. Wang, Y., Liu, S., Xu, D., Zhao, Y., Hao, S., Gao, X.: Development and application of wall-climbing robots. In: Proceedings of the 1999 IEEE International Conference on Robotics and Automation, Vol. 2, pp. 1207–1212. Detroit, AL, USA, 10–15 (1999)
7. Xu, Z., Ma, P.: A wall-climbing robot for labelling scale of oil tank's volume. Robotica **20**, 209–212 (2002)
8. Zhang, H., Zhang, J., Liu, R., Zong, G.: Mechanical design and dynamics of an autonomous climbing robot for elliptic half-shell cleaning. Int. J. Adv. Robot. Syst. **4**, 437–446 (2007)
9. Funatsu, M., Kawasaki, Y., Kawasaki, S., Kikuchi, K.: Development of cm-scale wall climbing hexapod robot with claws. In: Proceedings of the 3rd International Conference on Design Engineering and Science—ICDES, pp. 101–106. Pilsen, Czech Republic (2014)
10. Schmidt, D., Berns, K.: Climbing robots for maintenance and inspections of vertical structures—a survey of design aspects and technologies. Robot. Auton. Syst. **61**, 1288–1305 (2013). https://doi.org/10.1016/j.robot.2013.09.002
11. CMP Express, Periodic facade inspection. https://cmpexpress.com.sg/periodic-facade-inspection/

Chapter 14
Motion Planning of Humanoid Robots Walking in Any Direction on Plane Surfaces with Arbitrary Orientation

Brandon D. Salazar-Bravo, J. Alfonso Pamanes, and J. Eduardo Fierro-Proa

Abstract This chapter presents an approach for planning stable motions of a humanoid robot while walking. The proposed motions are based on cycloidal time functions of the pelvis and the free foot of the robot. The parameters employed by such functions must be appropriately specified by the user, considering the behavior of the ZMP given by the simulation of the humanoid dynamics. This approach considers the robot moving on a plane surface with arbitrary orientation, walking in a straight path in any direction. The method is applied to the Bioloid Premium Humanoid robot and its efficacy is experimentally validated.

Keywords Humanoid robot · Stable walking · ZMP

14.1 Introduction

This contribution is related to the efficient and stable biped walking of humanoid robots. The kinematic behavior of this type of mechanism was one of the areas of interest of Prof. Carlos López-Cajún. Because of the inherent instability of a biped while walking, slow perturbations could cause it to fall. Accordingly, a relevant criterion in designing, motion planning, and control of such machines must be to keep the stability as high as possible while the robot walks or runs. The stability criterion of biped robots applied in most research works is based on the Zero Moment Point (ZMP) [1]. Shi et al. [2], for instance, minimize the distance between the center of the stable region and a set of ZMP of the support foot, by specifying a suitable trajectory for the hip of a humanoid. They analyzed a 12 degree of freedom (DOF) robot and specified its pelvis motion by using a sinusoidal function. This kind of function, however, produces impact forces during the landing of the free foot. On the other hand, Huang et al. [3] maximize the stability margin of a biped walking with

B. D. Salazar-Bravo · J. A. Pamanes (✉) · J. E. Fierro-Proa
División de Posgrado e Investigación del TecNM, Instituto Tecnológico de La Laguna, Torreón, Coah. CP 27000, México
e-mail: japamanesg@correo.itlalaguna.edu.mx

M. Ceccarelli and J. C. Jauregui-Correa (eds.), *State-of-the-Art and Innovations in Mechanism and Machine Science*, Mechanisms and Machine Science 150,
https://doi.org/10.1007/978-3-031-47040-0_14

motions based on third-order spline functions. The authors use two parameters as optimization variables. Unfortunately, the applied spline functions don't even allow for avoiding the impact forces. In other work [4], an *ideal* trajectory of the ZMP, based on a human walking, is specified. The pelvis motion is such that the real trajectory of the ZMP is as near as possible to the ideal one. Other criteria were applied in more recent studies for synthesizing gaits. Hirabayashi et al. [5] introduce an approach that compensates for the yaw moment of the robot during walking. In [6] a method was presented to generate walking patterns that require the lowest friction forces. Alternatively, a method based on cycloidal functions was presented in [7] for moving the pelvis and the free foot. This kind of function produces walking without impact forces of the free foot during the landing. Indeed, during the second half of a step time, the velocity and acceleration of such a foot are smoothly reduced until they become zero when the foot touches the floor. An optimization process based on the approach in [7] was carried out to maximize the stability margin of a humanoid [8].

The previously referred approaches have been developed by considering that the walking plane is horizontal. In such a case, the postures of the robot's legs during walking are symmetrical and its kinematic analysis is relatively simpler than that required when the humanoid walks on an inclined surface. In this chapter, the approach in [7] is extended in order to consider the humanoid walking on a straight path in any direction on a plane surface with arbitrary orientation. Thus, asymmetric robot postures must be taken into account in the new formulation.

The geometric parameters of the considered architecture for the humanoid studied here are described in the next section. The position analysis of the robot is presented in Sect. 14.2. The basic notions about the applied motion of the pelvis and the free foot are considered in Sect. 14.3. The formulation to specify the desired motion of the robot in arbitrary directions on a sloping surface is developed in Sect. 14.4. The equations to solve the inverse kinematics of velocity and acceleration are presented in Sect. 14.5. Then, our approach is applied to an experimental walking of the Bioloid Premium Humanoid robot for a rectilinear path on a sloping surface. Finally, the results are discussed, and the conclusion of the work is presented.

14.2 Position Analysis of the Legs of a Humanoid

The architecture of the legs of the humanoid considered here is shown in Fig. 14.1. One orthonormal reference frame is attached to each link. The frames are numbered from *0* (support foot) to *12* (free foot). The geometric parameters defining the pose of each frame (beginning with the *1*) with respect to the precedent one are given in Table 14.1. Besides, supplementary orthonormal frames Σ_W, Σ_P and Σ_F, attached with the floor (or *world*), the pelvis (link 6) and the free foot (link 12) respectively, can be also identified in Fig. 14.1. The frame Σ_W is employed as the main reference of the robot's motion. The posture in Fig. 14.1 corresponds to a step with the right support foot (RSF). The frames of links associated with the posture using the left support foot (LSF) are assigned by applying the same criterion, beginning with frame 0 attached to the support foot. In both cases, the geometric parameters are those displayed in Table 14.1. Only ε_p switches from +1 for RSF to −1 for LSF. The elemental homogeneous matrices associated with the legs can be identified based on the geometric parameters in Table 14.1.

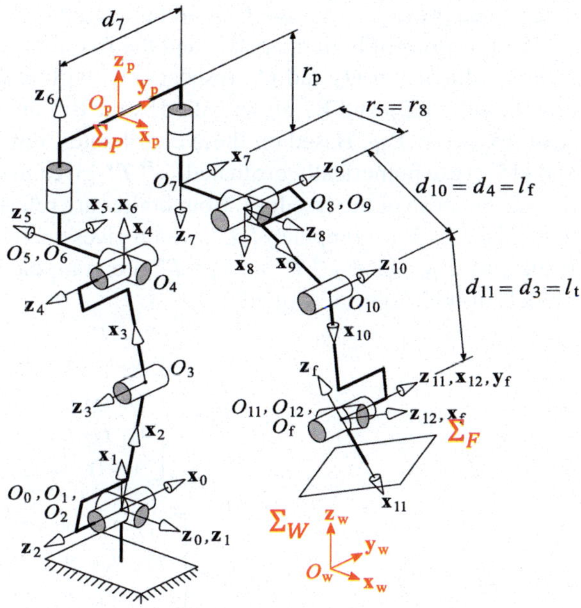

Fig. 14.1 Kinematic scheme of legs of the considered humanoid

Table 14.1 Modified Denavit-Hartenberg parameters of legs [9]

i	α_i	d_i	θ_i	r_i
1	90°	0	θ_1	0
2	−90°	0	θ_2	0
3	0°	d_3	θ_3	0
4	0°	d_4	θ_4	0
5	−90°	0	θ_5	0
6	90°	0	θ_6	0
7	180°	$\varepsilon_p d_7$	θ_7	0
8	−90°	0	θ_8	0
9	−90°	0	θ_9	0
10	0°	d_{10}	θ_{10}	0
11	0°	d_{11}	θ_{11}	0
12	−90°	0	θ_{12}	0

*$\varepsilon p = +1$ for right support foot, $\varepsilon p = -1$ for left support foot

14.2.1 Inverse Kinematics

Applying the Paul's method [10], the equations of the inverse kinematics model of legs can be obtained for specified poses of frames Σ_P and Σ_F attached to the pelvis and the free foot, respectively, related to the world frame Σ_W. Such equations were determined previously for the Bioloid humanoid [7].

The positions of both the pelvis and the free foot are specified through the cartesian coordinates of O_p and O_f, respectively, with respect to $\mathbf{\Sigma}_W$. Furthermore, the orientations of $\mathbf{\Sigma}_P$ and $\mathbf{\Sigma}_F$ are specified by the Bryant angles λ, μ and ν of each frame with respect to $\mathbf{\Sigma}_W$. Based on these coordinates, the homogeneous matrices (14.1) and (14.2) are numerically evaluated. ${}_P{}^W\boldsymbol{T}^*$ gives the pose of the pelvis and ${}_F{}^W\boldsymbol{T}^*$ defines the pose of the free foot. Both bodies are referred to $\mathbf{\Sigma}_W$. These matrices are applied in (14.3) to compute ${}_F{}^P\boldsymbol{T}^*$, which gives the relative pose of the free foot with respect to $\mathbf{\Sigma}_P$. Once ${}_P{}^W\boldsymbol{T}^*$ and ${}_F{}^P\boldsymbol{T}^*$ are computed, then the inverse kinematics models of both legs are applied.

$$ {}_P{}^W T^* = \begin{bmatrix} t_{11} & t_{12} & t_{13} & x_p \\ t_{21} & t_{22} & t_{23} & y_p \\ t_{31} & t_{32} & t_{33} & z_p \\ 0 & 0 & 0 & 1 \end{bmatrix} \tag{14.1} $$

$$ {}_F{}^W T^* = \begin{bmatrix} s_{fx} & n_{fx} & a_{fx} & p_{fx} \\ s_{fy} & n_{fy} & a_{fy} & p_{fy} \\ s_{fz} & n_{fz} & a_{fz} & p_{fz} \\ 0 & 0 & 0 & 1 \end{bmatrix} \tag{14.2} $$

$$ {}_F{}^P T^* = {}_W{}^P T^* \, {}_F{}^W T^* \tag{14.3} $$

In (14.3) ${}_W{}^P T^*$ is the inverse of ${}_P{}^W T^*$, given in (14.2). The entries of ${}_F{}^P T^*$ are designated as:

$$ {}_F{}^P T^* = \begin{bmatrix} s_{fpx} & n_{fpx} & a_{fpx} & p_{fpx} \\ s_{fpy} & n_{fpy} & a_{fpy} & p_{fpy} \\ s_{fpz} & n_{fpz} & a_{fpz} & p_{fpz} \\ 0 & 0 & 0 & 1 \end{bmatrix} \tag{14.4} $$

The entries of matrices (14.1) and (14.4) are data for inverse kinematic model of the robot.

Table 14.2 Time interval at each phase in a walking process

Phase	Number of step (n_{p_i})	Time interval	Time variable used in motion equations
Starting	$n_{pi} = 1$	$0 \leq t \leq T_1$	t
Cruising	$n_{pi} = 2, \ldots, n_p + 1$	$T_1 \leq t \leq (T_1 + n_P T_2)$	$t\prime = t - T_1 - (n_{Pi} - 2)T_2$
Stopping	$n_{pi} = n$	$(T_1 + n_P T_2) \leq t \leq (T_1 + n_P T_2 + T_3)$	$t'' = t - T_1 - (n_P T_2)$

14.3 Gait of a Humanoid Based on Cycloidal Motions

The basic notions concerning the considered walking pattern are presented in this Section, which is based on [7]. We consider here that the humanoid walks on a horizontal plane. The extension to a walking on an arbitrarily orientated plane is accomplished in Sect. 14.4.

For a walking process of the humanoid, the pelvis begins from the resting position, and it accelerates with a cycloidal motion in a first phase (*starting phase*), during one step in a time T_1, until a specified cruise speed is reached in direction of x_W. This speed is kept constant during the next n_p steps, completing a second stage (*cruising phase*). Each step in this phase needs a time T_2. Finally, the pelvis slow down its speed to a stop in the last step during the third phase in a time T_3 (*stopping phase*). Thus, the total number of steps for the walking process is $n_p + 2$, which is completed in the total time given by $T_1 + n_p\ T_2 + T_3$. The time intervals of the phases are summarized in Table 14.2.

The time functions that describe the poses of Σ_P and Σ_F, attached respectively to the pelvis and the free foot, are presented in Appendix A. The geometric meaning of parameters in these functions corresponding to the linear motion are indicated in Fig. 14.2. In Sect. 14.6, the complete list of parameters is given in Table 14.3 for a specific walking assigned to a Bioloid Premium humanoid robot. By specifying suitable values of these parameters in the time functions a stable walking of the robot could be obtained. The paths of O_f and O_p of a generic humanoid are shown in Fig. 14.2, corresponding to a given set of parameters. The Table 14.3 includes all parameters required for position and orientation of the pelvis and feet.

14.4 Sloping Surface for Walking

To formulate a solution of the problem of walking on a sloping surface, the arbitrarily oriented plane QRS is considered, as shown in Fig. 14.3. This plane is defined in Σ_W by the following Equation:

$$Ax + By + Cz + D = 0 \tag{14.5}$$

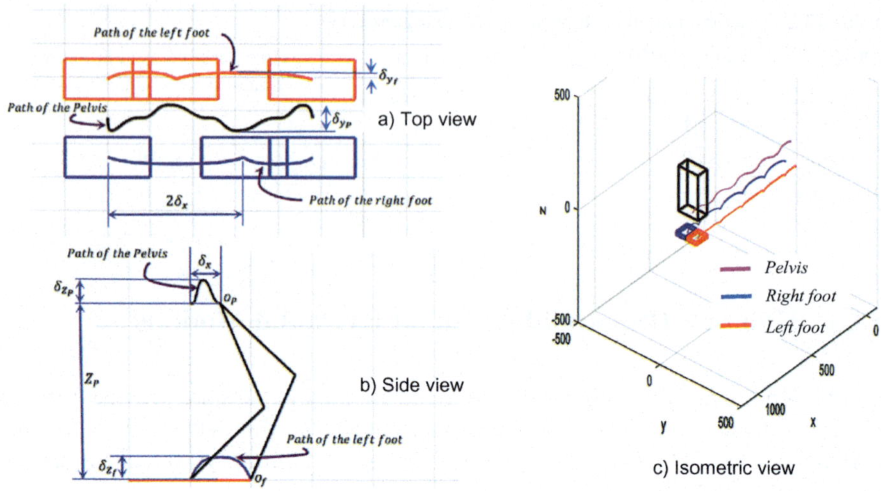

Fig. 14.2 Parameters of a typical gait based on cycloidal motions, and paths of O_p and O_f

where the parameters *A, B, C* and *D* determine the orientation of the surface as shown in [11]. Then, an orthonormal frame Σ_U is allocated such that the plane defined by the axis $\boldsymbol{x}_U$ and $\boldsymbol{y}_U$ lies on the walking surface. The origin O_U of Σ_U has coordinates r_{ux}, r_{uy} and r_{uz} in the world frame Σ_W. Then, the pose of Σ_U is described with respect to the frame Σ_W by the homogeneous matrix (14.11).

$$
{}_U{}^W T = \begin{bmatrix} x_{ux} & y_{ux} & z_{ux} & r_{ux} \\ x_{uy} & y_{uy} & z_{uy} & r_{uy} \\ x_{uz} & y_{uz} & z_{uz} & r_{uz} \\ 0 & 0 & 0 & 1 \end{bmatrix} \tag{14.6}
$$

The entries of this matrix can be expressed as functions of parameters *A, B, C* and *D*.

On the other hand, an orthonormal frame Σ_C is defined by a rotation of the frame Σ_U by an angle of δ about the axis z_U, and then by a translation determined by the vector $\boldsymbol{r}_C$ whose components in Σ_U are x_{OC}, y_{OC} and z_{OC}. The pose finally obtained by Σ_C with respect to Σ_U is shown in Fig. 14.4. The homogeneous matrix describing the pose of Σ_C with respect to Σ_U is

$$
{}_C{}^U T = \begin{bmatrix} c\delta & -s\delta & 0 & x_{OC} \\ s\delta & c\delta & 0 & y_{OC} \\ 0 & 0 & 1 & z_{OC} \\ 0 & 0 & 0 & 1 \end{bmatrix} \tag{14.7}
$$

where $s\delta = sin\delta$ and $c\delta = cos\delta$.

Table 14.3 Parameters of the walking considered in the case study

Parameter	Magnitude
T_1	2 s
T_2	2 s
T_3	2 s
T_t	20 s
n	10 steps
n_p	8 steps
$x_{P_{ini}}$	0 mm
$y_{P_{ini}}$	38.5 mm
$z_{P_{ini}}$	122 mm
δx_p	50 mm
δy_p	−28.5 mm
δz_p	2 mm
λ_{Pini}	5°
μ_{Pini}	0°
ν_{Pini}	0°
$\delta\lambda_P$	0°
$\delta\mu_P$	0°
$\delta\nu_P$	0°
x_{fini}	0 mm
y_{fini}	77 mm
z_{fini}	0 mm
δx_f	50 mm
δy_f	5 mm
δz_f	25 mm
λ_f	0°
μ_f	0°
ν_f	0°
$\delta\lambda_f$	0°
$\delta\mu_f$	0°
$\delta\nu_f$	0°
ϵ	±1
r	250 mm
ϕ	85°
δ	90°

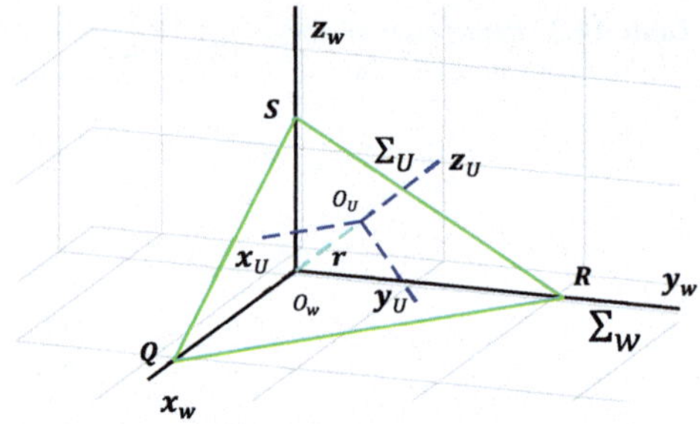

Fig. 14.3 Orthonormal frame Σ_U attached to the sloping plane

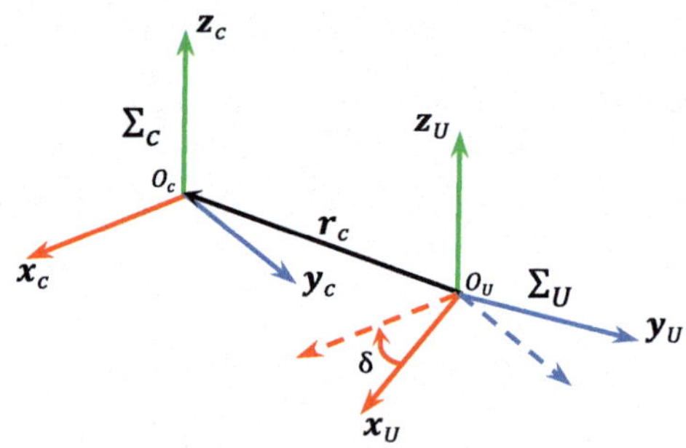

Fig. 14.4 Pose of the supplementary orthonormal frame Σ_C on the sloping plane with respect to Σ_U

Consequently, the pose of Σ_C with respect to Σ_W is obtained by

$$ {}_C^W T = {}_P^W T {}_C^P T \tag{14.8} $$

Therefore

$$ {}_C^W T = \begin{bmatrix} t_{wc11} & t_{wc12} & t_{wc13} & t_{wc14} \\ t_{wc21} & t_{wc22} & t_{wc23} & t_{wc24} \\ t_{wc31} & t_{wc32} & t_{wc33} & t_{wc34} \\ 0 & 0 & 0 & 1 \end{bmatrix} \tag{14.9} $$

where:

$t_{wc11} = x_{ux} c\delta + y_{ux} s\delta$	$t_{wc23} = z_{uy}$
$t_{wc12} = -x_{ux} s\delta + y_{ux} c\delta$	$t_{wc24} = x_{uy} x_{Oc} + y_{uy} y_{Oc} + z_{uy} z_{Oc} + r_{uy}$
$t_{wc13} = z_{ux}$	$t_{wc31} = x_{uz} c\delta + y_{uz} s\delta$
$t_{wc14} = x_{ux} x_{Oc} + y_{ux} y_{Oc} + z_{ux} z_{Oc} + r_{ux}$	$t_{wc32} = -x_{uz} s\delta + y_{uz} c\delta$
$t_{wc21} = x_{uy} c\delta + y_{uy} s\delta$	$t_{wc33} = z_{uz}$
$t_{wc22} = -x_{uy} s\delta + y_{uy} c\delta$	$t_{wc34} = x_{uz} x_{Oc} + y_{uz} y_{Oc} + z_{uz} z_{Oc} + r_{uz}$

As a result of the precedent analysis, the frame Σ_C is described with respect to Σ_W by the entries of the matrix ${}_C^W T$, which are determined by the parameters A, B,

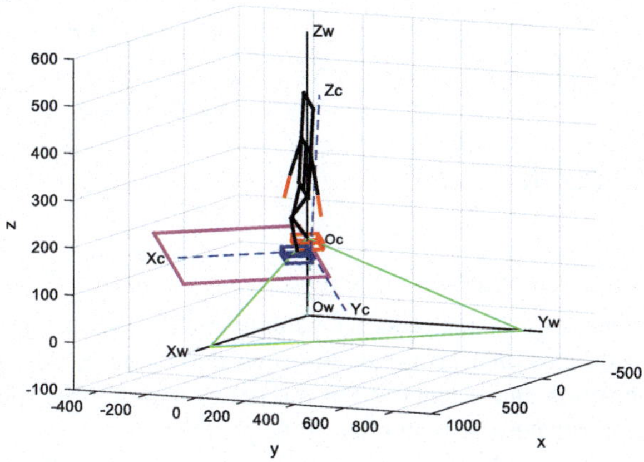

Fig. 14.5 Simulation of a humanoid walking on a carpet of a sloping plane

$C, D, \delta, x_{OC}, y_{OC}$ and z_{OC}. These parameters specify the pose of the walking plane, the start point and the direction of walking of the humanoid. In other words, such parameters specify the desired pose of a hypothetic *carpet* where the humanoid have to walk on the sloping plane. When the frame Σ_C has been defined, the gait of the humanoid should be specified with respect to this frame by applying the equations of the walking pattern as usually applied to a horizontal walking surface. In Fig. 14.5 a posture of a humanoid in a simulation is observed walking on the carpet defined on a sloping plane by applying the formulation introduced here.

14.5 Inverse Kinematics of Velocity and Acceleration

The velocity state $\dot{\boldsymbol{x}}_P$ of the pelvis, and the relative velocity state $\dot{\boldsymbol{x}}_{F/P}$ of the free foot with respect to the pelvis, are given respectively by (14.10) and (14.11):

$$\dot{\boldsymbol{x}}_P = J_P \dot{\boldsymbol{q}}_P \tag{14.10}$$

$$\dot{\boldsymbol{x}}_{F/P} = J_F \dot{\boldsymbol{q}}_F \tag{14.11}$$

where J_P is the Jacobian matrix of the support leg and $\dot{\boldsymbol{q}}_P$ the corresponding 6-dimensional array of joint rates. Besides, J_F is the Jacobian matrix of the oscillating leg and $\dot{\boldsymbol{q}}_F$ the corresponding 6-dimensional array of joint rates. Symbolic expressions for entries of both matrices can be obtained by applying the *SYMORO* (*SYmbolic MOdeling of RObots*) software [12] based on the geometric parameters of the humanoid's legs. Solving (14.10) and (14.11) to determine the joint rates of both legs associated with the specified states of velocities $\dot{\boldsymbol{x}}_P$ and $\dot{\boldsymbol{x}}_{F/P}$, respectively, we obtain (14.12) and (14.13), corresponding to the inverse model of velocity of the robot.

$$\dot{\boldsymbol{q}}_P = J_P^{-1} \dot{\boldsymbol{x}}_P \tag{14.12}$$

$$\dot{\boldsymbol{q}}_F = J_F^{-1} \dot{\boldsymbol{x}}_{F/P} \tag{14.13}$$

Clearly

$$\dot{\boldsymbol{x}}_{F/P} = \dot{\boldsymbol{x}}_F - \dot{\boldsymbol{x}}_P \tag{14.14}$$

where $\dot{\boldsymbol{x}}_F$ is the state of velocity of the free foot. $\dot{\boldsymbol{x}}_P$ and $\dot{\boldsymbol{x}}_F$ are given by the 6-dimensional arrays of the time derivatives of the corresponding coordinates of position and orientation of the pelvis and the free foot. The time functions of such coordinates of are defined in Appendix A.

On the other hand, the state of acceleration $\ddot{\boldsymbol{x}}_P$ of the pelvis, and the state of relative acceleration $\ddot{\boldsymbol{x}}_{F/P}$ of the free foot with respect to the pelvis, are given respectively by (14.15) and (14.16):

$$\ddot{\boldsymbol{x}}_P = J_P \ddot{\boldsymbol{q}}_P + \dot{J}_P \dot{\boldsymbol{q}}_P \tag{14.15}$$

$$\ddot{\boldsymbol{x}}_{F/P} = J_F \ddot{\boldsymbol{q}}_F + \dot{J}_F \dot{\boldsymbol{q}}_F \tag{14.16}$$

where $\dot{J}_P$ is the time derivative of J_P, and $\ddot{\boldsymbol{q}}_P$ the corresponding 6-dimensional array of time derivatives of $\dot{\boldsymbol{q}}_P$. Besides, $\dot{J}_F$ is the time derivative of J_F, and $\ddot{\boldsymbol{q}}_F$ the corresponding 6-dimensional array of time derivatives of $\dot{\boldsymbol{q}}_P$.

Solving (14.15) and (14.16) we obtain the inverse model of acceleration of the robot, expressed by (14.17) and (14.18), that give the time derivatives of joint rates for both legs:

$$\ddot{\boldsymbol{q}}_P = J_P^{-1}\left[\ddot{\boldsymbol{x}}_P - \dot{J}_P \dot{\boldsymbol{q}}_P\right] \tag{14.17}$$

$$\ddot{\boldsymbol{q}}_F = J_F^{-1}\left[\ddot{\boldsymbol{x}}_{F/P} - \dot{J}_F \dot{\boldsymbol{q}}_F\right] \tag{14.18}$$

Note that

$$\ddot{\boldsymbol{x}}_{F/P} = \ddot{\boldsymbol{x}}_F - \ddot{\boldsymbol{x}}_P \tag{14.19}$$

where $\ddot{\boldsymbol{x}}_F$ is the state of acceleration of the free foot. The states of acceleration of the pelvis and the free foot are given by the 6-dimensional arrays of time derivatives of the corresponding linear and angular velocities. The solution of the inverse kinematics of velocity and acceleration is essential for solving the inverse dynamics of the robot.

14.6 Analysis of a Walking of the Bioloid Robot on a Sloping Surface Using Cycloidal Functions

To analyze the stability of a specific humanoid walking by using cycloidal motions, we apply such a gait to the Bioloid Premium robot (Fig. 14.6). The analysis include the resolution of the inverse dynamics of the robot, which is applied in a numeric simulation. The generalized forces actuating on the robot during the walking allow us to determine the behavior of the ZMP on the feet. A suitable performance of the ZMP will mean that the walking is feasible and consequently, an experimental validation could be achieved.

The algorithm applied to solve the inverse dynamics is based on the Newton-Euler formulation. For the case considered here, the angle of the axis normal to the surface where the robot walks is $\phi = 85°$ with respect to the horizontal plane and the direction of the walking carpet on the plane is defined by $\delta = 90°$. The set of proposed parameters of the gait is given in Table 14.3.

The inertial parameters of links of the robot are specified in Appendix B. The architecture of Bioloid's legs is similar to that one appreciated in Fig. 14.1. The corresponding magnitudes of parameters in Table 14.1 are $d3 = d4 = d7 = d10 = d11 = 77$ mm.

As a result of our analysis, the behavior of the generalized forces applied to the 12 joints of the legs was obtained, as appreciated in plots of Fig. 14.7a, b. Furthermore, the paths of the ZMP on the support feet were identified. They are presented in Fig. 14.8. Even if some of the ZMP in the paths are close to the edges of the footprints, it can be observed in Fig. 14.8 that such points never leaves the footprints. Consequently, it can be claimed that the stability of the robot will be assured during the real walking. Hence, taking into account this observation, we proceed to an experimental validation of the walking. A sequence of postures of the humanoid during the experiment is appreciated in Fig. 14.10 that confirm the viability of its motions.

Fig. 14.6 Bioloid Premium humanoid robot considered in the case study

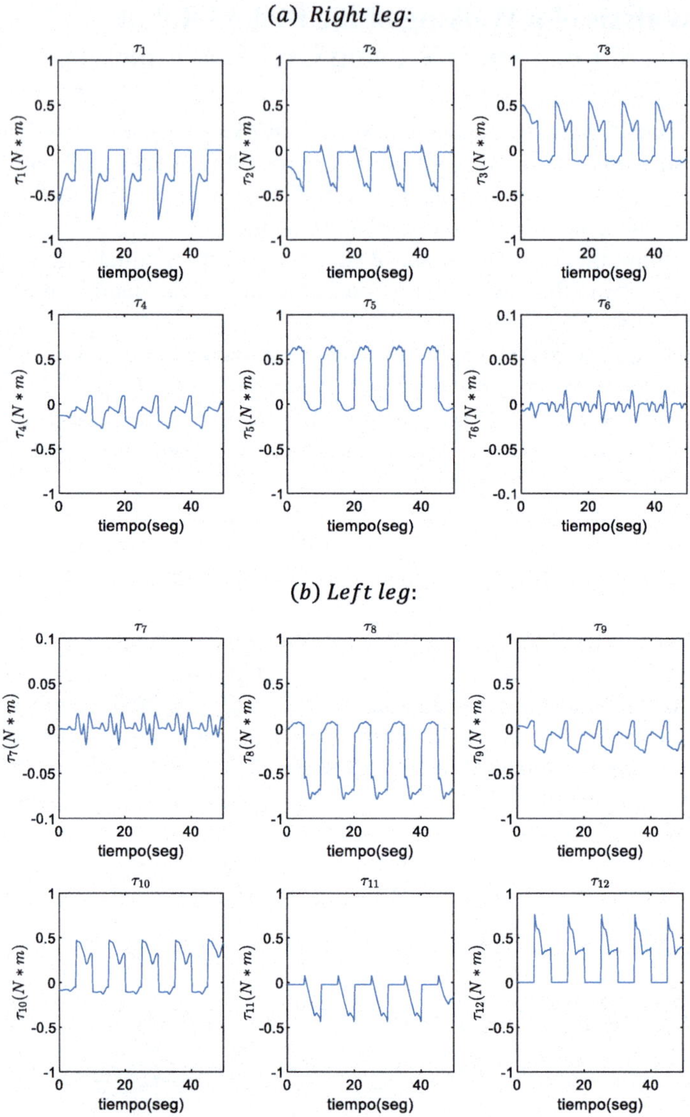

Fig. 14.7 Behavior of the generalized forces obtained by simulation

14.6.1 Stability Index

To numerically assess the performance of the robot's postures during the walking we introduce a stability index based on the behavior of the ZMP while a single support phase is accomplished. We consider, for one step, a sample of *npm* zero

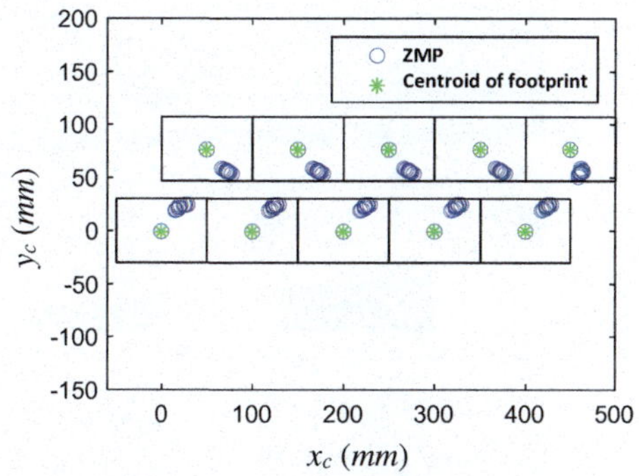

Fig. 14.8 Paths of the *Zero Moment Points* inside the footprints at each step

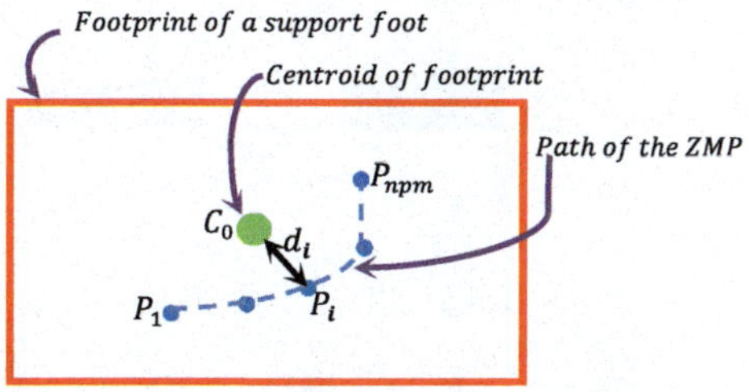

Fig. 14.9 Sample of *Zero Moment Points* P_i for a step of a walking

moment points Pi ($i = 1, 2, \ldots, npm$). These points are represented in Fig. 14.9 inside a polygonal footprint. The centroid C_0 of this polygon is also shown in the Figure.

It is observed that any ZMP, Pi, is associated to a posture $\boldsymbol{qi}$ of the robot. Besides, based on the theory in [1], when a posture $\boldsymbol{qi}$ of a robot during a walking is such that Pi is outside the footprint of the support foot, then the robot falls. So, based on the stability of the robot, a posture $\boldsymbol{q0}$ which locate its ZMP on C_0 can be considered as an ideal posture. Therefore, the corresponding distance d_i from the ZMP P_i to C_0 can be interpreted as a *distance* from the posture $\boldsymbol{q_i}$ to the ideal one $\boldsymbol{q_0}$. Such a distance is inversely proportional to the stability of the robot. For the sample of n_{pm} zero moment points, we compute a typical large distance d^* of the set by:

$$d^* = \overline{d} + d_\sigma \tag{14.20}$$

where $\overline{d}$ is the average distance and d_σ is the standard deviation of the sample. The inverse of d is an index of the stability. The smaller the distance d^* is, the greater the stability is of the corresponding posture of the humanoid. In the case study considered in this Section, the average of d^* for the 10 steps of the walking was 30.7936 mm.

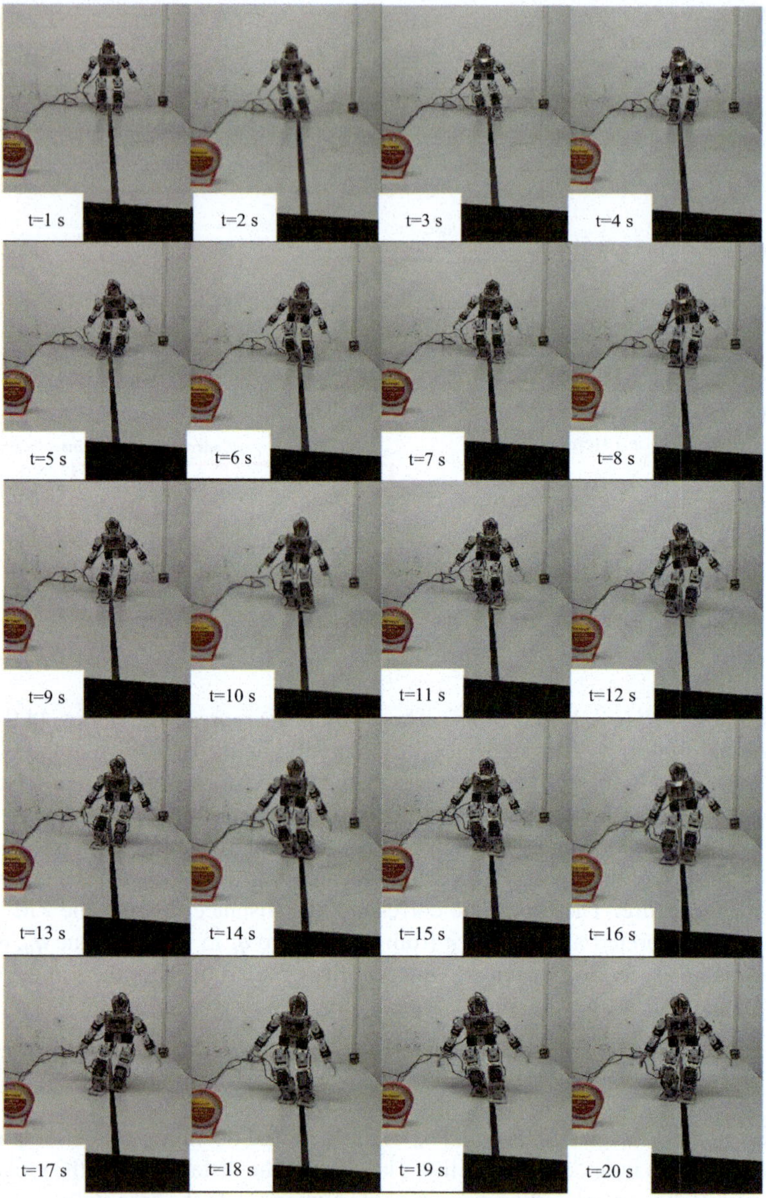

Fig. 14.10 Sequence of postures of the robot during the walking [13]

14.7 Conclusion

An integration of the different aspects of motion planning for walking of humanoids was achieved in this work. This one includes the application of our approach previously presented [11] for specifying the desired motions of a humanoid robot that walks in an arbitrary direction on a sloping surface. Such approach is extended here by incorporating the dynamics of the robot. Furthermore, this work includes a summary of the formulations applied to solve the whole inverse kinematics and dynamics of the humanoid. The solution of such problems allows us to create an interesting toolbox of software for simulation studies of walking of humanoids. Such a toolbox was applied to recognize the behavior of the ZMP in the support foot at each step during the walking in order to validate the feasibility of a given set of parameters employed to produce a stable gait of the robot. Such parameters generate a stable walking in a direction perpendicular to that one of the maximal slope on an inclined surface. The employed walking pattern was proposed in an early work [7] to specify motions of the pelvis and the free foot based on cycloidal functions. These equations provide smooth movements to the humanoid, which avoid impacts of the free foot during the landing in a walking. The proposed gait was successfully validated by applying it to the Bioloid Premium humanoid robot. A photographic sequence of 20 postures of the humanoid during the experimentation are presented [13].

The case study considered here considers the experimentation of only one set of walking parameters, which was previously chosen by testing several sets. The results of such previous tests are not included here because we don't have enough space. However, more extensive studies will be published in future works that will include more details of the process for choosing the amounts of walking parameters. One of the employed techniques will involves an optimization process addressed to minimize the distance d^* for a walking. Clearly, several sloping angles of the walking plane should be taken into account for strongly supporting our approaches.

Acknowledgements This work was supported by the CONACyT (National Council of Research and Technology), the TecNM (National Institute of Technology of Mexico) and the ITLag (Institute of Technology of la Laguna) of the Ministry of Public Education of Mexico.

Appendix A. Time Cycloidal Functions for Motion of Pelvis and Free Foot

The proposed cycloidal motion laws for the pelvis are characterized by equations (A14.1)–(A14.6), and those for the free foot are (A14.7) to (A14.12)

$$x_p(t) = \begin{cases} x_{p_{ini}} + \frac{\delta x_p}{2}\left(\frac{t}{T_1} - \frac{1}{\pi}sin\left(\frac{t\pi}{T_1}\right)\right), 0 \le t \le T_1 \\ x_{p_{ini}} + \delta x_p\left(j - \frac{1}{2} + \frac{t'}{T_2}\right), j = 1, \ldots, n, 0 \le t' \le T_2 \\ x_{p_{ini}} + n(\delta x_p) + \frac{\delta x_p}{2}\left(1 + \frac{t''}{T_3} + \frac{1}{\pi}sin\left(\frac{t''\pi}{T_3}\right)\right), 0 \le t'' \le T_3 \end{cases} \quad \text{(A14.1)}$$

$$y_p(t) = \begin{cases} y_{p_{ini}} + \epsilon\delta y_p\left(\frac{t}{T_1} - \frac{1}{2\pi}sin\left(\frac{2t\pi}{T_1}\right)\right), 0 \le t \le T_1 \\ y_{p_{ini}} + 2\epsilon\delta y_p\left(\frac{t\prime}{T_2} - \frac{1}{2\pi}sin\left(\frac{2t\prime\pi}{T_2}\right)\right), 0 \le t\prime \le T_2 \\ y_{p_{ini}} + \epsilon\delta y_p\left(\frac{t''}{T_3} - \frac{1}{2\pi}sin\left(\frac{2t''\pi}{T_3}\right)\right), 0 \le t'' \le T_3 \end{cases} \quad \text{(A14.2)}$$

$$z_p(t) = \begin{cases} z_{p_{ini}} + 2\delta z_p\left(\frac{t}{T} - \frac{1}{4\pi}sin\left(\frac{4t\pi}{T}\right)\right), 0 \le t \le \frac{T_i}{2} \\ z_{p_{ini}} + 2\delta z_p\left(\frac{1}{2} - \frac{t}{T} + \frac{1}{4\pi}sin\left(\frac{4t\pi}{T}\right)\right), \frac{T_i}{2} \le t \le T_i \end{cases} \quad \text{(A14.3)}$$

$$\lambda_p(t) = \begin{cases} \lambda_{p_{ini}} + \epsilon\delta\lambda_p\left(\frac{t}{T_1} - \frac{1}{2\pi}sin\left(\frac{2t\pi}{T_1}\right)\right), 0 \le t \le T_1 \\ \lambda_{p_{ini}} + 2\epsilon\delta\lambda_p\left(\frac{t\prime}{T_2} - \frac{1}{2\pi}sin\left(\frac{2t\prime\pi}{T_2}\right)\right), 0 \le t\prime \le T_2 \\ \lambda_{p_{ini}} + \epsilon\delta\lambda_p\left(\frac{t''}{T_3} - \frac{1}{2\pi}sin\left(\frac{2t''\pi}{T_3}\right)\right), 0 \le t'' \le T_3 \end{cases} \quad \text{(A14.4)}$$

$$\mu_p(t) = \begin{cases} \mu_{p_{ini}} - 2\delta\mu_p\left(\frac{t}{T} - \frac{1}{4\pi}sin\left(\frac{4t\pi}{T}\right)\right), 0 \le t \le \frac{T_i}{2} \\ \mu_{p_{ini}} - 2\delta\mu_p\left(\frac{1}{2} - \frac{t}{T} + \frac{1}{4\pi}sin\left(\frac{4t\pi}{T}\right)\right), \frac{T_i}{2} \le t \le T_i \end{cases} \quad \text{(A14.5)}$$

$$\nu_p(t) = \begin{cases} \nu_{p_{ini}} + \epsilon\delta\nu_p\left(\frac{t}{T_1} - \frac{1}{2\pi}sin\left(\frac{2t\pi}{T_1}\right)\right), 0 \le t \le T_1 \\ \nu_{p_{prec}} + 2\epsilon\delta\nu_p\left(\frac{t\prime}{T_2} - \frac{1}{2\pi}sin\left(\frac{2t\prime\pi}{T_2}\right)\right), 0 \le t\prime \le T_2 \\ \nu_{p_{prec}} + \epsilon\delta\nu_p\left(\frac{t''}{T_3} - \frac{1}{2\pi}sin\left(\frac{2t''\pi}{T_3}\right)\right), 0 \le t'' \le T_3 \end{cases} \quad \text{(A14.6)}$$

$$x_f(t) = \begin{cases} x_{f_{ini}} + \frac{\delta x_f}{2}\left(\frac{2t}{T_1} - \frac{1}{\pi}sin\left(\frac{2t\pi}{T_1}\right)\right), 0 \le t \le T_1 \\ x_{f_{ini}} + \delta x_f\left(\frac{2t\prime}{T_2} - \frac{1}{\pi}sin\left(\frac{2t\prime\pi}{T_2}\right)\right), 0 \le t\prime \le T_2 \\ x_{f_{ini}} + \frac{\delta x_f}{2}\left(\frac{2t''}{T_3} - \frac{1}{\pi}sin\left(\frac{2t''\pi}{T_3}\right)\right), 0 \le t'' \le T_3 \end{cases} \quad \text{(A14.7)}$$

$$y_f(t) = \begin{cases} y_{f_{ini}} + 2\delta y_f\left(\frac{t}{T} - \frac{1}{4\pi}sin\left(\frac{4t\pi}{T}\right)\right), 0 \le t \le \frac{T_i}{2} \\ y_{f_{ini}} + 2\delta y_f\left(\frac{1}{2} - \frac{t}{T} + \frac{1}{4\pi}sin\left(\frac{4t\pi}{T}\right)\right), \frac{T_i}{2} \le t \le T_i \end{cases} \quad \text{(A14.8)}$$

$$z_f(t) = \begin{cases} z_{f_{ini}} + 2\delta z_f\left(\frac{t}{T} - \frac{1}{4\pi}sin\left(\frac{4t\pi}{T}\right)\right), 0 \le t \le \frac{T_i}{2} \\ z_{f_{ini}} + 2\delta z_f\left(\frac{1}{2} - \frac{t}{T} + \frac{1}{4\pi}sin\left(\frac{4t\pi}{T}\right)\right), \frac{T_i}{2} \le t \le T_i \end{cases} \quad \text{(A14.9)}$$

$$\lambda_f(t) = \begin{cases} \lambda_{f_{ini}} - 2\delta\lambda_f\left(\frac{t}{T} - \frac{1}{4\pi}sin\left(\frac{4t\pi}{T}\right)\right), 0 \le t \le \frac{T_i}{2} \\ \lambda_{f_{ini}} - 2\delta\lambda_f\left(\frac{1}{2} - \frac{t}{T} + \frac{1}{4\pi}sin\left(\frac{4t\pi}{T}\right)\right), \frac{T_i}{2} \le t \le T_i \end{cases} \quad \text{(A14.10)}$$

$$\mu_f(t) = \begin{cases} \mu_{f_{ini}} - 2\delta\mu_f\left(\frac{t}{T} - \frac{1}{4\pi}sin\left(\frac{4t\pi}{T}\right)\right), 0 \le t \le \frac{T_i}{2} \\ \mu_{f_{ini}} - 2\delta\mu_f\left(\frac{1}{2} - \frac{t}{T} + \frac{1}{4\pi}sin\left(\frac{4t\pi}{T}\right)\right), \frac{T_i}{2} \le t \le T_i \end{cases} \quad \text{(A14.11)}$$

$$v_f(t) = \begin{cases} v_{f_{ini}} - 2\delta v_f\left(\frac{t}{T} - \frac{1}{4\pi}sin\left(\frac{4t\pi}{T}\right)\right), 0 \le t \le \frac{T_i}{2} \\ v_{f_{ini}} - 2\delta v_f\left(\frac{1}{2} - \frac{t}{T} + \frac{1}{4\pi}sin\left(\frac{4t\pi}{T}\right)\right), \frac{T_i}{2} \le t \le T_i \end{cases} \quad \text{(A14.12)}$$

Appendix B. Inertial Properties of Links of the Bioloid Robot

Properties defined with respect to frames attached to the links as shown in Fig. 14.1.

Link	Mass (g)	Center of mass	Inertia tensor (kg m^2)
Link $_0 i^1$	91.16	$x = -2.33$ $y = 0.02$ $z = -21.99$	$\begin{bmatrix} 0.00006689408 & 0.00000003736 & -0.0000239406 \\ 0.00000003736 & 0.00009945628 & -0.0000000197 \\ -0.0000239406 & -00.0000000197 & 0.0000558316 \end{bmatrix}$
Link $_1 i^2$	68.66	$x = 12.79$ $y = 2.66$ $z = -0.79$	$\begin{bmatrix} 0.00002289274 & 0.00000068491 & 0.00000014994 \\ 0.00000068491 & 0.00001999689 & 0.00000014061 \\ 0.00000014994 & 0.00000014061 & 0.00002544922 \end{bmatrix}$
Link $_2 i^3$	80.39	$x = -11.99$ $y = -13.46$ $z = -0.55$	$\begin{bmatrix} 0.000025985 & 0.000001375 & -0.00000131725 \\ 0.00000137 & 0.00005144842 & -0.00000019712 \\ -0.0000013172 & -0.00000019712 & 0.00004966677 \end{bmatrix}$
Link $_3 i^4$	84.92	$x = -23.49$ $y = 0.4$ $z = -0.33$	$\begin{bmatrix} 0.0000227885 & -0.00000215017 & -0.00000096655 \\ -0.0000021501 & 0.00005969328 & 0.0000001671 \\ -0.0000009665 & 0.00000016719 & 0.0000485586 \end{bmatrix}$
Link $_4 i^5$	68.66	$x = 0.02$ $y = 12.79$ $z = 28.87$	$\begin{bmatrix} 0.00003066058 & 0.00000001364 & 0.00000000911 \\ 0.00000001364 & 0.00002556439 & 0.00000045146 \\ 0.00000000911 & 0.00000045146 & 0.00001745718 \end{bmatrix}$
Link $_5 i^6$	84.92	$x = 9.92$ $y = 3.39$ $z = 42.32$	$\begin{bmatrix} 0.000037991 & -0.0000023473 & 0.00000487244 \\ -0.000002347 & 0.0000321353 & -0.00000561942 \\ 0.000004872 & -0.0000056194 & 0.0000324199 \end{bmatrix}$
Link $_6 i^P$	229.58	$x = -38.15$ $y = -0.15$ $z = 50.1$	$\begin{bmatrix} 0.00048687615 & -0.0000000636 & 0.00002614566 \\ -0.00000006362 & 0.0004776422 & 0.00000032058 \\ 0.00002614566 & 0.00000032058 & 0.00027129261 \end{bmatrix}$
Link $_P i^7$	70.16	$x = -0.15$ $y = -38.15$ $z = -50.1$	$\begin{bmatrix} .00003806738 & -0.00000248054 & 0.00000487133 \\ -0.00000248054 & 0.00003213482 & -0.0000569001 \\ 0.00000487133 & -0.0000569001 & 0.00003249317 \end{bmatrix}$

(continued)

(continued)

Link	Mass (g)	Center of mass	Inertia tensor (kg m^2)
Link $_7 i^8$	68.66	$x = 12.79$ $y = 0.02$ $z = -28.87$	$\begin{bmatrix} 0.0000255643 & 0.00000001364 & -0.00000045146 \\ 0.0000000136 & 0.00003066058 & -0.00000000911 \\ -0.0000004514 & -0.00000000911 & 0.00001745718 \end{bmatrix}$
Link $_8 i^9$	84.92	$x = 23.48$ $y = 0.57$ $z = -0.56$	$\begin{bmatrix} 0.00002279453 & 0.00000234787 & 0.00000070805 \\ 0.00000234787 & 0.00005966783 & 0.00000018155 \\ 0.00000070805 & 0.00000018155 & 0.00004857334 \end{bmatrix}$
Link $_9 i^{10}$	80.39	$x = 12.03$ $y = -13.46$ $z = -0.55$	$\begin{bmatrix} 0.00002598519 & -0.0000014009 & 0.00000131081 \\ -0.0000014009 & 0.00005136858 & -0.0000001971 \\ 0.00000131081 & -0.0000001971 & 0.00004958693 \end{bmatrix}$
Link $_{10} i^{11}$	68.66	$x = -12.79$ $y = 2.61$ $z = -0.79$	$\begin{bmatrix} 0.0000229102 & -0.00000065763 & -0.00000014994 \\ -0.0000006576 & 0.00001999689 & 0.00000014579 \\ -0.0000001499 & 0.00000014579 & 0.00002546677 \end{bmatrix}$
Link $_{11} i^{12}$	91.16	$x = -0.02$ $y = -2.33$ $z = -21.99$	$\begin{bmatrix} 0.00009945633 & -0.00000003736 & 0.00000001971 \\ -0.00000003736 & 0.00006689404 & -0.00002394052 \\ 0.00000001971 & -0.00002394052 & 0.0000558318 \end{bmatrix}$

References

1. Vukobratović, M., Juricić, D.: Contribution to the synthesis of biped gait. IEEE Trans. Bio-Med. Eng. **16**(1) (1969)
2. Shih, C., Zhu, Y., Gruver, W.: Optimization of the biped robot trajectory. In: Proceedings of the 1991 IEEE International Conference on Robotics and Automation, pp. 899–903 (1991)
3. Huang, Q., Yokoi, K., Kajita, S., Kaneko, K., Arai, H., Koyachi, N., Ta-nie, K.: Planning walking patterns for a biped robot. IEEE Trans. Robot. Autom. **17**(3), 280–289 (2001)
4. Shi, J., Li, Y., Chung, S., Lee, T., Gruver, W.: Trajectory synthesis and physical admissibility for a biped robot during the single support phase. In: Proceedings of the IEEE Inter-national Conference on Robotics and Automation, pp. 1646–1652 (1990)
5. Hirabayashi, T., Ugurlu, B., Kawamura, A., Zhu, C.: Yaw moment compensation of biped fast walking using 3D inverted pendulum. In: Proceedings of the AMC'08. 10th IEEE International Workshop on. Advanced Motion Control, pp. 296–300 (2008)
6. Khadiv, M., Moosavian, S.A.A., Yousefi-Koma, A., Sadedel, M., Mansouri, S.: Optimal gait planning for humanoids with 3D structure walking on slippery surfaces. Robotica, pp. 1–19, Cambridge University Press (2015)
7. Arias, L., Olvera L., Pamanes, J.A., Nunez, J.V.: Patrón de marcha 3D de tipo cicloidal para humanoides y su aplicación al robot Bioloid. Revista Iberoamericana de Ingeniería Mecánica (RIBIM), pp. 03–22 (2014)
8. Reyes, G., Pámanes, J.A., Fierro, J.E., Nunez, V.: Optimum walking of the bioloid humanoid robot on a rectilinear path. In: Zeghloul, S., Laribi, M.A., Gazeau, J.P. (Eds.) Proceedings of the

IFToMM CK 2017, the 7th IFToMM International Workshop on Computational Kinematics, pp. 143–151. © Springer International Publishing Switzerland, May 22–24. Capítulo 17. ISBN 978-3-319-63272 (2017)
9. Khalil, W., Kleinfinger: A new geometric notation for open and closed-loop robots. In: Proceedings of the IEEE Conference on Robotics and Automation (1986)
10. Gorla, B., Renaud, M.: Modèles des Robots Manipulateurs: Application à leur Commande, Cepadues Editions (1984)
11. Fierro, J.E., Pamanes, J.A., Aquino, J.A., Ollervides, E.J.: Analysis of a humanoid robot walking in an arbitrary direction on a sloping surface. In: 2021 XXIII Robotics Mexican Congress (ComRob), pp. 56–62 (2021). https://doi.org/10.1109/ComRob53312.2021.9628548. ISBN-13: 978-1-6654-3825-4
12. Khalil, W., Lemoine, P.: SYMORO+: Symbolic Modeling of Robots. User's Guide. Ed. Laboratoire des Sciences du Numérique de Nantes, France (2003)
13. Salazar-Bravo, B.D., Pamanes, J.A., et al.: SlopingSurface_ParametersCase3, YouTube. https://youtu.be/KuYu3qjE5h4, http://www.itlalaguna.edu.mx/ (2023)

Chapter 15
A Cable-Based Quadrotor Test Bench: Preliminary Results

Eusebio E. Hernández, Irandi Gutiérrez-Carmona, and Alejandro Arreola-Nepomuceno

Abstract This paper focuses on the design and construction of an experimental test bench aimed to verify the 6DOF quadrotor flight controllers. The system is based on a cable-driven parallel mechanism. Usually, to test quadrotor flight controllers require a long time development, including a simulation stage and experimental verification. Last procedure can be dangerous because propellers are moving parts at high speed. The cable based quadrotor test bench presented in the paper aims to fill the gap between controller simulations and real time flight. First, the mathematical model of general quadrotors is revised considering aerodynamic characteristics. Then, paper gives a detailed description of the proposed test bench, the hardware, operation features and its components. Finally, preliminary tests were implemented to indicate the feasibility and value of the test bench.

Keywords Test bed development · Quadrotor · Flight test system · Cable-driven parallel robot · Laboratory environment

15.1 Introduction

Quadrotors, as representative and popular UAVs (Unnamed Aerial Vehicles), still receiving significant attention not only in the academic and scientific field but also in the commercial field, with increasingly improved systems for a wide variety of appli-

E. E. Hernández (✉) · A. Arreola-Nepomuceno
ESIME Ticomán, Instituto Politécnico Nacional, Av. Ticoman 600, Del. Gustavo A. Madero, Ciudad de México 07340, Mexico
e-mail: euhernandezm@ipn.mx

A. Arreola-Nepomuceno
e-mail: aarreolan1200@alumno.ipn.mx

I. Gutiérrez-Carmona
Escuela de Ingeniería y Ciencias, Tecnológico de Monterrey, Ave. Eugenio Garza Sada 2501, Monterrey, N.L. 64849, México
e-mail: irandi_gutierrez@tec.mx

M. Ceccarelli and J. C. Jauregui-Correa (eds.), *State-of-the-Art and Innovations in Mechanism and Machine Science*, Mechanisms and Machine Science 150,
https://doi.org/10.1007/978-3-031-47040-0_15

cations in areas such as agriculture, inspection, monitoring, and security, terrestrial mapping, environmental preservation practices, or simply for video and photography [1]. In addition, the development of this type of system is still growing due to its inherent advantages over other aerial vehicles, including simple architecture, functionality, reliability, ease and economy in its construction, performance, agility, and increasing stability [2]. The idea is to take advantage of these characteristics so that the air vehicle goes to risky regions for humans or performs repetitive tasks. For these tasks, commercial versions offer remotely piloted vehicles or essential functions in semi-automatic mode with excellent performance in terms of stability and control. On the other hand, mainly in the research field, using vehicles with open architecture is encouraged to achieve tasks in automatic mode. In both versions, modern UAV systems include an aerial vehicle and complex hardware and software systems that allow them to operate in one of three modes: manual, semi-automatic, or automatic. Recent advances in new materials and more efficient manufacturing processes have led to lighter and more accurate vehicles. Coupled with the development of increasingly smaller electronic components, such as microcontrollers, sensors, drivers, cameras, and electronic components for telemetry, this has helped to obtain increasingly precise navigation and stabilization systems. These vehicles can be considered aerial robots; thus, their development process follows the stages of a typical mechatronic system. It begins with the system's modeling and design, control scheme design, simulation, and flight tests to evaluate the vehicle response [2, 3]. Last stage can be dangerous mainly due to failures in the controllers since the propellers, which are moving parts at high speed, can cause injuries to the operators and personnel involved in the tests. Also, a runaway vehicle can accidentally get lost or crash. Controllers are traditionally tested at the simulation level before being implemented in micro-controllers; however, the correct performance of the control schemes may differ from the real behavior, mainly due to simplifications or numerical errors in the simulation stage. Thus, a more appropriate option to verify that the controller works appropriately is to use test benches, which help test the vehicle's performance more realistically and safely.

A reasonable number of published papers have addressed proposals of test benches for multirotor vehicles. In closed environments, traditional evaluations of quadrotor flight have been done with camera systems. The most popular of these is the Vicon motion capture system [4, 5], but a Realsense trailing camera and a Hall sensor gimbal joystick can also be used [2]. In the literature, one can also find proposals based on mechanical arrangements that obtain the vehicle's position, combined with various sensors. These include IMU (Inertial Measurement Unit), accelerometers, and gyro arrays to simulate 6 DOF quadrotor flight [6], the combination of two IMUs and a gyro structure [7], or a 3-axis force sensor [8]. In fact, in the last reference, a hybrid test bench-simulation software strategy is used to test small UAVs. The test bench uses an arrangement of motors, a counterweight, ball joints, and a force sensor.

Recently, the use of cables to restrain the flight of quadrotors on test benches has begun to be proposed. In [1], a 3DOF test bench is proposed for a simple vertical take-off and landing UAV flight evaluation. They used a counterweight connected by a cable and a pulley that is installed to offset the UAV's gravitational acceleration error.

Wire-based systems can combine the free-flight advantages of vision camera-based systems with the ability to estimate vehicle thrust. In the area of robotics, cable-based systems have been extensively studied. There is a wide variety of systems for various applications, including astronomical observation systems [9], rehabilitation prototypes [10–14], aerial robots [15, 16], cable-suspended camera robots [17], or cable-based measurement systems [18, 19].

In this contribution to the book, considered a tribute to Dr. Carlos S. Lopez Cajun, a test bench is proposed to estimate the flight of small quadrotor UAVs. The system provides six degrees of freedom and is designed to estimate the necessary thrust and lift for quadrotor vehicles. In the last years before his retirement, Dr. Lopez Cajun was an enthusiast for developing this type of cable-based system. He introduced the first author on this interesting topic, explaining the inherent advantages of this type of parallel structure, among which are its large potential workspace, high operating speed, and good load-to-weight ratio. Some examples of collaborations during this stage in this area were [20–22]. The development of cable-based systems has been increasing in the last two decades; even today, for the so-called cable-driven parallel robots, one can frequently find calls for special issues of journals or conferences specialized on this topic (for instance, the biannual CABLECON Conference).

The rest of this contribution is organized as follows: Sect. 15.2 discusses the quadrotor vehicle dynamics principles that should accomplish the 6DOF flight prototype; Sect. 15.3 presents an overview of the main subject, describing the proposed test bench and its components; in Sect. 15.4, we present experiments and preliminary results of the cable-based test bench; and finally, the conclusion and future work are outlined.

15.2 Physical Foundations for Test Bench Design

Quadrotors arose from the idea of the helicopter, however, they are rotary-wing aircraft with a variant that uses two pairs of rotors to generate the vehicle's propulsion. These pairs of rotors have the characteristics of being counter-rotating and having a fixed pitch angle. Using rotors with fixed pitch angles allow these to be directly connected to four individual motors connected and placed in an "X" configuration, thus eliminating the need to use complex mechanical linkages to control the pitch angle, see Fig. 15.1. A battery and a micro-controller are used to propel and control the rotors, which are placed near the center of the aircraft. To achieve changes in the altitude and attitude of the vehicle, one must vary the rotors' rotation speed individually.

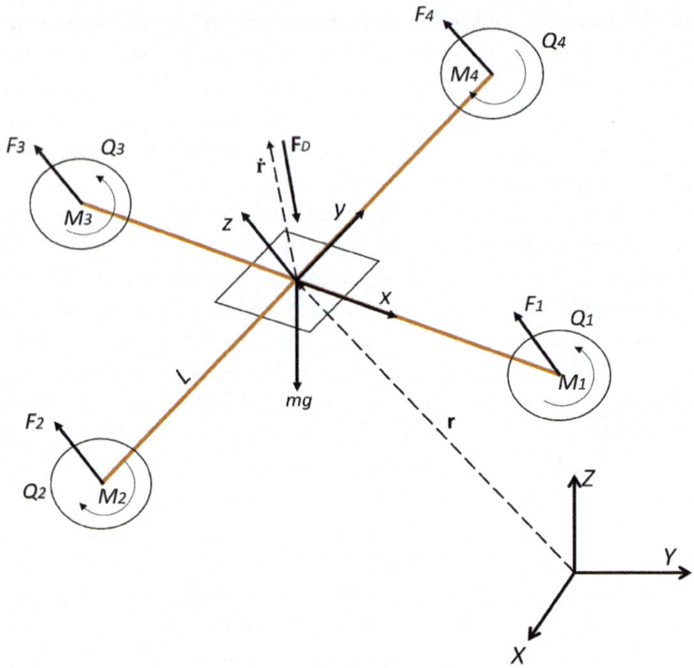

Fig. 15.1 Approximation of the desired trajectory (red) through the path optimization (precision points)

15.2.1 Quadrotor Dynamics

For the present work, it is proposed to monitor the behavior of a quadrotor, which is why it is necessary to understand some aspects of the flight dynamics of this type of vehicle, such as the matrix of rotation and angular velocities that will serve as a starting point for the development of theoretical models later. To understand the dynamics, one must derive the operation of the quadrotor in two operating frames. The frame inertial, which is defined by the ground, with gravity pointing in the negative Z direction and the frame of the body, which is defined by the orientation of the quadrotor, with the rotor axes pointing in the positive Z direction and the arms pointing in the X and Y directions. Position of the quadrotor in the inertial frame can be defined as $(x, y, z)^T$ and velocity in the same inertial frame as $(\dot{x}, \dot{y}, \dot{z})^T$. Similarly, roll, pitch and yaw are defined in the body frame as $Q = (\phi, \theta, \psi)^T$, with their corresponding angular velocities equivalent to $\dot{Q} = (\dot{\phi}, \dot{\theta}, \dot{\psi})^T$. The rotation matrix relates the rotations in the frame inertia and the body frame. It can be obtained by referencing the Euler angles in the body frame towards the inertial frame, which is conventionally applied to the study of the flight dynamics of quadrotors taking as sign convention the axes notation that can be seen in Fig. 15.1.

$$R(\phi,\theta,\psi)=\begin{bmatrix} c\psi c\theta & c\psi s\theta s\phi - s\psi c\phi & c\psi s\theta c\phi + s\psi s\phi \\ s\psi c\theta & s\psi s\theta s\phi + c\psi c\phi & s\psi s\theta c\phi - s\phi c\psi \\ -s\theta & c\theta s\phi & c\theta c\phi \end{bmatrix} \tag{15.1}$$

Now, considering $\omega = (p, q, r)^T$ as the absolute angular velocity vector of the quadrotor expressed in the rotation body-fixed frame, it is worth noting that the angular velocity vector $\omega \neq \dot{Q}$. Angular velocity is a vector pointing along the axis of rotation, while $\dot{Q}$ is just the time derivative of yaw, pitch, and roll angles. By using the rotation matrix, the relationship between the angular rates and the time derivative of the Euler angles can be written as:

$$\begin{bmatrix} p \\ q \\ r \end{bmatrix} = \begin{bmatrix} 1 & 0 & -S_\phi \\ 0 & C_\phi & C_\theta S_\phi \\ 0 & -S_\phi & C_\theta C_\phi \end{bmatrix} \begin{bmatrix} \dot{\phi} \\ \dot{\theta} \\ \dot{\psi} \end{bmatrix} \tag{15.2}$$

Then, the quadrotor rotational motion is given by the angular rate components in the three axis. These rotational velocities occur due to the moments applied on the system by four electrical motors. Thus, with reference Fig. 15.1, U_i (i = 1, 2, 3, 4) represents the input for each motor of the system, which can be written in the form:

$$U_1 = \sum_{i=1}^{4} M_i = b(Q_1^2 + Q_2^2 + Q_3^2 + Q_4^2) \tag{15.3}$$

$$U_2 = (-M_2 + M_4) = b * \iota(-Q_2^2 + Q_4^2) \tag{15.4}$$

$$U_3 = (M_1 + M_3) = b * \iota(Q_1^2 - Q_3^2) \tag{15.5}$$

$$U_4 = (-1)^i \sum_{i=1}^{4} M_{D_i} = d(-Q_1^2 + Q_2^2 - Q_3^2 + Q_4^2) \tag{15.6}$$

where
U_1 is the total thrust
U_2 is the pitch moment
U_3 is the roll moment
U_4 is the yaw moment
b is the hover thrust factor
d is the hover drag factor
l is the length of the arm that supports the propeller
Q_i is the rotational velocity of each propeller.

The first three equations describe the linear acceleration of the vehicle in each axis, while the rest are the equations of the angular velocities in each axis.

According to [4], and following the Newton-Euler formalism, the following equations can be considered to complete the quadrotor dynamics:

$$m\ddot{r} = R\begin{bmatrix}0\\0\\U_1\end{bmatrix} - \begin{bmatrix}0\\0\\mg\end{bmatrix} + F_D \tag{15.7}$$

$$I\ddot{q} = R\begin{bmatrix}U_2\\U_3\\U_4\end{bmatrix} - \dot{q} \times I\dot{q} \tag{15.8}$$

where m is the mass of the quadrotor, r denotes the position vector of the center of mass in the world frame, $\mathbf{q}$ is attitude in body frame, and $\mathbf{I}$ is the rotational inertia.

15.2.2 Rotor and Aerodynamic Characteristics

Here, the main rotor and aerodynamic properties are reviewed. Each rotor has an angular speed Q_i and produces a vertical force F_i according to

$$F_i = k_T Q_i^2 \tag{15.9}$$

The parameter k_T is frequently estimated by experimentation [4]. The rotors also produce a moment that can be calculated as

$$M_i = k_M Q_i^2 \tag{15.10}$$

The constant, k_M, can be estimated by matching the angular acceleration and the rotor speed from a step response of the yaw angle. Then, when quadrotor flights, each rotor contributes some torques about the body z axis. This torque is required to provide thrust; it creates the instantaneous angular acceleration and overcomes the frictional drag forces. The drag equation from fluid dynamics gives us this quantity

$$F_D = \frac{1}{2} C_D \rho A v^2 \tag{15.11}$$

where ρ is the air density, A is the propeller cross-section, and C_D is the drag force coefficient.

15.3 System Overview

Due to the quadrotors' simple design, building lighter vehicles than rotary-wing vehicles with traditional configurations is more convenient. However, despite having such a simple design, its construction also presents some issues that could be improved. Control in these vehicles is difficult because their mass is concentrated in a small area, so the vehicle is prone to tipping over. This condition forces the rotors to react quickly to counteract the overturning tendency. For this reason, they use an IMU composed of accelerometers and gyroscopes that continuously measure the vehicle's behavior. When the quadrotor begins to lean to one side, the micro-controller receives a signal from the UMI and calculates the necessary signals to counteract the overturning tendency. Subsequently, the signals are sent to the motor controller, which will adjust the speed of the required rotors. For this point, a corrective force returns the vehicle to a level position. This entire process must occur one hundred times per second to keep the vehicle stable. For this reason, there is a necessity to use test bench platforms allowing an easier transition between numerical simulations and a real behaviour analysis. The objective is to evaluate the performance of a controller under a safe environment both for user and aerial vehicle, without any risk of collision or tipping over. In next section, a description of the proposed test bench is depicted.

15.3.1 Structure of the Test Bench

The test bench consists of a quadrotor and a base. The base uses a set of six cables attached to an end effector, the aim is to place over this rigid triangle the quadrotor structure, by measuring the vehicle performance in six DOF, see Fig. 15.2. This cable-driven system has a 3-2-1 configuration that has been applied for a variety of measurements systems in different applications, such as for rehabilitation purposes or 3D pose measurements. The objective is to estimate the position and orientation by measuring the six cable lengths. These cable lengths are the inputs for a kinematic model that belongs the position and orientation estimations. It is well-known that the direct kinematics of the 3-2-1 and 3/2 in cable-based parallel manipulators can be solved by a sequence of trilaterations [23, 24]. On the other hand, the test bench will estimate the necessary quadrotor thrust that be able to hover the vehicle into the test bench workspace and to ensure all the cable are in tension. For this, six load cells have been placed under the test bench and they will belong a measure of the cable's tension for a given aerial vehicle lifting trajectory.

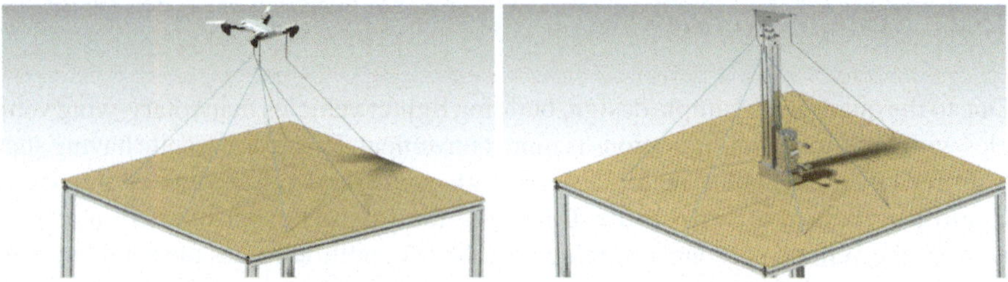

Fig. 15.2 Left image: a 3D CAD output for conceptual modelling of the proposed test bench. Right image: placing an electrical cylinder on the test bench to emulate quadrotor lifting

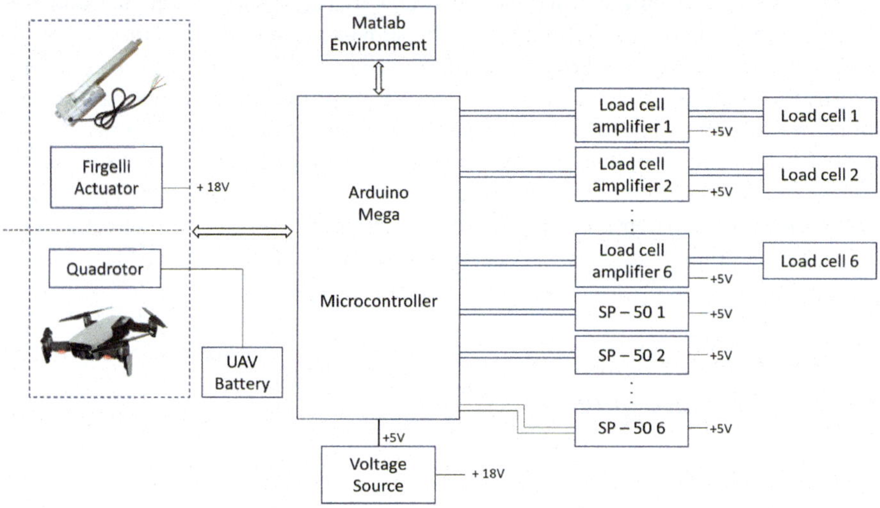

Fig. 15.3 A block scheme of the implemented hardware for the quadrotor test bench

15.3.2 Hardware and Software Implementation

The hardware system of the test bench is depicted in Fig. 15.3. The hardware implementation consists of a set of six string pot sensors that deliver the six cable's displacements, and six load cells to measure the cable tensions. The SP2 String Pot from Celesco is a compact, economical and water resistant device that utilizes a flexible cable, a spring-loaded spool and a potentiometer to detect and measure linear position. It has 50 in of full stroke range and accuracy of 0.25% of full stroke. A set of load cell amplifiers have been used to signal conditioning for the micro-controller. The power for the test bench is supplied by an external voltage source. Cable lengths and tension are measuring by an implemented code in Arduino IDE program. For obtaining the pose and the angular velocity, a processing code is developed using Matlab on a personal computer, see Fig. 15.4. The data obtained from sensors and load cells are transmitted to the computer through USB ports.

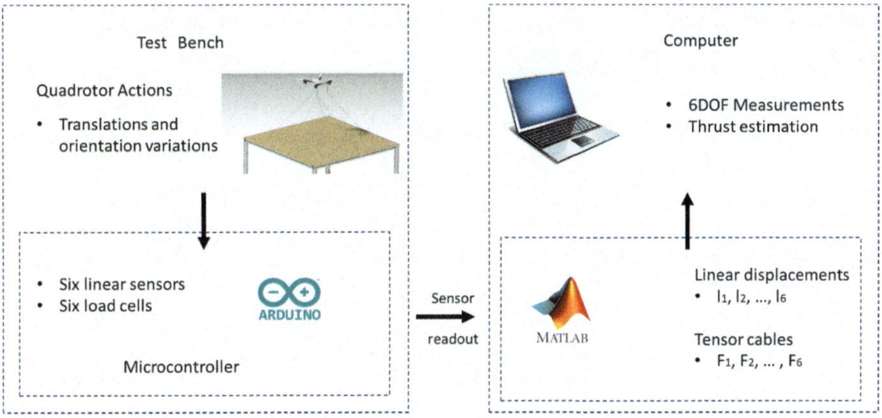

Fig. 15.4 A block diagram to describe the processing code implemented

15.4 Preliminary Experimental Results

In this section, the merits of the proposed test bench are analysed through a preliminary stage experiment. The first objective was to evaluate the kinematic model, in this experimental approach just translation coordinates and the roll angle were estimated. Even though measuring tension cables, using load cells and pulleys has been implemented, these measures have not been included to estimate the quadrotor thrust at this stage. As mentioned above, the processing code is developed using Matlab on a personal computer, this code includes an implementation of the direct kinematic formulation based on the Cayley-Menger determinants [24]. A set of repetitive movements was considered by placing an electrical cylinder on the test bench base, as is shown in Fig. 15.2. Two sequences were commanded to the linear actuator in order to analyse the kinematic model and test bench performances. The first one sequence deploys the actuator for a known distance and back to the original position (Seq 1). The second sequence just considers the actuator movement in one direction (Seq 2). Table 15.1 shows voltage and length values measured as inputs to mapfun function for the string pot sensors, which is a Matlab function that takes an input value mapped from a possible value and scales it to the desired output.

Some experiments were conducted to analyse kinematic characteristics from the implemented model, by executing two simple sequences, labeled as Seq 1 and Seq 2 in the Figs. 15.5, and 15.6, respectively. Figure 15.5 shows the resulting estimated coordinates and roll angle for Seq 1 as a repetitive movement, which was commanded when the rod actuator slides out a known distance and back to original position. From these graphs, one can note the system is more sensible and there are major changes when the rod actuator goes to original position. In the same way, Fig. 15.6 shows the resulting estimated coordinates and roll angle for Seq 2 as a repetitive movement, which was commanded when the rod actuator just slides out a known distance, i.e. emulating just a quadrotor lifting trajectory. From the resulting graphs, the system

Table 15.1 SP2 sensor's measured values to implement Mapfun function in matlab environment

Sensor	l_{max}	V_{max}	V_{min}
1	79.1	29.814	0.0391
2	79.16	29.521	0.0635
3	78.1	29.423	0.0684
4	78.6	29.179	0.044
5	79.5	30.010	0.0684
6	79.95	30.108	0.0635

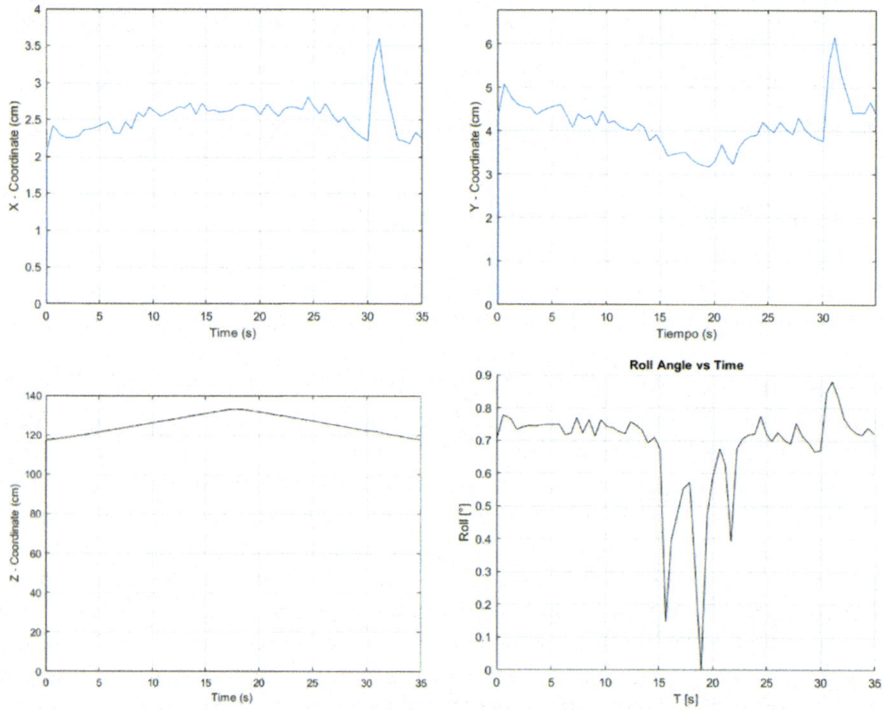

Fig. 15.5 Estimated x,y, and z coordinates and roll angle for Seq 1 movement

is more stable when the rod actuator goes up, major changes occur at the end of the commanded movement. According to [25], cable-based tracking systems are affected by dimension deviation due to manufacturing tolerances, cable length uncertainties and error measures by the abrasion in the holes due to cable friction. Then, these issues could be addressed for an improved system performance.

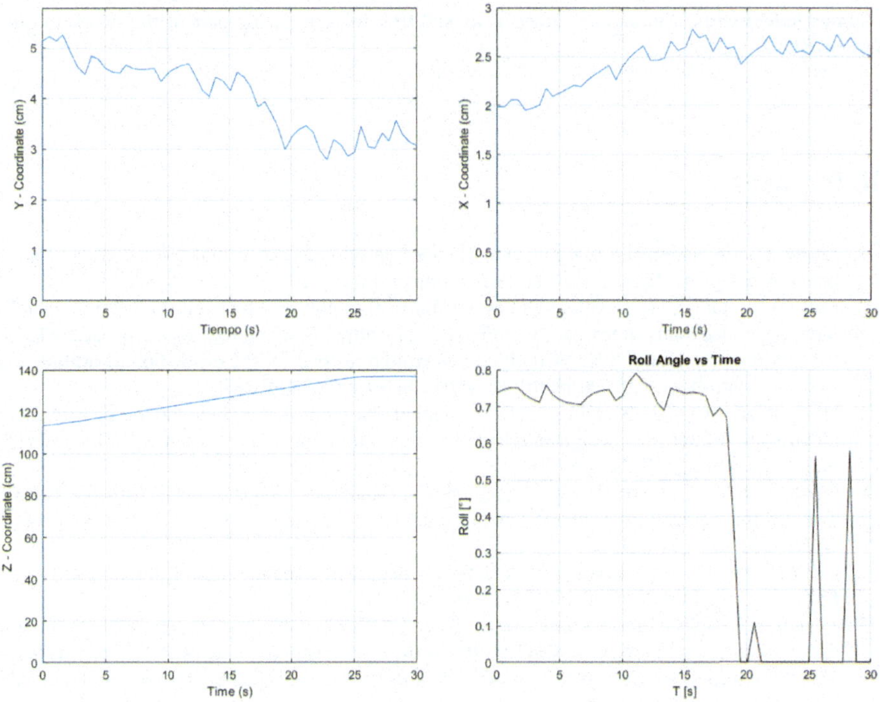

Fig. 15.6 Estimated x, y, and z coordinates and roll angle for Seq 2 movement

15.5 Conclusions and Future Work

In this paper, we presented the design of a quadrotor test bench with a cable-based structure. The aim is to validate controllers with minor risk and development time possible. In addition, it can be considered a valuable platform for evaluating the performance and operation verification of control systems without the necessity of expert quadrotor operators. Due to the hardware and software implemented in its construction, the prototype can be easily replicable for laboratory equipment and teaching. First, the mathematical model of general quadrotors has been revised considering the aerodynamic characteristics, as the physical foundations for the flight testing. The proposal includes the possibility of 6 DOF flight and thrust estimations, even though, as preliminary results, we estimated just the translation coordinates and roll angle for two repetitive movements depicting quadrotor lifting. These preliminary tests show the merits of this development, however some issues have been identified in order to improve performance and solve the current drawbacks. The direct future work is to finish the thrust estimation using the load cells implementation and include a trust model. In addition, we will conduct more research to reduce sources of uncertainty since cable abrasion, sensor noise, and structure vibration affect the system's performance.

Acknowledgements Authors are grateful to SIP-IPN for supporting part of this work through project SIP-20232083.

References

1. Jatsun, S., Emelyanova, O., Leon, A.S.M.: Design of an experimental test bench for a UAV type convertiplane **714**(1), 012009 (2020). https://doi.org/10.1088/1757-899X/714/1/012009
2. Mukras, S., Omar, H.: Development of a 6-dof testing platform for multirotor flying vehicles with suspended loads. Aerospace **8**, 355 (11 2021). https://doi.org/10.3390/aerospace8110355
3. Xuan Mung, N., Hong, S.K.: A multicopter ground testbed for the evaluation of attitude and position controller (2018). https://doi.org/10.14419/ijet.v7i4.39.23708
4. Dong, W., Gu, G., Zhu, X., Ding, H.: Development of a quadrotor test bed–modelling, parameter identification, controller design and trajectory generation. Int. J. Adv. Robot. Syst. **12**, 1 (2015). https://doi.org/10.5772/59618
5. Palacios, F.M., Quesada, E.S.E., Sanahuja, G., Salazar, S., Salazar, O.G., Carrillo, L.R.G.: Test bed for applications of heterogeneous unmanned vehicles. Int. J. Adv. Robot. Syst. **14**(1), 1729881416687111 (2017)
6. Yu, Y., Ding, X.: A quadrotor test bench for six degree of freedom flight. J. Intell. Robot. Syst. **68** (2012). https://doi.org/10.1007/s10846-012-9680-y
7. Veyna, U., Garcia-Nieto, S., Simarro, R., Salcedo, J.V.: Quadcopters testing platform for educational environments. Sensors **21**(12) (2021), https://www.mdpi.com/1424-8220/21/12/4134
8. Kim, S., Philip, O.A., Tullu, A., Jung, S.: Development and verification of a ros-based multi-dof flight test system for unmanned aerial vehicles. IEEE Access **11**, 37068–37081 (2023)
9. Nan, R.: Five hundred meter aperture spherical radio telescope (fast). Sci. China Ser. G: Phys. Astron. **49**, 129–148 (2006). https://doi.org/10.1007/s11433-006-0129-9
10. Ben Hamida, I., Laribi, M.A., Mlika, A., Romdhane, L., Zeghloul, S., Carbone, G.: Multi-objective optimal design of a cable-driven parallel robot for rehabilitation tasks. Mech. Mach. Theory 104141. https://doi.org/10.1016/j.mechmachtheory.2020.104141
11. Gallina, P., Rosati, G., Rossi, A.: 3-d.o.f. wire driven planar haptic interface. J. Intell. Robot. Syst. **32**, 23–36 (2001). https://doi.org/10.1023/A:1012095609866
12. Hernandez, E., Valdez Peña, S., Carbone, G., Ceccarelli, M.: Design optimization of a cable-driven parallel robot in upper arm training-rehabilitation processes 413–423 (2018). https://doi.org/10.1007/978-3-319-67567-1_39
13. Rosati, G., Gallina, P., Masiero, S.: Design, implementation and clinical tests of a wire-based robot for neurorehabilitation. IEEE Trans. Neural Syst. Rehabil. Eng.: Publ. IEEE Eng. Med. Biol. Soc. **15**, 560–569 (2008). https://doi.org/10.1109/TNSRE.2007.908560
14. Venkata Sai Prathyush, I., Ceccarelli, M., Russo, M.: Control design for cableankle, a cable driven manipulator for ankle motion assistance. Actuators **11**(2) (2022). https://doi.org/10.3390/act11020063, https://www.mdpi.com/2076-0825/11/2/63
15. Jiang, Q.: Determination and stability analysis of equilibrium configurations of objects suspended from multiple aerial robots. J. Mech. Robot. **4**, 021005 (2012). https://doi.org/10.1115/1.4005588
16. Ding L, Yao, Y., Ma, R.: Observer-based control for a cable-driven aerial manipulator under lumped disturbances. Comput. Model. Eng. Sci. **135**(2), 1539–1558 (2023). https://doi.org/10.32604/cmes.2022.023003, http://www.techscience.com/CMES/v135n2/50182
17. Spidercam. https://www.spidercam.tv/. Accessed 09 July 2023
18. Hernandez, E., Valdez Peña, S., Ceccarelli, M., Hernandez-Aguirre, A., Botello, S.: Design optimization of a cable-based parallel tracking system by using evolutionary algorithms. Robotica **33**, 1–12 (2014). https://doi.org/10.1017/S0263574714000484

19. Varela, M., Ceccarelli, M., Flores, P.: A kinematic characterization of human walking by using catrasys. Mech. Mach. Theory **86** (2015). https://doi.org/10.1016/j.mechmachtheory.2014.12.006
20. Hernandez, E., Carbone, G., López-Cajún, C.: Operation Features of Milli-CaTraSys, pp. 191–199 (2008). https://doi.org/10.1007/978-1-4020-8915-2_24
21. Hernandez, E., Ceccarelli, M., Carbone, G., López-Cajún, C., Jauregui, J.: Characterization of a cable-based parallel mechanism for measurement purposes. Mech. Based Des. Struct. Mach. **38**, 25–49 (2010). https://doi.org/10.1080/15397730903386101
22. Hernandez, E., López-Cajún, C., Jauregui, J.: Calibración de manipuladores paralelos y su aplicación a las máquinas herramienta. un estudio del estado del arte. Ingeniería, investigación y tecnología **11**, 141–154 (06 2010)
23. Ottaviano, E., Ceccarelli, M., Toti, M., Avila Carrasco, C.: Catrasys (cassino tracking system): a wire system for experimental evaluation of robot workspace. J. Robot. Mechatron. **14**, 78–87 (2002)
24. Thomas, F., Ros, L.: Revisiting trilateration for robot localization. IEEE Trans. Robot. **21**(1), 93–101 (2005). https://doi.org/10.1109/TRO.2004.833793
25. Thomas, F., Ottaviano, E., Ros, L., Ceccarelli, M.: Uncertainty Model and Singularities of 3-2-1 Wire-Based Tracking Systems, pp. 107–116. Springer Netherlands, Dordrecht (2002). https://doi.org/10.1007/978-94-017-0657-5_12

Chapter 16
Synchronization in Mechanical Systems

Mihir Sen and Juan Carlos Jauregui-Correa

Abstract This chapter compiles Carlos López-Cajún and the authors' work on synchronization. Synchronization is present in many mechanical systems, and its identification and analysis require an understanding of the complexity of the individual elements' interrelations. This chapter starts with a reference to previous works, then describes the general concepts of synchronization, presents some of the most commonly used methods for studying synchronization, and ends by summarizing the limitations and future work. The methods analyzed are frequency spectrum analysis, signal correlation, spectrograms, and Kuramoto's order parameter. The described methods are exemplified with experimental vibration data recorded from twenty-two rotor blades.

Keywords Vibration analysis · Synchronization · Frequency spectrum analysis · Signal correlation · Spectrograms · Experimental analysis

16.1 Introduction

Synchronization is a phenomenon that Carlos López-Cajún and the authors have studied for several years. Their interest began when one of them observed the behavior of thermal systems, originally the air conditioner equipment in a hotel, where they noticed that after a time, the control units started switching on and off at the same time. This opens up many questions, and they started studying practical applications where the phenomenon was present. The following list includes the authors publications concerning synchronization:

M. Sen (✉)
Department of Aerospace and Mechanical Engineering, University of Notre Dame, Notre Dame, IN 46556, USA
e-mail: Mihir.Sen.1@nd.edu

J. C. Jauregui-Correa
Universidad Autónoma de Querétaro, Santiago de Querétaro, México
e-mail: jc.jauregui@uaq.mx

M. Ceccarelli and J. C. Jauregui-Correa (eds.), *State-of-the-Art and Innovations in Mechanism and Machine Science*, Mechanisms and Machine Science 150,
https://doi.org/10.1007/978-3-031-47040-0_16

- Thermal–hydraulic networks [1]
- First-order systems [2]
- Van der Pol oscillators [3]
- Self-excited elastic beams [4]
- Blade vibration [5]
- Thermally-coupled rooms [6]
- Effect of walls [7, 8]
- Limit-cycle oscillators [9]
- Heated plates [10]
- Automobiles as complex systems [11]
- Analysis of experimental data [12–15]
- Non-local interactions [16]
- Thermostatically controlled adjacent cavities [17, 18].

Reviews on mechanical systems have been written [19] "Foreground and background components in separable complex systems" [20], vibrations of rotor blades [13]. Based on this, it is evident that various types of machines can synchronize.

In mechanical engineering, the system can be either thermal [6–8, 10, 17, 18] or mechanical [1, 4, 5, 13]. In addition to experimental studies [12–15], such systems can be mathematically modeled [2, 3, 9, 16]. Synchronization is thus a phenomenon that can occur in complex systems [11].

Practical examples of complex machinery may be washers/dryers, turbines, or automobiles. In each case, the complete system consists of several sub-systems, each composed of fundamental components such as masses, springs, or dampers. Each sub-system is coupled to another or others through mechanical coupling elements. In particular circumstances, it is possible to have isolated sub-systems within a machine that are completely disconnected from the rest and which can be safely excluded from the network, at least from the mechanical response point of view. There are many other characteristics of complex machines that are of interest, such as the interaction between foreground and background components [8, 9], multiple scales [10], and others [11], which will not be explored further here.

16.2 Overview

Synchronization is a phenomenon that is present in nature and can be identified in many biological environments, such as the wing motion of ducks and geese when they all move their wings in phase and at the same rhythm. Similar situations can be found in fish banks and the rhythm of people's steps walking on the street [21]. Mechanical systems also show this phenomenon. Synchronization can be defined as the coordinated motion of multiple components of different machines or systems that harmonize their dynamic responses.

In certain operations, synchronization is part of the system's design, assuring that the interrelationship among components results in a smooth and optimal performance. Historically, music boxes and clocks were designed to ensure that every element worked synchronously to produce harmonic music or display the hour accurately. Modern machinery also operates synchronously, mainly where multiple parts or processes have to interact coordinately. Examples are assembly lines and conveyors, the coordination of the spark in combustion engines, or any positioner that has to actuate in coordination with other elements. In an automobile gearbox, synchronization is crucial to assure the best performance and components' life. It is also essential when operating different robots in manufacturing or assembly lines.

This paper deals with another type of synchronization. It presents examples where two or more components couple their motion and move in harmony due to an external excitation. This phenomenon was first reported by Christiaan Huygens (1629–1695) in a letter sent to the Royal Society in 1664 [22]. He found that placing two clocks in the same table synchronized the pendulums at a certain period. Huygens was involved in determining the distance that sail boats traveled by measuring the time with pendulum clocks at sea. This idea was the first procedure for measuring the earth's longitude. This experiment brought out a novel concept that led to the analysis of complex systems. An individual pendulum will move at its own natural frequency and its phase angle is related only to a reference frame, whereas two or more pendulums behave totally differently. If one pendulum starts moving and the rest are standing still, after a certain time, the stationary pendulums will swing and then all of them will swing in phase. The interaction between the pendulums is uncertain and it might depend on the stiffness of the supporting structure and the interaction through the surrounding air. A similar situation is observed in rooms with multiple clocks, if they have different oscillating periods, with a small perturbation, they will synchronize in a short time. Another example can be observed in wind turbines farms, in general they start operating at different angles of rotation, but the aerodynamic interactions lead to synchronization and every turbine rotates in phase with the rest of them, the angular velocity depends on the wind speed, but all the rotors pass over the tower at the same time. A similar situation occurs in turbomachinery when rotor blades and stator vanes vibrate harmonically due to small external excitations.

Another type of synchronization of mechanical systems occurs in traffic flow, which can be affected by slight variations in the car density, a driver behaving differently from the others, or road conditions. These types of perturbations synchronize the vehicle speed, causing traffic jams.

The opening of the Millennium Bridge in London was worldwide news. The unexpected motion of the bridge is an excellent example of synchronization in mechanical structures. The flutter presented by the bridge, when many people walked harmonically, is an example of self-sustained synchronization. The people's foot impact harmonized due to the soft stiffness of the bridge, which preserved step frequency and synchronized the motion of both people and structure.

One of the aspects that needs to be clarified is the difference between synchronization and resonance. Synchronization, as described before, is the phenomenon that describes how two or more oscillating systems move in phase or harmonize with each other. It is caused by an identifiable or not coupling element that “allows” for transmitting a waveform. It can be a soft connecting structure, the air, or any other surrounding medium that facilitates this transmission. On the other hand, resonance is a phenomenon that occurs when the system is excited by a dynamic periodic force, which is composed of one or more periodic functions, and one of these periodic functions has a frequency that is almost equal to the system’s natural frequency.

In summary, synchronization means a coordinated motion among the system’s elements facilitated by the surroundings. After a certain period, this motion is in phase, and the synchronized elements have almost the same frequency. On the contrary, resonance is a phenomenon that results when the excitation frequency coincides or is close to one of the system’s natural frequencies and amplifies the system’s response.

16.3 Models of Analysis

Although synchronization can be understood with analytical models and simulations, the results are not easily interpreted because they depend on two-time scales: the oscillation period and the time until the elements move in phase. Additionally, synchronization occurs in systems having many components and complex configurations. When identifying synchronization using experimental data, the analysis is more complicated than only analyzing simulation results. Therefore, there are different approaches to synchronization analysis.

16.3.1 Spectral Analysis

Spectral analysis is very useful because it identifies the frequencies of each element. The most commonly used is the Fast Fourier Transform (FFT). Mathematically it is expressed as a convolution function of the form:

$$(f * g)(t) = \int_{-\infty}^{\infty} f(\tau)g(t-\tau)d\tau \tag{16.1}$$

where $f(\tau)$ is the motion data and $g(t-\tau) = e^{-i\omega(t-\tau)}$. Thus, the frequency spectrum is calculated as:

$$X(\omega) = \langle x|e^{-i\omega t}\rangle = \int_{-\infty}^{\infty} x(t)e^{-i\omega t}dt \tag{16.2}$$

This transformation provides the frequency components of the motion data. The Fourier transform provides the motion frequency of every element and it also computes the phase between the signals; thus, if the system’s element frequencies are

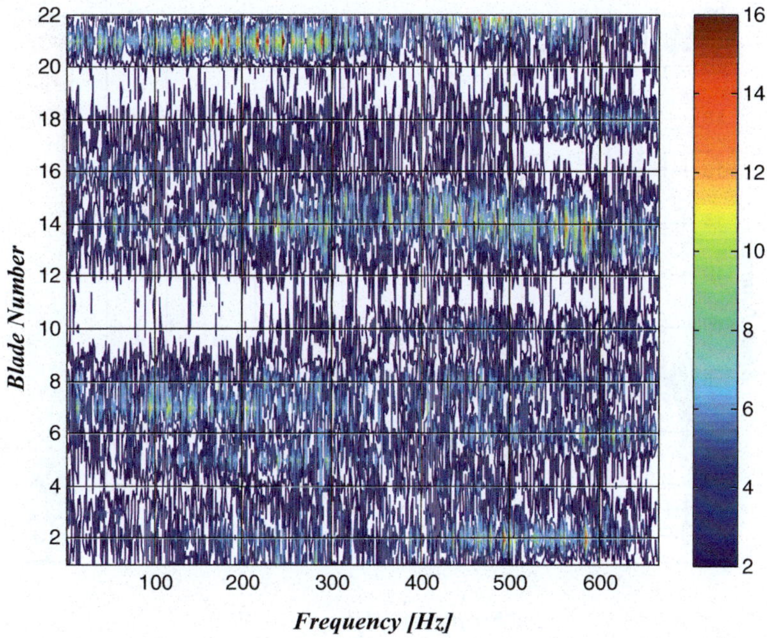

Fig. 16.1 Comparison of the frequency spectra of 22 blades. First measurement at 10:31 am

similar and in phase, it is possible to establish that the elements move synchronously. The limitation of this analysis is that it is applicable only when the signals are harmonic.

The present authors and Carlos López-Cajún analyzed the synchronization phenomenon in 22 static blades under a steady stream of wind [5]. It is possible to see how the blades synchronized by comparing the frequencies of the 22 blades at different measuring times. Figures 16.1, 16.2, and 16.3 compare the blade's frequency spectra measured at three different instants. It can be seen that after a particular time, most of the blades have a dominant frequency at 375 Hz, and only blades 8 and 21 have a broader frequency response. The fact that not all the elements have the same answer indicates that synchronization can have features "outside" the overall behavior.

16.3.2 Correlation

Correlation measures the statistical relationship between two variables. There are different correlation factors; but for analyzing synchronization it is recommended using the Pearson correlation coefficient.

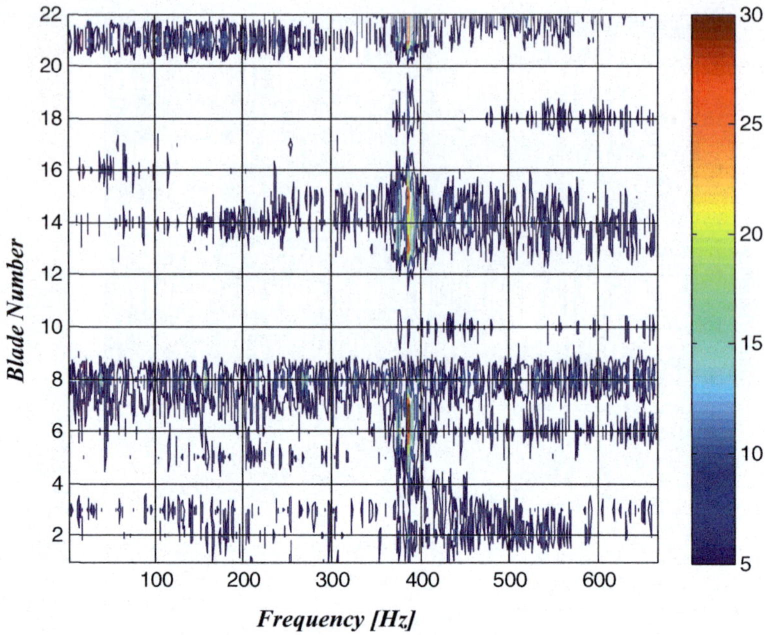

Fig. 16.2 Comparison of the frequency spectra of 22 blades. Intermediate measurement at 11:29 am

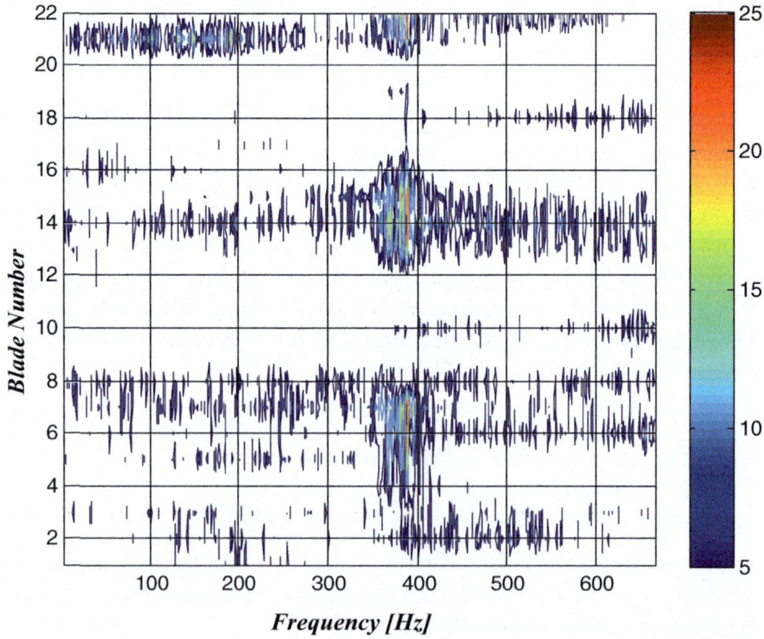

Fig. 16.3 Comparison of the frequency spectra of 22 blades. Last measurements at 12:16 pm

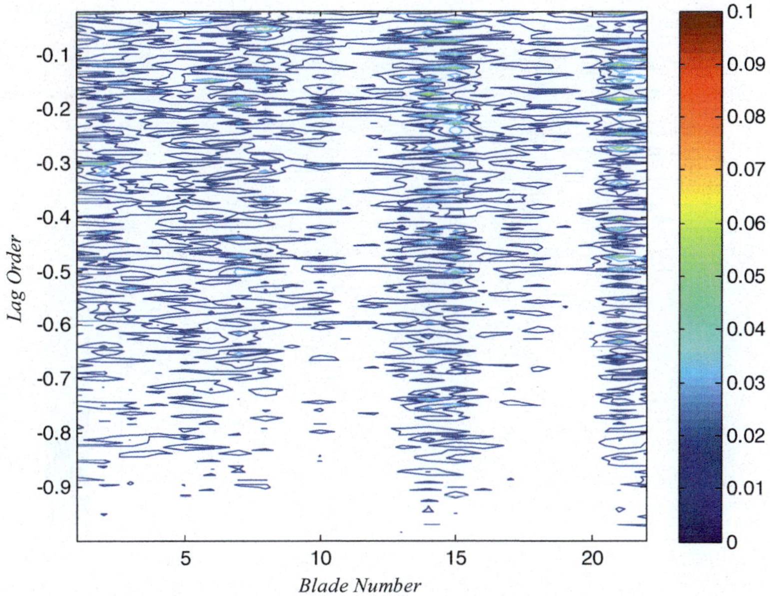

Fig. 16.4 Correlation coefficient of blades 1 and 8. First measurement at 10:31 am

$$c = \frac{P[(x - \overline{x})(y - \overline{y})]}{\sigma_x \sigma_y} \tag{16.3}$$

where c is the correlation coefficient, P is the expected value, $\overline{x}$ is the mean of one signal, and σ_x is the standard deviation. The numerator is the covariance. The correlation coefficient indicates the similarity between the two functions. The correlation coefficient of the 22 blades were calculated, in this chapter only the correlation coefficient between blades 1 and 8 are included. Figures 16.4, 16.5 and 16.6 show the correlation coefficient (lag order) for the measurements taken at 10:31, 11:29 and 12:16. The patterns are similar as those results analyzed with the frequency spectrum.

16.3.3 Spectrograms

Spectrograms, or time–frequency maps, are powerful tools for analyzing vibration signals. It separates the original signal into a set of individual frequencies and preserves the time when each frequency response occurs. Spectrograms can identify the instant when different elements move at the same frequency, showing if they are in phase, meaning that they move synchronously. The Continuous Wavelet Transform is the most appropriate procedure for analyzing vibration signals, and it provides the best results using Morlet's wavelet function.

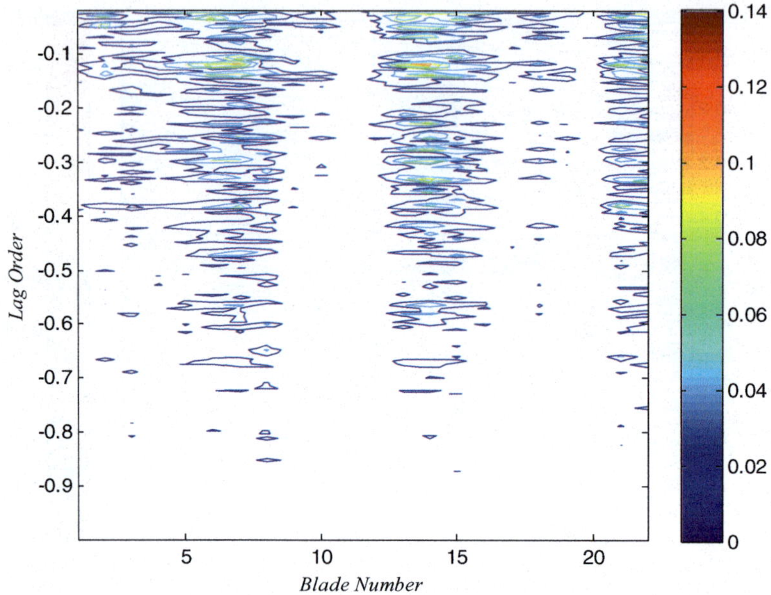

Fig. 16.5 Correlation coefficient of blades 1 and 8. First measurement at 11:29 am

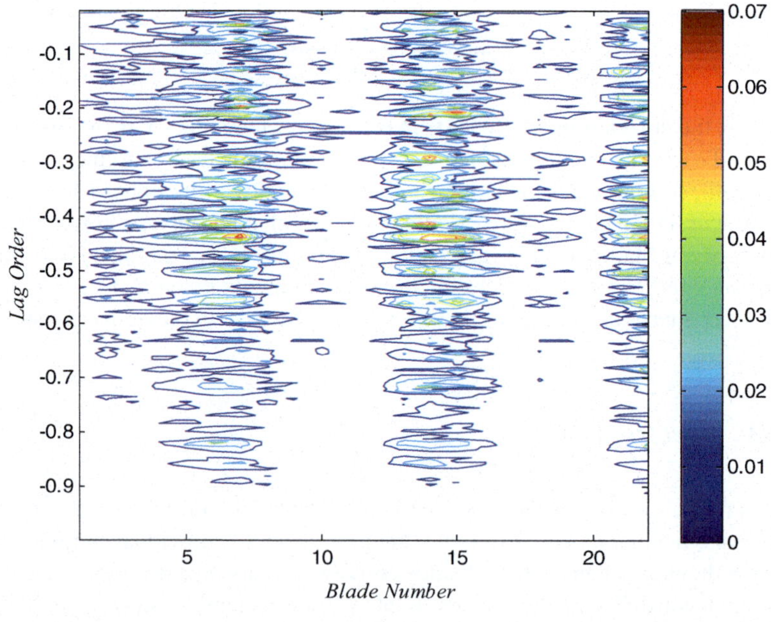

Fig. 16.6 Correlation coefficient of blades 1 and 8. First measurement at 12:16 pm

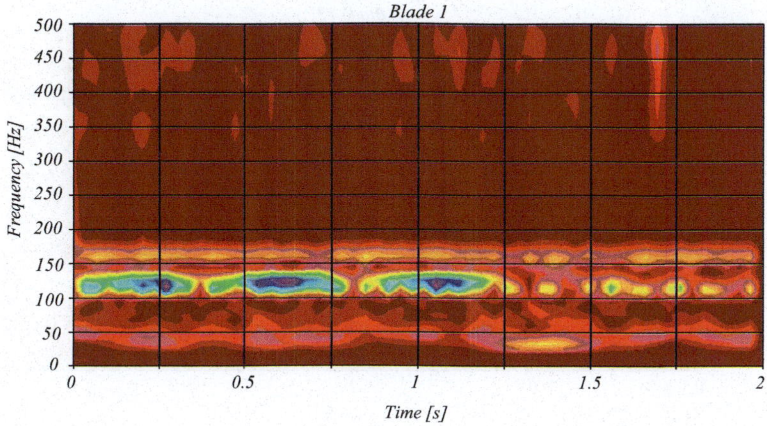

Fig. 16.7 Spectrogram of blade 1 measurement at 12:16 pm

The Continuous Wavelet Transform (CWT) is determined from the convolution of the mother wavelet and the signal data as:

$$X(s,\tau) = \langle x|\psi(s,\tau)\rangle = \frac{1}{\sqrt{s}}\int_{-\infty}^{\infty} x(t)\psi^*\left(\frac{t-\tau}{s}\right)dt \qquad (16.4)$$

where $s > 0$ is the scaling parameter and is inversely proportional to the frequency. ψ^* is the complex conjugated function of the mother wavelet.

The Morlet's mother wavelet is defined as:

$$\psi\left(\frac{t-\tau}{s}\right) = e^{i2\pi f_0\left(\frac{t-\tau}{s}\right)}e^{-\alpha\frac{(t-\tau)^2}{s^2\beta^2}} \qquad (16.5)$$

where f_0, α, β are constant.

The CWT was applied to the 22 blade signals, and the spectrograms show those intervals when the blades vibrate synchronously. Figures 16.7, 16.8 and 16.9 show spectrogram of different blades measured at the same time.

The three blades have the highest peaks at the same intervals, showing the instants when synchronization occurs.

16.3.4 Kuramoto's Parameter

The Kuramoto's parameter determines the phase between several signals. It is based on calculating the phase angle between two oscillators and sets a parameter from 0 to 1. If the parameter is equal to one, it means that the oscillators are in phase, and zero that they are out of phase. The parameter is calculated using the Hilbert's transform

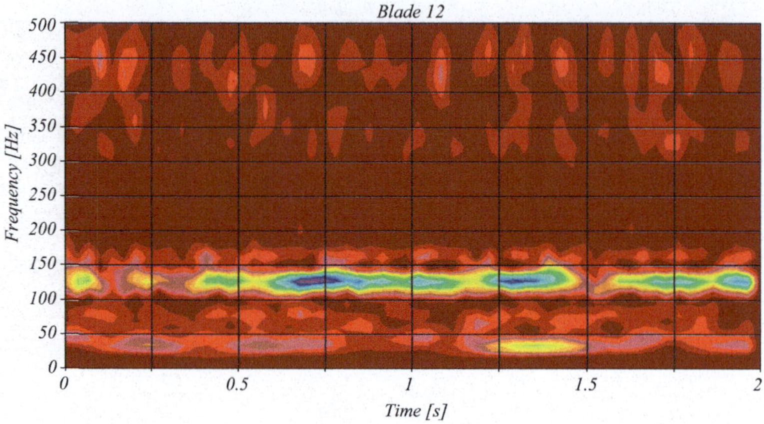

Fig. 16.8 Spectrogram of blade 12 measurement at 12:16 pm

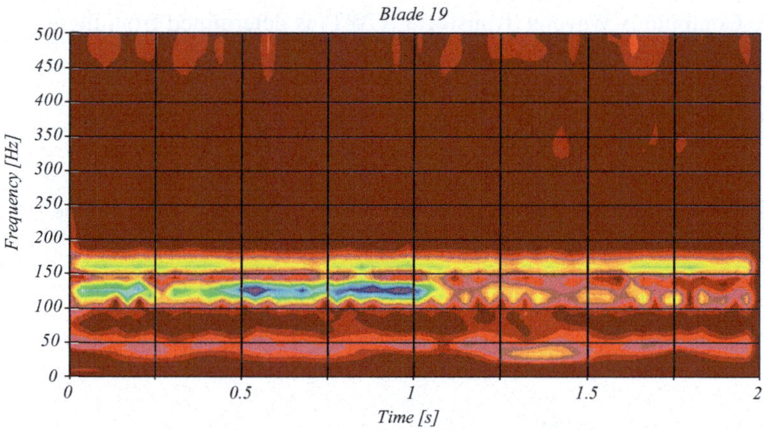

Fig. 16.9 Spectrogram of blade 19 measurement at 12:16 pm

as:

$$\overline{x}(t) = \frac{1}{\pi}\int_{-\infty}^{\infty} \frac{x(\tau)}{t-\tau} d\tau \tag{16.6}$$

The original data is transformed into a complex function

$$y(t) = x(t) + i\overline{x}(t) \tag{16.7}$$

And the phase angle is calculated as:

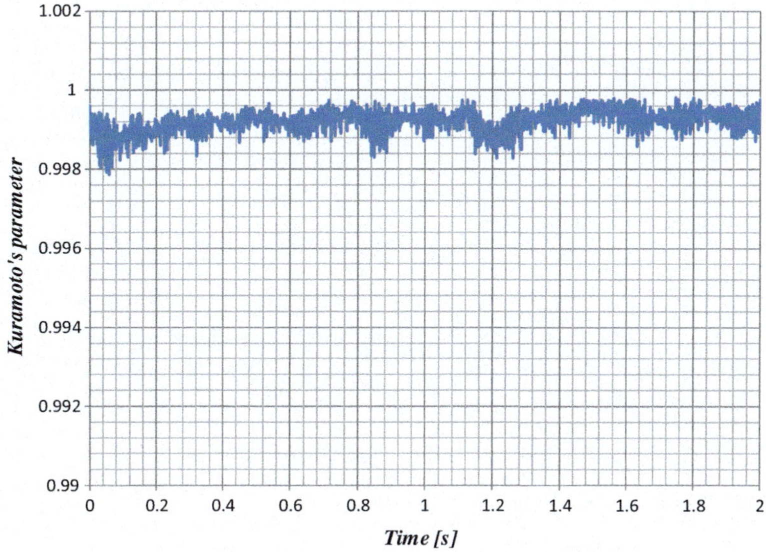

Fig. 16.10 Kuramoto's parameter for the measurements at 12:16 pm

$$\theta(t) = arctan\left(\frac{\overline{x}(t)}{x(t)}\right) \tag{16.8}$$

The Kuramoto's parameter that considers the phase angle of several oscillators is:

$$r(t) = \frac{1}{N}\left|\sum\nolimits_{k}^{N} e^{i\theta_k(t)}\right| \tag{16.9}$$

where $\theta_k(t)$ is the phase angle of each oscillator, N is the number of oscillators and $i = \sqrt{-1}$

The Kuramoto's parameter was applied to the 22 blade measurements. Figure 16.10 shows the parameter for the 22 blades as a function of time. Figure 16.11 shows the histogram of the parameter; this figure shows that most of the values are higher than 0.998, meaning that synchronization occurs almost all the time.

16.4 Recent Publications

After the last López-Cajún publication, several researchers have been working on the analysis and simulation of the synchronization of mechanical systems. Artyunin et al. [23] made an experimental study of the oscillations of a pendulum mounted on a rotating shaft. The rotating shaft was coupled to an electric motor. They showed that

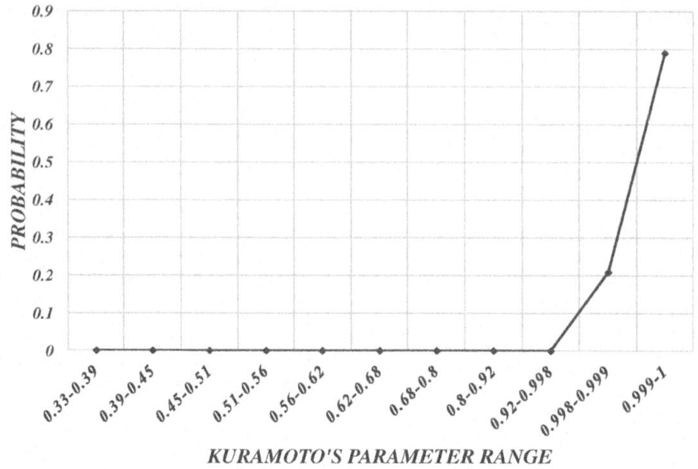

Fig. 16.11 Histogram of the Kuramoto's parameter

the pendulum's motion synchronized with one of the system's natural frequencies. The system's stability depended on each element's mass and moment of inertia.

A similar work was presented by Pena-Ramirez et al. [24]. They found limitations in the master–slave configuration of a control system since it cannot maintain synchronization. To overcome this problem, they proposed a new master–slave control scheme. They applied their method for synchronizing harmonic oscillators to a chaotic system. Once they validated their new scheme, they applied it experimentally to an electronic circuit that emulated a Rössler system. With their new scheme, they could synchronize unidirectional coupled elements.

Herrera et al. [25] proposed a methodology for studying the orbital synchronization of homogeneous mechanical systems with one degree of freedom that shows under actuation. They demonstrated their method with a numerical simulation.

Dudkowski et al. [26] analyzed the possible stability states of two self-exited pendulums suspended from an oscillating beam. They considered three DOFs for each pendulum. They showed the influence of the model parameters and how that synchronization depended on the beam's oscillation angle. They also analyzed the scenarios where bifurcation occurred using phase diagrams. Their results can be extrapolated to other classical models with identical and different pendulums.

Chen et al. [27] analyzed the synchronization of an even number of eccentric rotors evenly distributed and mounted on a rigid mass supported by four springs. Each rotor was moved by an electric motor and had an eccentric mass. They solved the model numerically and experimentally. For the model, they determine the synchronization condition by applying an indirect Lyapunov's method. With their model, they could set the design parameters of similar vibration systems [28, 29].

References

1. Cai, W., Sen, M., Yang, K.T., McClain, R.L.: Synchronization of self-sustained thermostatic oscillations in a thermal-hydraulic network. Int. J. Heat Mass Transf. **49**, 4444–4453 (2006)
2. Cai, W., Sen, M.: Synchronization of thermostatically controlled first-order systems. Int. J. Heat Mass Transf. **51**(11–12), 3032–3043 (2008)
3. Barrón, M.A., Sen, M.: Synchronization of four coupled van der Pol oscillators. Nonlinear Dyn. **56**(4), 357–367 (2009)
4. Barrón, M.A., Sen, M.: Synchronization of coupled self-excited elastic beams. J. Sound Vib. **324**, 209–220 (2009)
5. Jáuregui-Correa, J.C., Sen, M., López-Cajún, C.: Experimental characterization of blade vibration synchronization. ASME Turbo Expo, Paper No. GT2011-46105, Vancouver, BC, June 6–10 (2011)
6. O'Brien, J., Sen, M.: Temperature synchronization, phase dynamics and oscillation death in a ring of thermally-coupled rooms. In: ASME International Mechanical Engineering Congress and Exposition, Paper No. IMECE2011-63153, Denver, CO, 11–17 November (2011)
7. Sen, M.: Effect of walls on synchronization of thermostatic room-temperature oscillations. In: XIX Congreso de la SOMIM, Salamanca, Mexico, 19–21 September (2012)
8. Sen, M.: Effect of walls on synchronization of thermostatic room-temperature oscillations. Ingeniería Mecánica, Tecnología y Desarrollo (Journal of the Mexican Society of Mechanical Engineering) **4**(3), 81–88 (2012)
9. Vinod, V., Balaram, B., Narayanan, M.D., Sen, M.: Mutual synchronization of dissipatively coupled limit cycle oscillators in a ring. In: 8th Conference on Nonlinear Systems and Dynamics, Indian Institute of Technology, Indore, India (2013)
10. Barrón, M.A., Sen, M.: Synchronization of temperature oscillations in heated plates with hysteretic on-off control. Appl. Therm. Eng. **65**(1–2), 337–342 (2014)
11. González, C.A., Jáuregui-Correa, J.C., López-Cajún, C.S., Domínguez-González, A., Sen, M.: Experimental analysis of synchronization and dynamics in an automobile as a complex system. Mech. Syst. Signal Process. **60–61**, 472–484 (2015)
12. Jáuregui-Correa, J.C., López-Cajún, C.S., Sen, M.: Analysis of experimental data from complex multibody system. In: Ceccarelli, M., Hernández, E. (eds.) Multibody Mechatronic Systems, pp. 211–218. Springer (2015)
13. Jáuregui-Correa, J.C., López-Cajún, C.S., Sen, M.: Determining the coupling source on a set of oscillators from experimental data. Complexity 2017, Article 8017138, 10 pages (2017)
14. Jáuregui-Correa, J.C., López-Cajún, C.S., Sen, M.: Analysis of experimental data from complex multibody system. In: Fifth international symposium on multibody systems and mechatronics, Huatulco, Mexico, 21–24 October (2014)
15. Jáuregui-Correa, J.C., López-Cajún, C.S., Sen, M.: Analysis of experimental data from complex multibody system. In: Ceccarelli, M., Hernandez, E. (eds.) Multibody Mechatronic Systems, pp. 211–218. Springer (2015)
16. Vinod, V., Balaram, B., Narayanan, M.D., Sen, M.: Effect of configuration symmetry on synchronization in a Van der Pol ring with nonlocal interactions. Nonlinear Dyn. **89**(3), 2103–2114 (2017)
17. Chávez-Martínez, R., Sánchez-López, M., Solorio-Ordaz, F., Sen, M.: Synchronization of natural convection in thermostatically-controlled adjacent cavities. In: 70th Annual Meeting of the APS, 19–21 November, Denver, Colorado (2017)
18. Chávez-Martínez, R., Sánchez-López, M., Solorio, F.J., Sen, M.: Synchronization of natural convection in thermostatically controlled adjacent cavities: effect of common wall properties. In: 71st Annual Meeting of the APS Division of Fluid Dynamics, Atlanta, GA, 18–20 November (2018)
19. Sen, M., López-Cajún, C.S.: Review of synchronization in mechanical systems. In: Jáuregui-Correa, J.C. (ed.) Nonlinear Structural Dynamics and Damping, Vol. 69 of Mechanisms and Machine Science. Springer (2019)

20. Sen, M., Jáuregui-Correa, J.C., López-Cajún, C.S.: Foreground and background components in separable complex systems. Systems **4**(3), 1–5, Article 27 (2016)
21. Ma, Y., Lee, E.W.M., Shi, M., Yuen, R.K.K., Kit, R.K.: Spontaneous synchronization of motion in pedestrian crowds of different densities. Nat. Hum. Behav. **5**, 447–457 (2021)
22. Huygens, C.: (translated and edited by R.J. Blackwell), The Pendulum Clock or Geometrical Demonstrations Concerning the Motion of Pendula as Applied to Clocks. Edinburgh Books, Edinburgh, U.K. (1986)
23. Artyunin, A.I., Eliseev, S.V., Sumenkov, O.Y.: Experimental studies on influence of natural frequencies of oscillations of mechanical system on angular velocity of pendulum on rotating shaft. In: Proceedings of the 4th International Conference on Industrial Engineering, ICIE 2018, Lecture Notes in Mechanical Engineering, Springer (2019)
24. Pena-Ramirez, J., Garcia, E., Alvarez, J.: Master-slave synchronization via dynamic control. In: Communications in Nonlinear Science and Numerical Simulation, vol. 80 (2020)
25. Herrera, L., Rodríguez-Liñán, M., Meza-Sánchez, M., Clemente, E.: Orbital synchronization of homogeneous mechanical systems with one degree of underactuation. Int. J. Robust Nonlinear Control **32–8**, 1049–8923 (2022)
26. Dudkowski, D., Czołczyński, K., Kapitaniak, T.: Multistability and synchronization: the coexistence of synchronous patterns in coupled pendula. Mech. Syst. Signal Process. **166** (2022)
27. Chen, X., Liu, J., Zhang, J., Li, L.: Synchronization of four axisymmetrically distributed eccentric rotors in a vibration system. Machines **10**, 457 (2022)
28. Jáuregui-Correa, J.C., Sen, M., López-Cajún, C.S.: Simultaneous vibrations of rotor blades (in Spanish). In: Mexican Society of Mechanical Engineering (SOMIM), 22–24 September, Monterrey, Mexico (2010)
29. Jáuregui-Correa, J.C., Sen, M., López-Cajún, C.S.: Experimental characterization of blade vibration synchronization. ASME Turbo Expo, Paper No. GT2011–46105, Vancouver, BC, 6–10 June (2011)

Chapter 17
Designer's Perspective on Applying Mechanisms for Biomechanics Solutions: Unlocking the Future of Healthcare

C. R. Torres-SanMiguel

Abstract In the Biomechanics Applied Lab at the National Polytechnic Institute of Mexico, several multidisciplinary projects in Biomechanics have been developed that combine principles of biology, physics, engineering, and medicine to analyze and understand the mechanics of biological systems. This research aims to show the study of movement, forces, and interactions within the human body to enhance performance, prevent injuries, and develop innovative solutions for various trades through three cases of study in the context of biomechanics, especially in the direction of medical robots, to design and optimize the machines for safe and efficient interaction with patients.

Keywords Biomechanics · Mechanisms · Prosthesis · Soft and hard tissue · Passive safety · Medical robots

17.1 Introduction

In recent years, the field of Biomechanics has witnessed remarkable improvements, revolutionizing healthcare practices around the world. These solutions offer remarkable potential to assist healthcare professionals in numerous tasks, including surgery, diagnostics, rehabilitation, and patient care. While the functionality and capabilities of mechanisms keep improving, biomechanics integration has propelled these machines to the forefront of modern medicine. By applying principles of biomechanics, medical robots can mimic human movement, enhance precision, patient recovery, and ultimately revolutionize the healthcare sector. Also, the knowledge of Professor Carlos S. López Cajún transmitted in his book "Mecanismos: Fundamentos cinematicos para el diseño y optimizacion de maquinaria" overview of the fundamentals and application of different mechanisms to solve problems [1]. The following

C. R. Torres-SanMiguel (✉)
Instituto Politécnico Nacional, Escuela Superior de Ingeniería Mecánica y Eléctrica Unidad Zacatenco, Sección de Estudios de Posgrado e Investigación, 07320 Mexico City, México
e-mail: ctorress@ipn.mx

M. Ceccarelli and J. C. Jauregui-Correa (eds.), *State-of-the-Art and Innovations in Mechanism and Machine Science*, Mechanisms and Machine Science 150,
https://doi.org/10.1007/978-3-031-47040-0_17

background summarizes the best articles in the field of the four research lines: prosthesis, soft and hard tissue analysis, designs for passive safety solutions, and medical robots; since 2015, Dr. Christopher René Torres San Miguel has developed and supervised at the National Polytechnic Institute of Mexico (IPN).

A prosthesis is an artificial device that replaces a missing body part. Prostheses are used to restore function and appearance to people who have lost limbs or biological body parts due to injury, illness, or congenital disabilities. The design of a prosthesis is a complex process involving many different factors, including the patient's needs, the type of prosthesis being designed, and the available materials and technologies. The design process for a prosthesis typically begins with an assessment of the patient's needs. This includes establishing the patient's aspirations for the prosthesis, such as walking again, returning to work, or re-joining sporting activities.

Customized prostheses are becoming progressively more available, and technology is constantly advancing. As technology improves, customized prostheses will become more affordable and accessible to a broader range of patients. The customized prosthesis requires a lot of designer experience. One real case resolved was a partial jaw restitution in a Mexican patient who underwent cancer. A multidisciplinary solution was developed by applying engineering mechanical tools, advanced medical surgery techniques, X-Ray tomography data, finite element method (FEM) simulations, and a fused deposition modeling technique to optimize a forging manufacturing process to produce customized lower jaw prosthesis. The titanium mandibular prosthesis (Fig. 17.1a) provides bone implant consolidation, hygiene, and easiness [2]. With all this experience, an Intramedullary Telescopic Nail (ITN) for Osteogenesis Imperfecta (IO) was designed for child patients with minimal external anchorage on the distal section and screw fixation on the proximal section. The experimental test was made with the regulation ASTM F1264-16a four-point bending tests (Fig. 17.1b). The device works exceptionally under flexion loads until one-third of the device's total extension is reached. After that length extension, the nail only supports half of the whole admissible original load [3]. In Addition, a numerical-experimental development of the mechanical behavior of an osteosynthesis implant for rib fractures (Fig. 17.1c) where subjected to loading conditions within the human body during deep breathing. The operating conditions of the thorax and the action produced on the ribs during deep breathing were analyzed. By performing the experimental and numerical analysis, it was possible to determine that the implant fulfilled the function of providing the fractured rib with adequate stiffness, thus preventing loosening at the junctions of the implant with the bone [4]. The design of a prosthesis for a short, long, or flat bone depends on the patient's needs and the severity of the injury. However, all prostheses are designed to provide the patient with the best function and comfort. The following report develops the design of a flexible mechanism as a transmission system for its application in a new human sternum prosthesis (Fig. 17.1d). To carry out the design of this mechanism, it was first necessary to study the functioning, structure, and movements that follow in the human thorax. The designed device can generate movements between the ribs during the respiratory process to preserve the ventilatory dynamics naturally generated in the thorax. A finite element analysis was performed to describe the mechanical behavior of the mechanism [5].

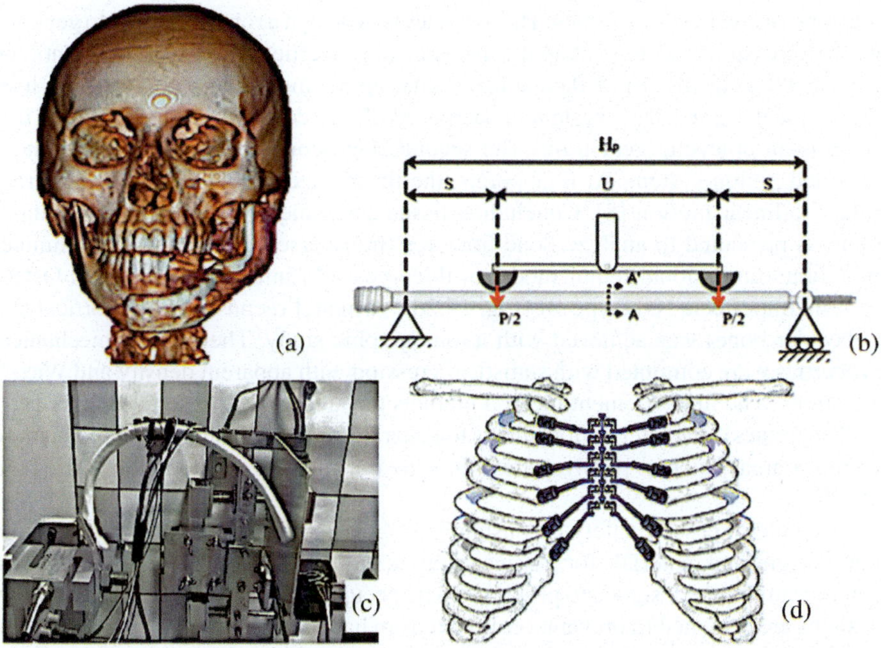

Fig. 17.1 **a** Jaw prosthesis[2], **b** Intramedullary telescopic nails [3], **c** Experimental artificial rib with osteosynthesis prosthesis [4], **d** *S*ternum prosthesis [5]

In previous case studies, tissue characteristics are vital for prosthesis design. The analysis of soft and hard tissue is another line of research Dr. Christopher Torres carries out at IPN.

Biomechanics' analysis of mechanical properties in soft and hard tissue describes their function. It is a complex field. Some of the critical factors considered in the biomechanics of soft and hard tissue include the structure of the tissue, the composition of the tissue, and the properties of the fluid in the tissue. A CAD-CAE design of the scalpel clamp was elaborated to compress soft tissues during surgical procedures (Fig. 17.2a). The normativity of the design of medical instruments and the anthropometric measures of human hands were considered. The device controls a rack and pinion mechanism with a micro servomotor, limit switches, and LabVIEW® software. The results show the experimental evaluation of the viscoelastic properties found in hydrolyzed collagen specimens [6]. Similarly, an aortic blunt trauma analysis during a frontal impact was analyzed using the Finite Element Method to estimate injuries when the three-point seat belt generates a compressive load on the thorax (Fig. 17.2b). Ogden's mathematical model was used to represent hyperelastic and viscoelastic characteristics in the arterial wall. Matlab pseudo-code was used to select specific boundary conditions that ensure the natural movement of the artery within the rib cage [7]. Moreover, some reports aim to get the mechanical properties of soft tissues using methodologies that go from imageology to universal testing machines using in vivo samples. This work aims to design a robotic system to generate the compression load and obtain the mechanical properties to verify a computational liver model (Fig. 17.2c). A device for evaluating the mechanical properties of liver tissue was designed. This system is composed of a three-stage mechanism. The first stage is an epicyclic gear train. The second stage considered a bevel gear and a rack and pinion system. At four points, the third stage generates the change from rotary to linear motion [8]. Concluding tissue assessment, a non-invasive methodology is presented to analyze bone fractures (transverse, oblique, and comminute) in a three-dimensional femur model of three-year-old infants affected with Osteogenesis Imperfecta (OI) type III (Fig. 17.2d). A model reconstruction (cortical and trabecular bone) was achieved with a tomographic study. The tissues' mechanical properties were computed with statistic regression with apparent density and Wirtz's equations. The finite element method predicts the areas susceptible to fracture [9].

The tissues' characterization allows to generate knowledge that is applied in more complex areas, such as passive safety. It is the third line of investigation carried out at IPN.

The field of passive safety is constantly progressing. Researchers are working on new designs and materials for passive safety features to make vehicles safer. Passive safety features are designed to protect vehicle occupants during a crash. Active safety systems are designed to prevent a collision from happening in the first place. Several

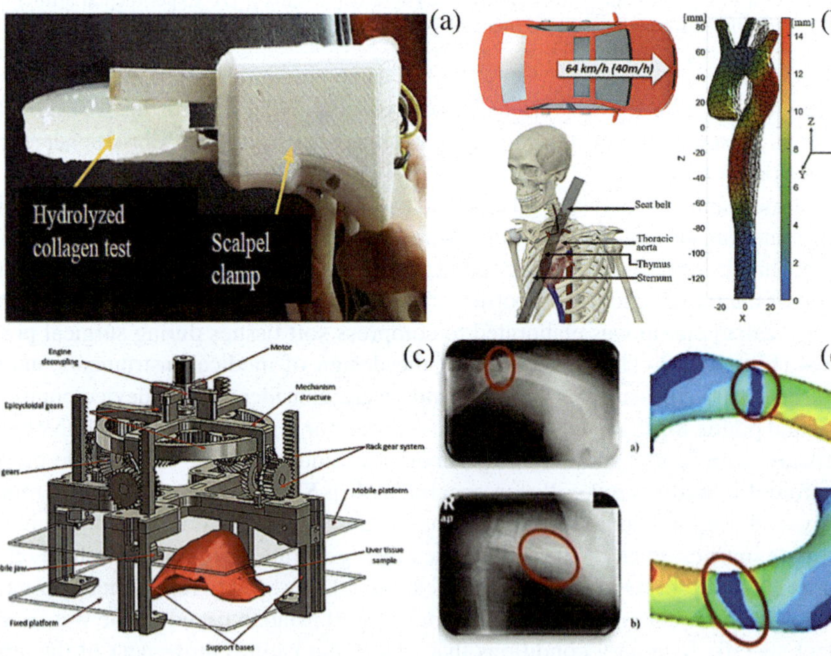

Fig. 17.2 **a** Scalpel clamp for soft tissue [6], **b** Aortic blunt Trauma [7], **c** Testbed for liver tissue compression [8], **d** Transverse and oblique bone fractures [9]

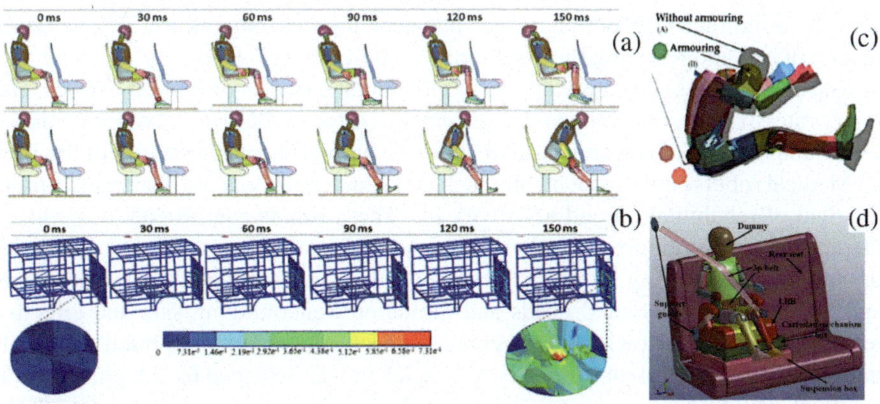

Fig. 17.3 **a** Simulation of the dummy in front and rear collision [10], **b** front impact scenario [11], **c** Dummy kinematics in a frontal collision [12], **d** Mechanical coupling with low back booster [13]

results of numerical analyses, experimental tests, and mathematical models have been developed in IPN. For instance, an Intercity Bus Occupant Safety assessment has been carried out, looking for the possibility of head and chest injury criteria in a 50th percentile dummy during a frontal and rear-end collision at 32 km/h (Fig. 17.3a). Biomechanical analysis using HIC (Head Injury Criterion) and CSI (Chest Severity Index) helps to create effective safety mechanisms to mitigate the level of injury. These simulations were carried out using the finite element method by LS-DYNA software [10].

Furthermore, a bus superstructure was evaluated according to European regulation No. 66, which allows knowing if the superstructure is able to prevent penetrations into the survival compartment (Fig. 17.3b). This includes the analysis of the movement by the occupants, given by the results obtained from the forces and loads experienced during the crash. The data obtained from these tests help to assess the risk of injury and determine the safety devices efficient in buses, such as passive safety systems like seat belts or airbags [11]. Likewise, an assessment of retention system effects of level III armored vehicles evaluated a frontal impact of an SUV with level III vehicle armor (Fig. 17.3c). The work was carried out under the National Highway Traffic Safety Administration (NHTSA) Federal Motor Vehicle Safety Standard (FMVSS) 208 regulation. The injuries comparison shows that armored increases significantly impact energy that a human being supports [12].

In the same way, a coupling mechanism was designed for Child Restraint System according to the R129 standard for reducing injuries during a frontal and lateral collision. This coupling device uses springs and dampers that allow displacements in the three Cartesian axes (x, y, z) to dissipate some of the energy produced by a traffic accident. Head, neck, and thorax with a six-year Hybrid III during a frontal impact were evaluated (Fig. 17.3d). Overall, the injury rates are compared when using mechanical coupling with LBB and only LBB to analyze the system's efficiency, showing a significant reduction in head and neck injuries, obtaining a 24% variation in the HIC36, and reducing the neck range motion by 19.3° [13].

The study of passive safety allows us to calculate: ranges of motion, speeds, and injury rates. These data help the design and development of medical robots. It is the fourth and last line of research carried out in the biomechanics laboratory of the IPN.

Medical robots equipped with biomechanical understanding are also transforming the field of rehabilitation and assistive care. These robots can provide customized therapy treatments tailored to individual patients, assisting in recovery from injuries or surgeries. Integrating biomechanics allows the robots to adapt their movements and forces to meet patients' needs and limitations, encouraging safe and effective rehabilitation. The following proposes are biomechanics solutions applied to rehabilitation through medical robot design. A wheelchair was designed for adopting several positions using a four bars mechanism attached to the mainframe that balances the resistance and stability required for an optimal position in the person. The design was carried out under the Theory of Inventive Problem Solving (TRIZ) concepts (Fig. 17.4a). The main components of the wheelchair design are analyzed by the finite element method. A proportional–integral–derivative controller (PID) is developed using multibody simulation. The four-bar mechanism is integrated to assist the position of the backrest, and a microcontroller controls an electromechanical actuator fixed to the chair's backrest [14]. In Addition, a modeling and simulation process of upper limb prostheses with seven degrees of freedom was carried out with 3DOF in the shoulder, 1DOF in the elbow, and 3DOF in the palm (Fig. 17.4b), through a static structural analysis using the ANSYS Workbench® software considering a force of 5 N.

The model's mass is 0.96 kg, including the arm, forearm, and wrist, while the shoulder mechanism and motors are 0.96 kg, with a total mass of 1.92 kg. Topological

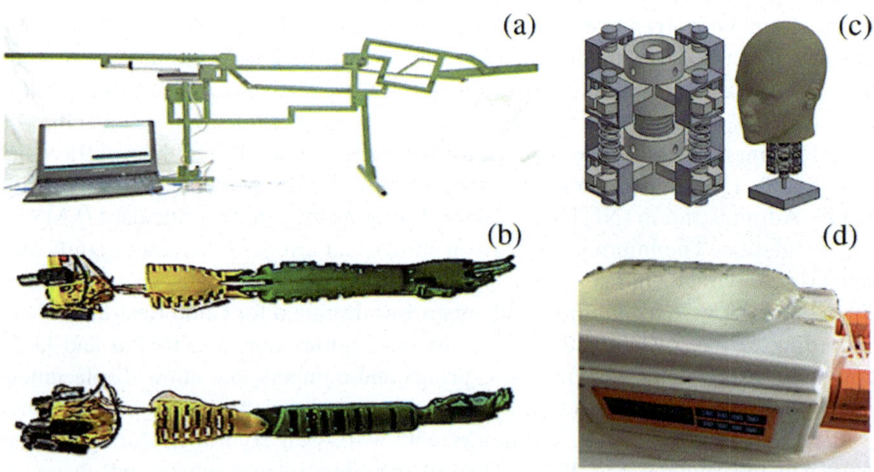

Fig. 17.4 **a** Automated multiposition dynamic wheelchair [14], **b** Upper limb prostheses with 7 DOF [15], **c** Articulated neck with sensor [16], **d** Automated peritoneal dialysis equipment [17]

optimization of the mechanisms located in the transhumeral zone of the prosthesis was carried out according to the Schwartz classification. The parts were manufactured using the Feature CAM® and Ultimaker Cura® [15]. Furthermore, an articulated low-cost artificial neck with sensors to assess the effects of head impacts was designed (Fig. 17.4c). Several tests are worked out considering the biomechanics involved in the most common accidents of head-neck crashes [16]. On the other hand, 3D Low-Cost Equipment for Automated Peritoneal Dialysis Therapy was developed for chronic kidney diseases (Fig. 17.4d). The manufacturing cost is reduced due to the materials used and the automation of the elements, thus creating three main subsystems, mechanical, electronic, and control [17].

This research aims to show that simple movements leading to complex solutions can be generated using a suitable methodology through the kinematic analysis of basic mechanisms such as 4 bars, cams, and gears applied to biomimetic observation. Three biomechanical problems are presented and solved using computational tools and analytical models.

17.2 Design Requirement

Solving Biomechanics problems depends on the engineering design experience and the capacity to synthesize inventions by incorporating simple mechanisms [1] that mimic human beings' natural biological circumstances. Mechanical design methods for biomechanics problems typically involve identifying the problem, understanding the field of biomechanics, generating design concepts, analyzing and optimizing the designs, prototyping, testing, and manufacturing. Mechanical design methods for biomechanics problems are a promptly evolving field. The following proposed methodology has been used to solve three different biomechanics problems.

17.2.1 Case I. Portable Mechanical Ventilator Design for Patients with COVID-19

This project proposes a low-cost, functional portable mechanical ventilator design for emergencies. The mechanical ventilator contemplates systems and subsystems for correct operation through a reliable design guided by safety standards. It is essential to highlight that one of the main areas of opportunity presented in this case is the lack of portable mechanical ventilators for transferring patients in the hospital areas and from their homes to the hospital. Therefore, this research shows the airflow patterns during inspiration and expiration in patients affected by COVID-19.

17.2.1.1 Design Requirements for Case I

During breathing, the lungs expand and retract cyclically, forcing atmospheric air into the alveoli and vice versa. This movement results from the forces between the respiratory muscles and the mechanical properties of the lung and the rib cage, especially the elastic properties. The pressure–volume curve of the lung and the respiratory system is practically linear, which makes it possible to consider, under physiological conditions, that the lung has a linear elasticity. Consequently, static compliance quantifies the lungs and the respiratory system [18]. Therefore, mechanical ventilation is a fundamental treatment in which a device that provides ventilation and oxygenation support, gas exchange, and breathing in patients with respiratory problems is assisted. The mechanical ventilator generates a pressure gradient between two points (upper airway: all anatomical structures above the alveolus) and produces a flow for a specific time, which generate pressure that must overcome resistance to flow and the elastic properties of the respiratory system. The behavior profiles of the theoretical tidal volumes supply a patient 500 ml at a rate of 40 ml/min to obtain a maximum inspiratory pressure of 20–40 cm H_2O. The expiratory phase maintains it at eight cm H_2O (basal state). The ventilator's cycling depends on the valve's opening and closing sequence, called inspiratory and expiratory. The inspiratory cycle valve can be programmed according to the respiratory rate set in the ventilator parameters. The inspiratory valve regulates the speed of the inspiratory flow by opening the outlet hole. The end of inspiration is usually cycled by time (Figs. 17.5 and 17.6).

A new design was proposed with the SOLIDWORKS® computer program, which comprises 23 pieces, most of which can be 3D printed or manufactured in sheet metal. The mechanical ventilator is created with gears with a follower of a crank-type mechanism. This design stands out because it is portable. The most essential element is the Ambubag, which is low-cost, supported by a motor. It compresses it by moving the gripping pliers of the proposed mechanism, generating pressure and velocity patterns (Fig. 17.7).

1. Metal sheet part support1, 2. Engine base support, 3. Bearing, 4. Rocker, 5. Cam, 6. Tie rod, 7. 12V DC Motor, 8. Sheet metal-Base vertical separation, 9. Short arrow, 10. Arrow shaft base motor 11. Base metal sheet, 12. AMBU-BAG, 13. Sheet metal-Base horizontal separation, 14. Metal sheet part support, 15. Pressure arm, 16. Pressure arm plate, 17. Flange with bearing, 18. Gear, 19. Screen 5", 20. 4 bars mechanism protector, 21. Gears mechanism protection, 22. Metal sheet cover, 23. Hinge. The P&ID for the design of the mechanical ventilator is described by a control algorithm, which has an intuitive graphical user interface (GUI). The primary data requested is elementary for any user (sex, age, weight). The algorithm calculates the initial flow parameters, which are monitored in real time and can be modified manually or self-regulated by the patient's needs. It has an autonomous operation, which the user can monitor and adjust (Fig. 17.8).

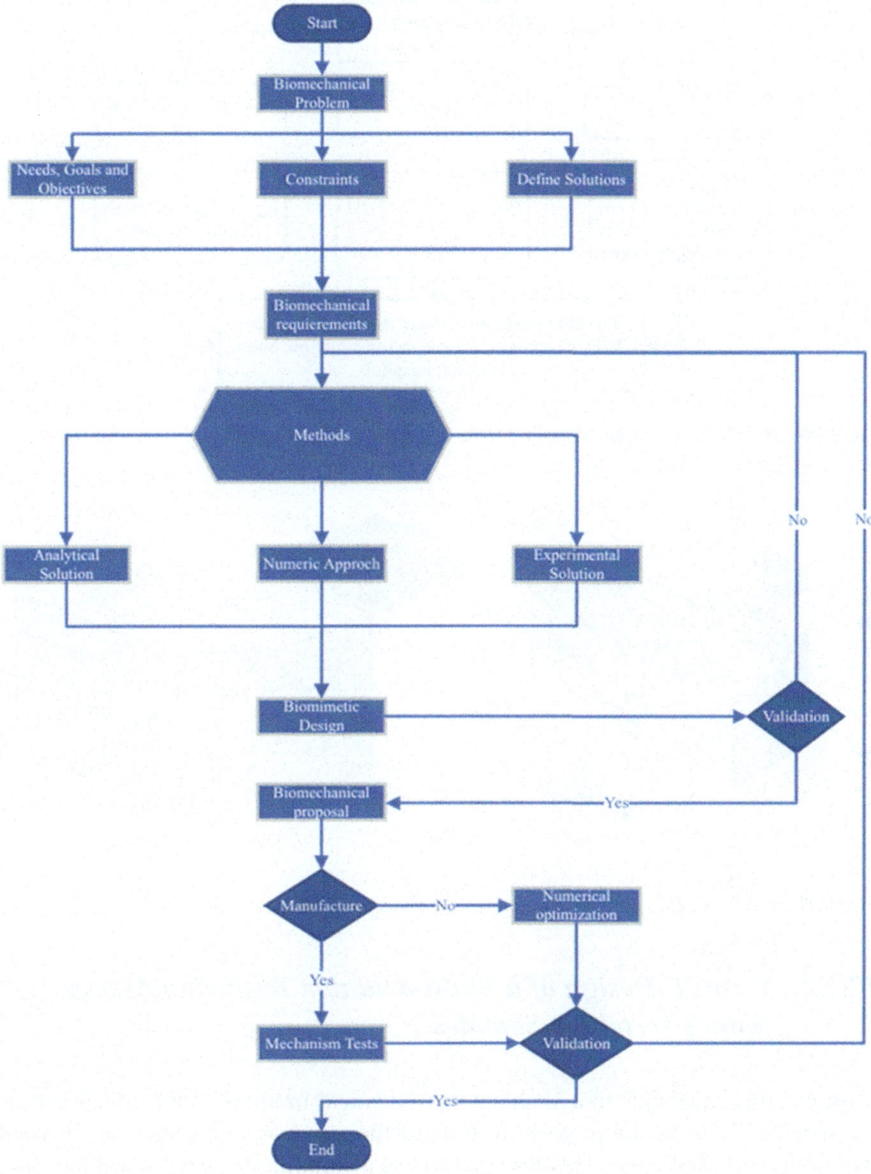

Fig. 17.5 Methodology to solve biomechanics and biomimetics issues

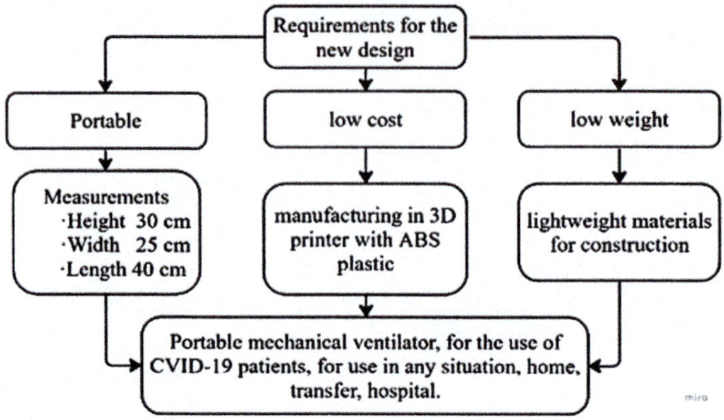

Fig. 17.6 Mechanical ventilator requirements

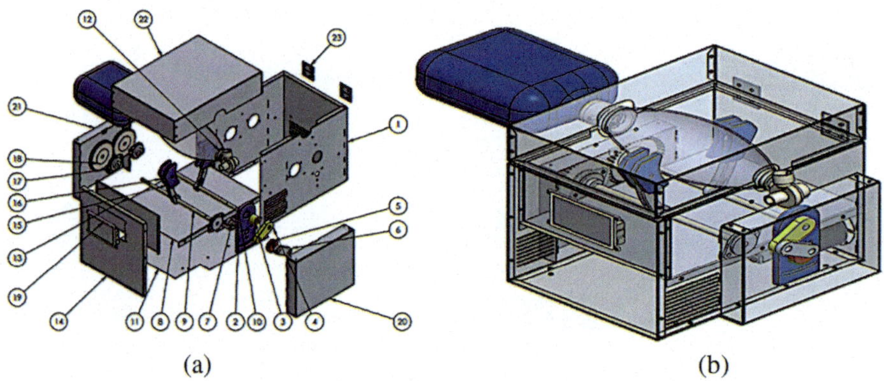

Fig. 17.7 **a** Mechanical ventilator assembly, **b** 3D Mechanic ventilator prototype

17.2.2 Case II. Design of a Multi-actuator Testbed to Assess Spinal Vertebrae Samples

Most existing machines that simulate the loads and motions of the spinal column are specifically designed to perform tests of the superior spine segments, particularly in the cervical zone. This research shows a simplified design of a spine simulator gadget that allows analyzing and evaluating the biomechanical behavior of the orthopedic devices used in the lumbar area. The calculations for correctly simulating movements and payloads on the lumbar segments L4 and L5 were considered. This design methodology of a testbed provides a comprehensive test of the lumbar spine, including flexo-extension, lateral tilt (inclination), rotation, and compression.

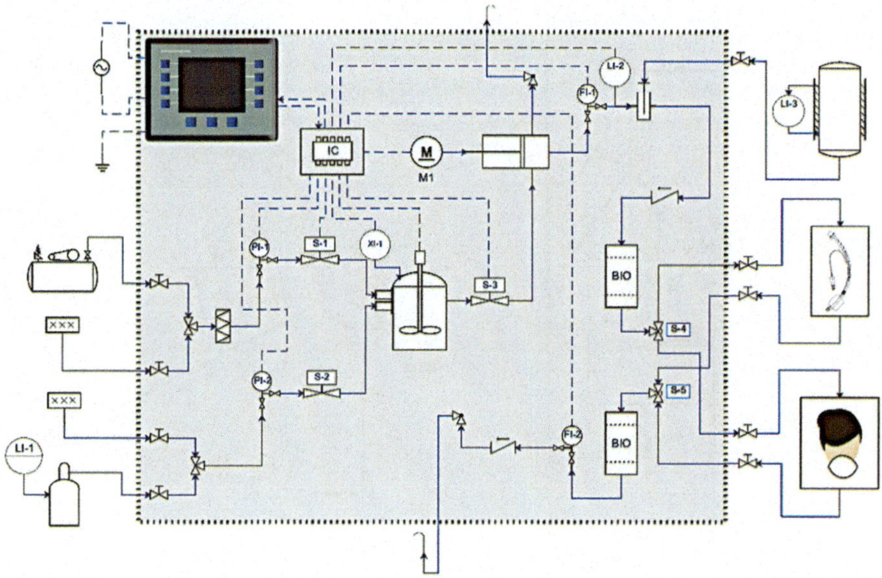

Fig. 17.8 P&ID for the mechanical ventilator

17.2.2.1 Design Requirements for Case II

The spine is a mechanical structure extending from the cranium base to the pelvis. It consists of a series of irregular bones that are more significant in the lower spine. Vertebrae join with ligaments, separated by intervertebral disks. A human adult's spine comprises 24 overlapped vertebrae, a sacrum, and the coccyx. Vertebrae are divided into seven cervical, twelve thoracic, and five lumbar. Forces in the spine produce combinations of motion that simultaneously encompass the six degrees of freedom. The spinal lumbar region possesses excellent flexibility and can transmit loads four times greater than the corporal weight.

The mechanical analysis of the spinal column is carried out with the 2D biomechanical model shown in (Fig. 17.9). A model is formulated to compute the lumbar stress considering the weight of the body trunk (W_{tr}), head and neck (W_{hn}), and arms (W_a). The distance from the load's center of gravity to lumbar vertebrae L_p, the distance from the trunk, head, neck, and arms set to the lumbar segments L_W. The force M that lumbar muscles exert to hold such a posture, the distance from the lumbar muscle's insertion point to lumbar vertebrae L_m, and the angle α of the back concerning the vertical. The higher body weight W, it is considered that $W_{tr} = 0.5$ W_{total}, $W_{hn} = 0.084$ W_{total}, and $W_a = 0.051$ W_{total}, according to [19]. The gravity centers of the corporal segments intervening in the action are calculated depending on the L_W posture and the gravity centers, as shown in (Fig. 17.9b).

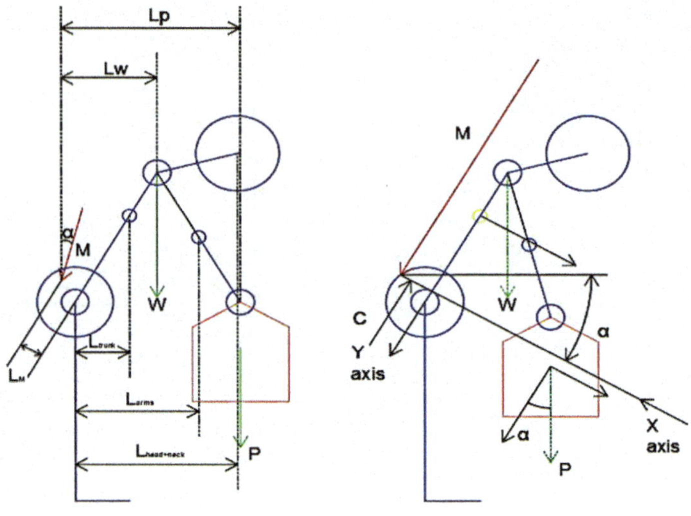

Fig. 17.9 Biomechanical model of the spinal column

$$L_W = \frac{W_{tr}L_{tr} + W_{hn}L_{hn} + W_a L_a}{W_{tr} + W_{hn} + W_a} \tag{17.1}$$

Medical studies [19] estimate the distance from the lumbar muscle's insertion point to lumbar segments at $L_M = 5$ cm. The lengths and weights for the biomechanical model were analyzed in three cases: percentile 5 (women), percentile 50 (men), and percentile 95 (men) of the Mexican population. A free-body diagram per motion can be used to calculate forces and loads exerted on lumbar vertebrae due to physiological motions and external loads (see Fig. 17.9). The maximal values of the involved forces obtain the extreme cases of each motion. The statistical results are shown in Table 17.1.

Flexion and extension are treated as separated motions because an external load is typically carried out in flexion, while in extension, not. For flexion in the extreme case when $\alpha = \pi/3$, translational and rotational equilibrium conditions should be imposed:

Table 17.1 Considered weights and lengths of the biomechanical model

Percentile	Wt	Height	Wtr	Whn	Wa	LP	Ltr	Lhn	La	LW	W
	(kg)	(m)	(kg)	(kg)	(kg)	(cm)	(cm)	(cm)	(cm)	(cm)	(kg)
5	50.0	1.58	25.0	4.20	2.5	31.1	23.3	46.7	31.1	27.0	31.7
50	77.0	1.68	38.5	6.47	3.9	48.0	36.0	72.0	48.0	41.7	48.9
95	100	1.88	50.0	8.40	5.1	62.3	46.7	93.5	62.3	54.1	63.5

$$\sum F_x = (W + P)\sin\alpha - R_x = 0 \tag{17.2}$$

$$\sum F_y = M + (W + P)\cos\alpha - R_y = 0 \tag{17.3}$$

$$\sum M = (WL_w + PL_P) - ML_M = 0 \tag{17.4}$$

Which is a closed system of equations for three unknowns: M, R_x, and R_y. The normal force represents the compression load, and the tangential force represents the shear stress [19]. The resultant forces from solving the system of equations are given in Table 17.2. The lumbar forces are obtained by dividing R_x by the lumbar segments. The pressure is distributed and multiplied by the segments to analyze the device developed.

As extension without external loads is considered and in the extreme case of $\alpha = \frac{\pi}{6}$, Eqs. (17.2–17.4) can be solved for extension motion considering $P = 0$ and the appropriate use of signs. These results are shown in Table 17.3.

Lateral tilt motion has an extreme case of $\alpha = \pi$, so considering that an external load of $P = 5$ kg may be carried in hand, Eqs. (17.2–17.4) solved lateral tilt. These results are shown in Table 17.4.

Finally, due to the nature of rotational motion, no muscle force M is exerted when occurring, considering a maximal amplitude of $\alpha = \pi$ and with an external load of $P = 20$kg, Eqs. (17.2–17.4) solved rotation motion. These results are shown in Table 17.5.

The methodology used in the research is based on Fig. 17.5, and to develop the control of the motors and the degrees of freedom was based on [20], considering those proposed to implement them in the machine, which simulates the motions and loads that the lumbar spine undergoes. The device is divided into three subsystems to perform an independent motion: flexion–extension, rotation, and lateral tilt. Its main parts are observed in Fig. 17.10: (A) a stepper motor, (B) the lateral tilt stepper motor

Table 17.2 Compression and shear forces for flexion motion, with an external load of $P = 20/textkg$

Percentile	M (N)	Rx (N)	Ry (N)	Lumbar force (N)
5	−2910.92	−439.65	2657.04	1389.22
50	−5886.39	−585.31	5548.44	1849.48
95	−9197.46	−709.36	8787.90	2241.55

Table 17.3 Compression and shear forces for extension motion, with no external loads

Percentile	M (N)	Rx (N)	Ry (N)	Lumbar force (N)
5	−1687.83	−155.73	1418.13	472.70
50	−4002.86	−239.83	3587.42	1195.82
95	−6751.35	−311.47	6211.89	2070.61

Table 17.4 Compression and shear forces for lateral tilt motion, with an external load of $P = 5\text{kg}$

Percentile	M (N)	Rx (N)	Ry (N)	Lumbar force (N)
5	−1993.60	−123.30	1654.85	551.61
50	−4473.74	−180.83	3976.88	1325.64
95	−7362.85	−229.83	6731.43	2243.80

Table 17.5 Compression and shear forces for rotation motion, with an external load of $P = 20\,\text{kg}$

Percentile	Rx (N)	Ry (N)	Lumbar force (N)
5	−58.91	−505.71	−168.58
50	−44.25	−673.26	−224.43
95	−71.39	−816.00	−272.01

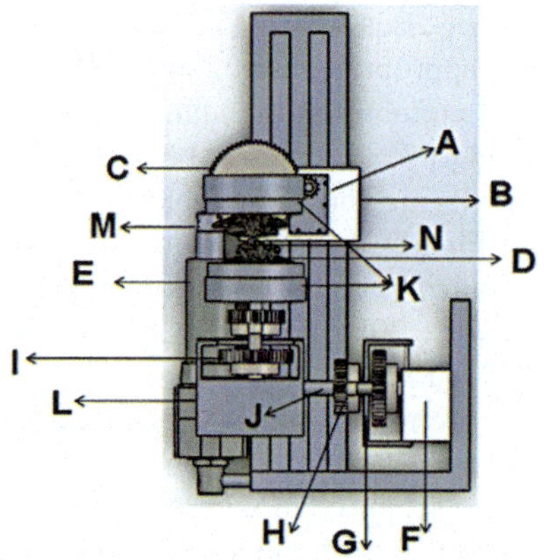

Fig. 17.10 Lumbar-spine simulator

box, (C) the lateral tilt gear system, (D) lumbar test vertebrae, (E) the linear actuator to generate compressing loads, (F) the motor box of flexion–extension subsystem, (G) the flexion–extension gear system, (H) the rotation gear system, (I) the second rotation gear system, (J) is the central axis, which supports the whole rotation mechanism, (K) are the bases to hold the lumbar vertebrae, (L) the rotation stepper motor box, (L) is the stepper motor box (M) is the linear actuator piston, and (N) is the intersomatic screw.

17.2.2.2 Gear Train Assessment

The required torque by each motor was calculated as follows:

$$M = Fr \tag{17.5}$$

Table 17.6 Required torques

Variable	Flexion	Extension	Lateral tilt	Rotation
F (N)	585.31	239.83	180.83	1908.39
r (m)	0.15	0.15	0.10	0.07
$M(Nm)$	87.21	35.73	18.083	133.5873

where F is the lumbar force previously calculated in Table 17.2, and r is the arm lever, which consists of the distance from the motor to the clamping mechanism of the vertebral segments. The results for each motion are summarized in Table 17.6.

For the three subsections, the gear ratios are as follows:

- For flexo-extension motion, a maximum torque of 7 Nm is considered to perform the movement, so the ratio required is about 4:1, where the first element is a pinion of 1/2" with 12 teeth and a diametral pitch of 0.75" and a gear of 48 teeth with a pitch diameter of 16; the second torque considers the same pinion as the first and a gear of 36 teeth with a diametral pitch of 16.
- For rotation motion, the powertrain requires an 18:1 ratio with two pinion-gear torques with two ratios, 6:1 and 3:1; the first has a pinion of 12 teeth and a gear of 72 teeth; the second has a pinion of 12 teeth and a gear of 36 teeth, all the elements have a diametral pitch of 16.
- For lateral tilt, the required torque is lower than in previous cases. Therefore, a stepper motor of 3.5 Nm is used; The powertrain needs a 5:1 ratio; the pinion has 12 teeth, and the gear has 60 teeth. Besides, both have a diametral pitch of 16.

17.2.3 Case III. Analysis of an Embedded Child Restraint System for Groups I, II, and III in a Vehicle

This research focuses on the design of a mechanism embedded in the rear seat of a sedan-type vehicle. A four bars mechanism was modeled in the SOLIDWORKS® computer program. The embedded child restraint system (E-CRS) facilitates its use and handling to all users who have children from 9 kg up to 36 kg or until they reach a height of 150 cm. The E-CRS was exported to MSC ADAMS® software to evaluate the mechanism. The results obtained from the analyses allow the mechanism to be validated based on standard N.129.

17.2.3.1 Design Requirements for Case III

Child restraint systems are divided into categories according to the children's weight. However, this weight is related to the age of the infant. Therefore, this study has presented the design for groups I to III. The parameters used in the device design are:

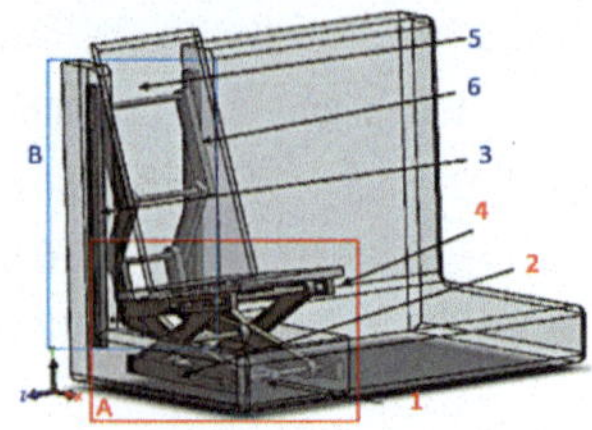

Fig. 17.11 Embedded child restraint system into the backseat

1. Piston linear actuator
2. Seat fastening base
3. Support fastening base
4. booster seat base
5. Backup basis.
6. Clamping Bolts

- The system elevation to reach the height for groups I to III of the child restraint systems.
- The weight that will support the design that, in this case, is from 9 to 36 kg. [21].
- Continue using the conventional rear seat without significantly modifying the existing system.

The design of the mechanism embedded in the child restraint system in the rear seat of the sedan vehicle is replaced in a conventional car since this mechanism is incorporated into the rear seat. Figure 17.11 shows the E-CRS outside the seat showing the folded and deployed tool. The mechanism comprises 16 bars, 9 bolts, and a piston actuator allowing displacement.

Two interlocking mechanisms of four bars with a follower are used to adopt the positions of group I to group III. It is divided into two sections. Section A comprises a pair of 4-bar mechanisms that interlock through a linear actuator manufactured by LINAK.

The LA20 is a compact actuator with a load capacity of 2500 N.

- Compact and highly efficient drive.
- High load capacity: 2,500 N push and 900 N pull.
- Speed 8.9 mm/s.
- Waterproof with medical approvals for the most demanding healthcare environments.
- Positioning signal and limit switch.
- Easy cable replacement.

The linear actuator makes the system rise. The kinematic analysis was carried out with the analytical method. It was defined to solve the problem with the inverse kinematics methodology. A mathematical study was carried out to achieve these two equations that reflect points C and D considering the initial data referring to the lengths of the mechanism (Fig. 17.12).

The unknowns are defined for points C and D. Points C and D start from the coordinate axis. Point C Eqs. (17.7) and (17.8). Point D Eqs. (17.11) and (17.12).

$$RC = O2C * e^{iq} \tag{17.6}$$

$$\mathrm{XC} = \mathrm{O2Ccos(q)} \tag{17.7}$$

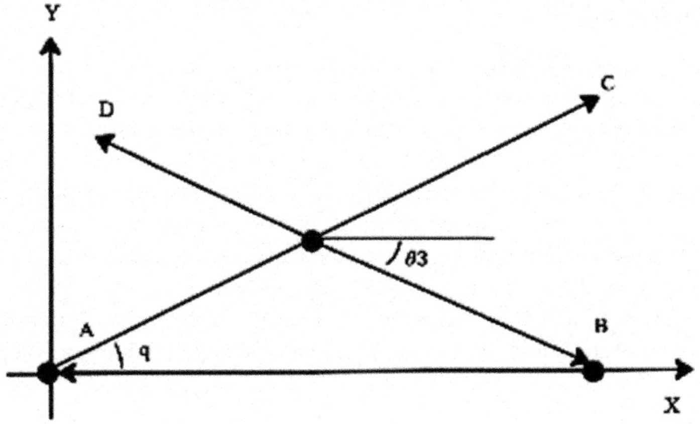

Fig. 17.12 Four bars position diagram

$$\mathrm{YC} = \mathrm{O2Csin(q)} \tag{17.8}$$

$$\mathrm{RD} = \mathrm{R2} + \mathrm{R3}\overrightarrow{\mathrm{AD}} \tag{17.9}$$

$$\mathrm{R3}\overrightarrow{\mathrm{AD}} = \overrightarrow{\mathrm{AD}} * \mathrm{r3}^{\mathrm{i}\theta 3} \tag{17.10}$$

$$\mathrm{XD} = \mathrm{r2cos(q)} - \overrightarrow{\mathrm{AB}}\mathrm{cos}(\theta 3) \tag{17.11}$$

$$\mathrm{YD} = \mathrm{r2sin(q)} - \overrightarrow{\mathrm{AB}}\mathrm{sin}(\theta 3) \tag{17.12}$$

The computational tools are a reinforcement to corroborate and have a greater certainty of the position that the seat mechanism must reach, and it is shown in Fig. 17.13.

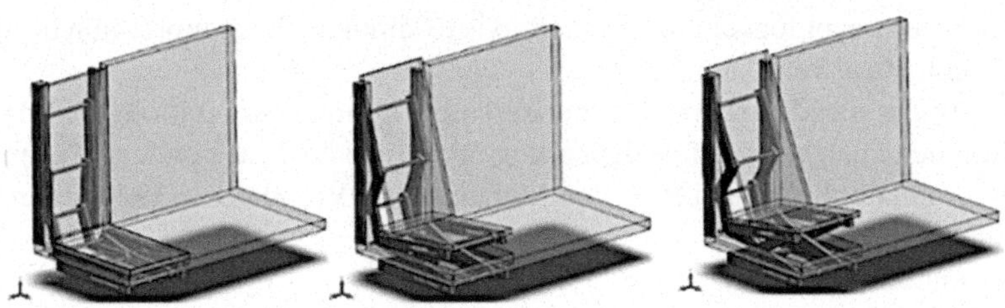

Fig. 17.13 The embedded system positions

17.3 Overview

Mechanism design is a rapidly growing field with many potential applications in biomechanics. Mechanisms will likely be used to design even more complex and sophisticated mechanisms, such as self-driving cars and robots that can interact with humans safely and reliably. The synthesis of basic mechanisms is a challenging task. However, using mathematical, geometrical, physical, and computer-aided methods has made designing more efficient, reliable, and safe mechanisms possible. The following section shows an overview of the solutions proposed for each attached problem mentioned in this article, showing specific simulation tools that support the correct design of the equipment to improve the quality of the human being.

17.3.1 Results Case I

The parameters for the proposed mechanical ventilator design can provide a numerical simulation of the compression action of the Ambubag. It performs a Fluid–Structure Interaction (FSI) analysis where through the contact of the pressure arms in the Ambubag, the pressure of the contained fluid increases and is expelled through the front connector. For this analysis, two different meshes were made, one to restrict the volume and properties of the CFD (Computational Fluid Dynamics) type fluid and another for the discretization of the structural part, joined from the FSI analysis.

It is necessary to define the boundary conditions that restrict its movements. In the case of the CFD mesh, there are two conditions:

- Pressure at the front end of Ambubag is equal to zero.
- The fluid's speed on the bag's walls is negligible (non-slip condition).

The fluid in the Ambubag is air at a temperature of 20 °C and a relative humidity of 60%. The properties of the air on 20 °C-h = 60% with these parameters are flow density = 1.20 kg/m^3 and a dynamic viscosity = 1.810×10^{-5} N-s/m^2.

In the case of the structural mesh, the movement of the front and rear connectors of the bag is restricted. In the case of the pressure arms, their movement is restricted in all directions except for rotation on their axis to simulate the compression movement of the mechanical ventilator.

The angular speed movement of the pressure arms in the simulation is 0.8816 rad/s. This parameter can be modified depending on the needs of each patient. The material used to simulate the behavior of the Ambubag is PVC Shore A35 by commercial specifications.

This material has a density = 1.12 kg/m^3, E = 938 kPa, and Poisson ratio = 0.4950. The results obtained from the simulations are shown in Fig. 17.14. It is shown the pressure and velocity of the flow at the outlet of the Ambubag, which are the units KPa and mm/s, respectively.

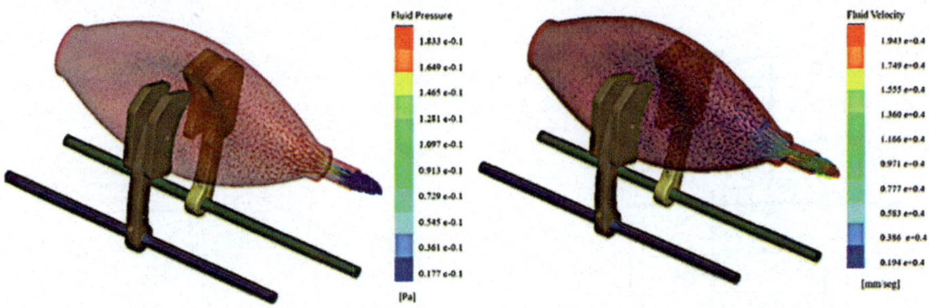

Fig. 17.14 Pressure and velocity of the fluid at the Ambubag outlet

Table 17.7 Comparison [Pa]

Ambubag fluid pressure steps						
Portable ventilator	1.83 e−0.1	1.64 e−0.1	1.09 e−0.1	0.91 e−0.1	0.54 e−0.1	0.17 e−0.1
El-Hadj [18]	2.18 e−0.1	1.75 e−0.1	1.39 e−0.1	1.08 e−0.1	0.60 e−0.1	0.77 e−0.1

The Ambubag has numerical stress, deformation, and displacement evaluations, but few CDF analyses were carried on. El-Hadj et al. developed a CFD analysis of a low-cost mechanical ventilator [18]. A comparison of our research and above mentioned is shown in Table 17.7. The main difference between both studies is related to boundary conditions that deform the Ambubag. Nevertheless, the main difference between both analyses is the claw grips involving the compression area.

17.3.2 Results Case II

Controlling the system's actuators is to make adequate movements to reach the desired points considering the time, position, velocity, and acceleration. In Fig. 17.15, the system is presented to appreciate the movements produced on the x, y, and z.

Calculating those parameters along an isochronous trajectory (which means that the motors all move to reach the point simultaneously) is presented, knowing that all the motions are rotational. A rotational equation of motion was used. A stepper motor can exert a torque that increases linearly with time $\tau = b_1 + b_2 t$, where b_1 and b_2 are constants. The rotational equation of motion is given by:

$$I\ddot{\theta} = b_1 + b_2 t \tag{17.13}$$

where I, is the moment of inertia. By redefining the constants in the latter equation and integrating twice, the angular speed (Eq. 17.15) and position (Eq. 17.16) for the

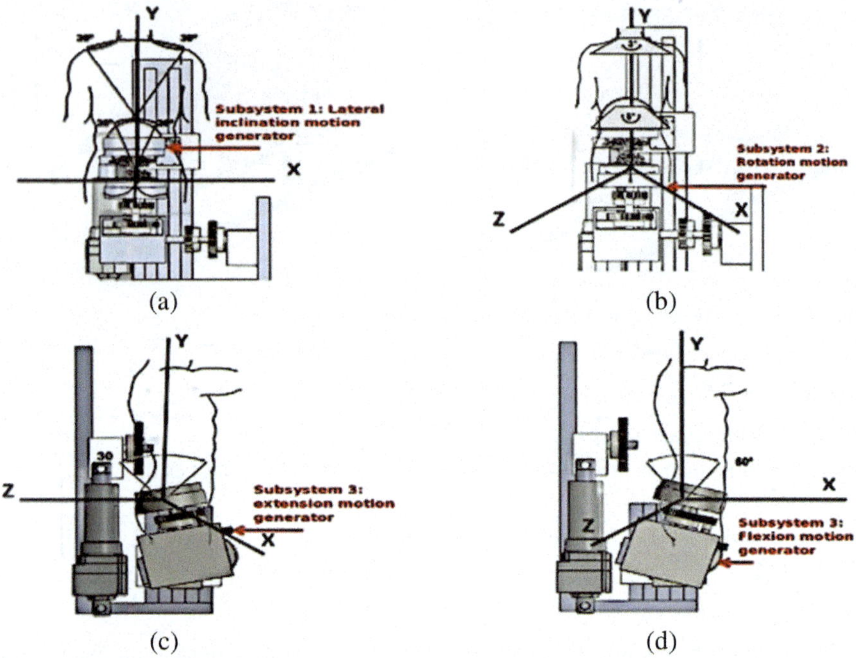

Fig. 17.15 **a** Lateral tilt subsystem, **b** Rotation subsystem, **c** Extension in flexion–extension subsystem, **d** Flexion in flexion–extension subsystem

time were obtained:

$$\ddot{\theta}(t) = 2a_2 + 6a_3 t \tag{17.14}$$

$$\dot{\theta}(t) = a_1 + 2a_2 t + 3a_3 t^2 \tag{17.15}$$

$$\theta(t) = a_0 + a_1 t + a_2 t^2 + a_3 t^3 \tag{17.16}$$

Thus, constants a_0, a_1, a_2 and a_3 ought to be determined according to initial and final conditions to determine the kinematic parameters which will determine the required trajectories. For instance, the calculation for the flexion motion should be done as follows: as the pinion actuated by the motor has 12 teeth and the flexion motion encompasses an angle of $\frac{\pi}{3}$, the motor requires a turn of 4π. Thus, considering that the motion starts and ends at rest, setting the initial time $t_i = 0$ s and considering that the motion should occur in $t_f = 5s$, initial and final conditions are: $\theta_i = \theta(t_i) = 0, \theta_f = \theta(t_f) = 4\pi$, $\dot{\theta}_i = \dot{\theta}(t_i) = 0$ and $\dot{\theta}_f = \dot{\theta}(t_f) = 0$. With such values, the coefficients for flexion motion should be: $a_0 = 0\,rad, a_1 = 0\,\frac{rad}{s}$, $a_2 = 0.48\,\pi rad/s^2$ and $a_3 = -0.064\,\pi rad/s^2$. Therefore, now the final angular acceleration, which is subject to the flexion motion, can be estimated, which is $\ddot{\theta}_f = -0.96\pi$. The calculation is straightforward for extension, rotation, and lateral tilt motions; the results are shown in Table 17.8.

The stepper motors, A4988 controllers, an Arduino One® platform, and a linear electric actuator were used. The chosen linear actuator was CAHB-21 by SKF, with push and pull loads of 2300 N and a 45–65 mm/s speed. It works at 22 A with 12–24 VDC, a work cycle of 25%, a working temperature of −40 to +85 C, and an IP rating of 66. For synchronization, a program was created in LabVIEW® software that generates the trajectories for each motor, considering the parameters mentioned before. The diagrams are presented as follows (Figs. 17.16 and 17.17).

The synchronization was made using a program in Arduino® that allows control of each motor's displacements by manipulating the steps. The gear ratio was used to activate each motor at a specific time. For example, when the gear ratio of 12:1 has

Table 17.8 Kinematics parameter of trajectories for each motion/subsystem

Motion	$\theta_i = a_0$ (rad)	$\ddot{\theta}_i = a_1$ (rad/s)	θ_f (rad)	$\dot{\theta}_f$ (rad/s)	a_2 (rad/s^2)	a_3 (rad/s^2)	$\ddot{\theta}_f$ (rad/s^2)
Flexion	0	0	4π	0	0.48π	-0.064π	-0.96π
Extension	0	0	2π	0	0.24π	-0.032π	-0.48π
Rotation	0	0	2π	0	$\frac{3\pi}{900}$	$-\frac{2\pi}{4500}$	$-\frac{24\pi}{900}$
lateral tilt	0	0	2π	0	$\frac{12\pi}{900}$	$-\frac{8\pi}{4500}$	$-\frac{6\pi}{900}$

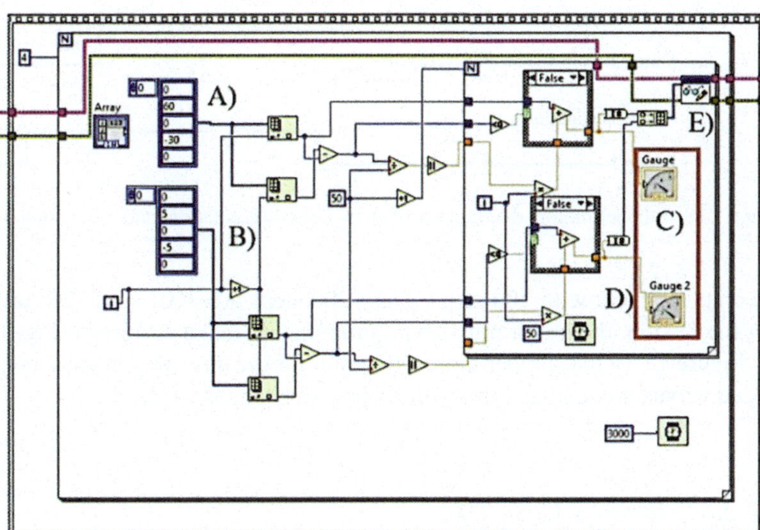

Fig. 17.16 Motor 1 and 2 synchronization. **a** Motor 1 Input data, **b** Motor 2 input data, **c** Motor 1 output signal, **d** Motor 2 output signal, **e** Output for following process

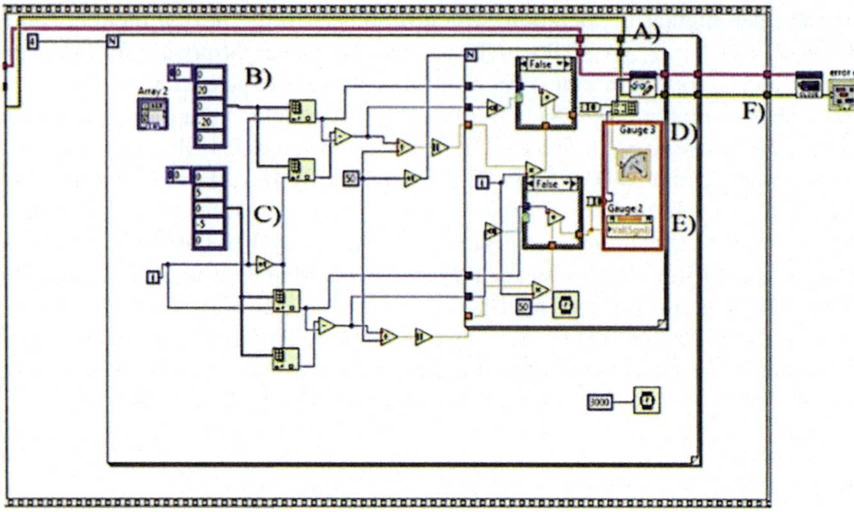

Fig. 17.17 Motor 2 and 3 synchronization. **a** Output for comparison of motor 2 and 3, **b** Motor 3 input data, **c** Motor 2 input data, **d** Motor 3 output signal, **e** Motor 2 Output signal, **f** Output final comparison

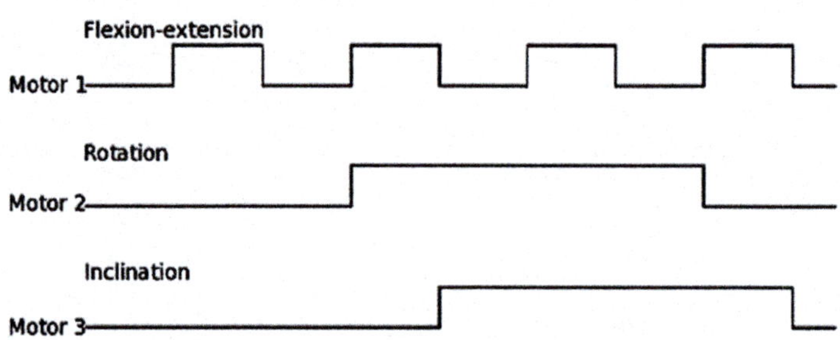

Fig. 17.18 Stepper motors diagram for flexion–extension, rotation, and inclination

a complete cycle with 200 steps, every 200 steps are 360°, and 720° are necessary for the flexion, the motor must be programmed to make 400 steps (Fig. 17.18).

Figure 17.19 the stepper motors diagram for the three trajectories, comparing the speed versus the load and the voltage consume versus the load.

17.3.3 Results Case III

The mechanism analysis determines the trajectory and the angles for the design in a simulation program such as Adams View® (Figs. 17.20, 17.21, and 17.22). A practical and straightforward solution to reproduce the movement to fold and position of a

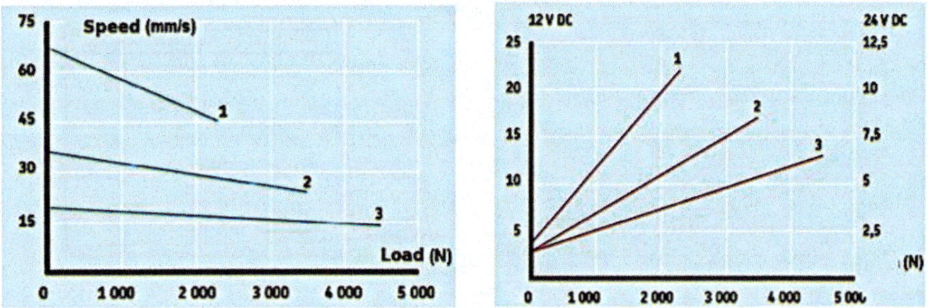

(a) (b)

Fig. 17.19 Linear actuator diagrams, **a** speed versus the load, **b** consume versus the load

child restraint system embedded in a rear seat of a sedan-type vehicle is shown. It is enough to simulate the movements with the weight and the force the child exerts when being on the mechanism.

The mechanism has a final displacement of 180 mm, covering groups I, II, and III. according to (CEPE) in regulation N.129 and an angular velocity of 9.5 deg/sec. The mechanism has no interference and does not go into a singularity.

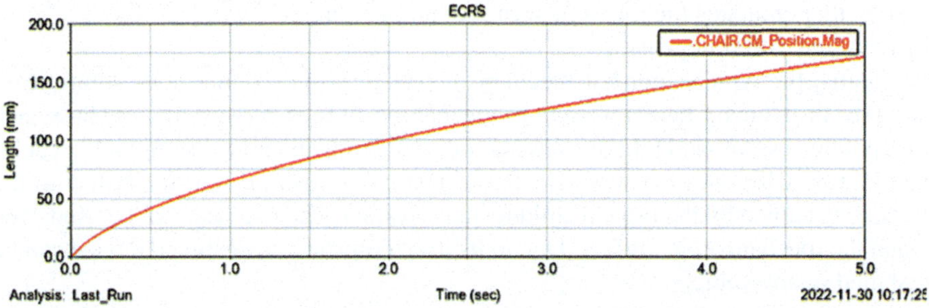

Fig. 17.20 E-CRS displacement graph

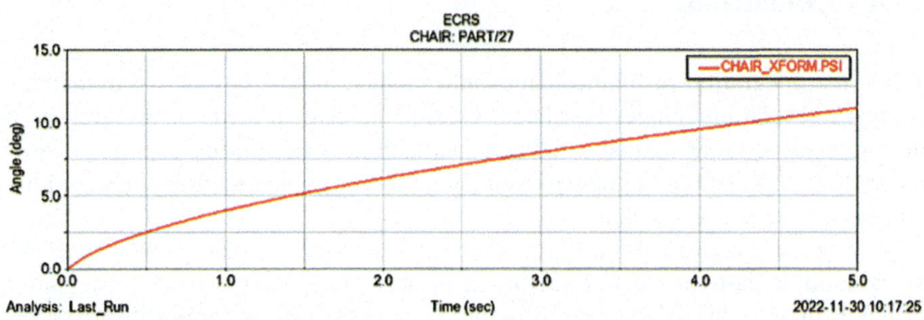

Fig. 17.21 E-CRS angle graph

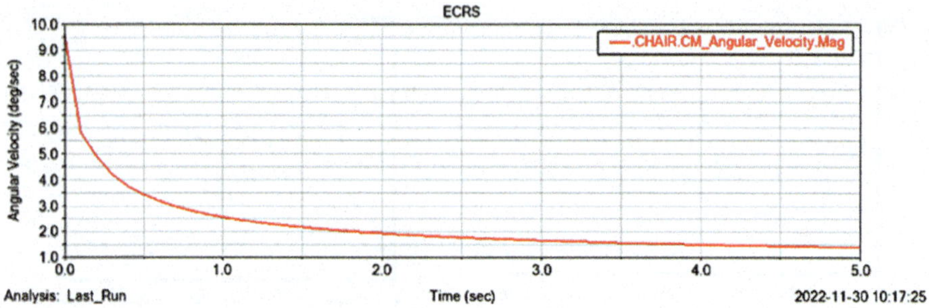

Fig. 17.22 E-CRS angular velocity graph

Professor Carlos S. López Cajún's literature shows that complex solutions can be synthesized by applying simple mechanisms or integrating different ones. Chapter 7 is about the transmission of potency or information which generate periodical movements. In this way, to mimic the breathing process, combining cams and an Ambubag under compression loads allow airflow.

The cinematic of a body is achieved by integrating a mechanical system mounted by gears. Chapter 8 shows how different types of gear and the technique can be coupled [1], allowing the motion and the potency between 2 axes; mathematical analysis and the biomechanical perspective fund, which system can be used. Thus, a system that emulates the movement of the lumbar zone of the spine was developed by coupling different gears.

Finally, the typical four-bar mechanism system provides enormous solutions to complex problems. Chapter 4 explains the features of four-bar systems and how each link point is essential [1]. In this case, a four-bar mechanism for the embedded child restraint (E-CRS) system was selected to resolve a position seat for reducing the volume occupied by the typical child restains system. The rear seat can be used even when the mechanism is folding. Due to its low footprint, the designed device can be installed in any vehicle.

17.4 Conclusion

While the integration of biomechanics into medical robots has shown tremendous promise, several challenges lie ahead. One of the critical areas of research is the development of more sophisticated and adaptable robotic systems capable of simulating the complexities of human movement with even greater accuracy. This requires advances in actuation mechanisms, sensor technologies, and control algorithms.

Additionally, addressing safety concerns and ensuring regulatory compliance will be essential to gain the trust of healthcare professionals. Biomechanics is a guiding principle in developing medical robots, unlocking possibilities for healthcare professionals and patients. By leveraging biomechanical principles, medical robots can enhance surgical precision, improve rehabilitation and assistive care, and optimize human–robot interaction.

While challenges remain, continued research and innovation in biomechanics will undoubtedly lead to even more sophisticated medical robots, revolutionizing healthcare practices and improving patient outcomes in the years to come.

Acknowledgements The authors acknowledge the financial support for the realization of this work to the Mexico Government by Consejo Nacional de Ciencia y Tecnología (CONAHCYT), the Instituto Politécnico Nacional (IPN). The authors also thank the support of projects SIP 2023113 and SIP 20231625 and the EDI grant, all by SIP/IPN.

References

1. López-Cajún, C.S., Ceccarelli, M.: Mecanismos: fundamentos cinemáticos para el diseño y optimización de maquinaria. Publishing house Trillas (2008)
2. Torres-Sanmiguel, C.R., Hernández-Gómez, J.J., Urriolagoitia-Sosa, G., Romero-ángeles, B., Martínez-Sáez, L.: Design and manufacture of a customised temporomandibular prosthesis. Revista Internacional de Métodos Numéricos para Calculo y Diseño en Ingenieria **35**(1) (2019). https://doi.org/10.23967/j.rimni.2019.02.001
3. Aguilar-Pérez, L.A., Sánchez-Cruz, J.I., Flores-Campos, J.A., Torres-Sanmiguel, C.R.: Numerical and experimental assessment of a novel anchored for intramedullary telescopic nails used in osteogenesis imperfecta fractures. Appl. Sci. (Switzerland) **11** (12) (2021). https://doi.org/10.3390/app11125422
4. Ramirez, O., Torres-San-Miguel, C.R., Ceccarelli, M. and Urriolagoitia-Calderon, G.: Numerical and experimental validation of a rib implant using and artificial rib. J. Mech. Med. Biol. https://doi.org/10.1142/S0219519422500361
5. Ramírez, O., Torres-San-Miguel, C.R., Ceccarelli, M., Rueda-Arreguin, J.L., Urriolagoitia-Calderon, G.: A compliant mechanism as a sternum prosthesis. In: New Trends in Medical and Service Robotics. MESROB Italy 2020. Mechanisms and Machine Science, vol. 93, pp. 143–151. Springer, Cham (2020). https://doi.org/10.1007/978-3-030-58104-6_17
6. Silva, K.N., Torres-Sanmiguel, C.R., Jimenez-Ponce, F., Urriolagoitia-Calderón, G.M.: Design of a scalpel clamp for viscoelastic properties assessment in soft brain tissues. J. Phys.: Conf. Series **1723**(1) (2021). https://doi.org/10.1088/1742-6596/1723/1/012015
7. Grave-Capistrán, M.A., Prieto-Vázquez, A.Y., Torres-Sanmiguel, C.R.: Aortic Blunt Trauma analysis during a frontal impact. Appl. Bionics Biomech. (2021). https://doi.org/10.1155/2021/5555218
8. Prieto-Vázquez, A.Y., Guerrero-Hernández, L.A., Gomez-Apo, E., Torres-San Miguel, C.R.: Numerical analysis of a testbed used for liver tissue of biomechanical behavior. Mech. Mach. Sci. **133** MMS, 251–259 (2023). https://doi.org/10.1007/978-3-031-32446-8_27
9. Ramírez-Vela, V., Aguilar-Pérez, L.A., Paredes-Rojas, J.C., Flores-Campos, J.A., Ortiz-Hernández, F.E., Torres-Sanmiguel, C.R.: Bone fractures numerical analysis in a femur affected by osteogenesis imperfecta. Children **8**(12) (2021). https://doi.org/10.3390/children8121177
10. Mexquititla, V.M.C., Rivera-Hernández, M.A., Torres-San Miguel, C.R.: Intercity bus occupant safety. Mechanisms Machine Sci. **108** MMS, 182–192 (2022). https://doi.org/10.1007/978-3-030-87383-7_20
11. Rivera-Hernández, M.A., Ramírez-Juárez, O., Cuautle-Estrada, A., Cantor-Mexquititla, V.M., Torres-SanMiguel, C.R.: Front impact simulation of urban bus. Adv. Struct. Mater. **172**, 15–25 (2022). https://doi.org/10.1007/978-3-030-97925-6_2

12. Islas-Lara, E., Torres-Sanmiguel, C.R., Paredes-Rojas, J.C., Cuautle-Estrada, A., De La Cruz-Alejo, C., Urriolagoitia-Calderón, G.M.: Assessment of retention system effects of level III armored vehicles. Energies **15**(6). https://doi.org/10.3390/en15062247
13. Cruz-Jaramillo, I.L., Martínez-Sáez, L., Torres-Sanmiguel, C.R.: Numerical simulation of mechanical coupling for low-Back Booster with a 6-year-old child during a Crash test. Appl. Sci. (Switzerland) **12** (11) (2022). https://doi.org/10.3390/app12115350
14. Aguilar-Pérez, L.A., Paredes-Rojas, J.C., Sanchez-Cruz, J.I., Leal-Naranjo, J.A., Oropeza-Osornio, A., Torres-San Miguel, C.R.: Design of an automated multiposition dynamic wheelchair. Sensors **21**(22) (2021). https://doi.org/10.3390/s21227533
15. Torres-Sanmiguel, C.R.: Modeling and simulation process via incremental methods of a production-aimed upper limb prosthesis. App. Sci. (Switzerland) **12**(6) (2022). https://doi.org/10.3390/app12062788
16. Rueda-Arreguín, J.L., Ceccarelli, M., Torres-Sanmiguel, C.R.: Design of an articulated neck to assess impact head-neck injuries. Life **12**(2) (2022). https://doi.org/10.3390/life12020313
17. Rivero-Urzua, S., Paredes-Rojas, J.C., Méndez-García, S.R., Ortiz-Hernández, F.E., Oropeza-Osornio, A., Torres-Sanmiguel, C.R.: 3D low-cost equipment for automated peritoneal dialysis therapy. Healthcare (Switzerland) **10**(3) (2022). https://doi.org/10.3390/healthcare10030564
18. Abdellah, E., Mohamed, K., Hijaz A., Houari A., Zamree B., Abdelhakime Y., Hanaa, A.: Design and simulation of mechanical ventilators. Chaos, Solitons Fractals **150**, paper 111169 (2021)
19. Izzo, R., Guarnieri, G., Guglielmi, G., Muto, M.: Biomechanics of the spine. Part I: spinal stability. Eur. J. Radiol. **82**(1), 118–126 (2013). ISSN 0720-048Xhttps://doi.org/10.1016/j.ejrad.2012.07.024
20. Perrusquia, A., Flores-Campos, J.A., Torres-Sanmiguel, C.R., Gonzalez, N.: Task space position control of Slider-Crank mechanisms using simple tuning techniques without linearization methods. IEEE Access **8**, 58435–58442 (2020). Article number 9037353. https://doi.org/10.1109/ACCESS.2020.2981187
21. UNECE. Un Regulation No. 129, Increasing the safety of children in vehicles, for policymakers and concerned citizens. United Nations Economic Commission for Europe, Ginebra, Suiza (2019)

Chapter 18
REST: A REmote Skeleton Telerehabilitation System

Daniele Cafolla and Betsy Dayana Marcela Chaparro-Rico

Abstract The system presented in this paper allows rehabilitation and/or physical-motor exercise sessions, usually carried out at the healthcare facility, directly at the home of each patient. The proposed system combines the convenience of performing sessions remotely with a quantitative biomechanical analysis, operated by a system that is guided and supported by an artificial intelligence. Thanks to the help of the latter, the system is not limited to a simple reproduction of the movements but is able to carry out a biomechanical quantitative analysis of these movements, immediately detecting and highlighting any inaccuracies and error. Each patient, using a camera, can carry out a rehabilitation session with a healthcare professional from his home and the latter from home or hospital can follow the patient in real time. The healthcare professional can evaluate the patient through a special graphic interface that shows a reconstructed virtual skeleton of the patient, and all the data detected and calculated by an artificial intelligence can be used for diagnosis and assistance purposes.

Keywords Biomechanics · Pose detection · Telerehabilitation · Rehabilitation robotics

18.1 Introduction

Applications for telemedicine have grown over the past few years as a result of the development of new computer science technologies and more sophisticated telemedicine equipment. A few methods of long-distance communication include videoconferencing, email, and texting. Today, it is possible to remotely control drones, robotic arms, and robots. The way that humans behave has significantly

D. Cafolla (✉) · B. D. M. Chaparro-Rico
Faculty of Science and Engineering, Department of Computer Science, Swansea University, Swansea, UK
e-mail: daniele.cafolla@swansea.ac.uk

B. D. M. Chaparro-Rico
e-mail: b.d.m.chaparrorico@swansea.ac.uk

M. Ceccarelli and J. C. Jauregui-Correa (eds.), *State-of-the-Art and Innovations in Mechanism and Machine Science*, Mechanisms and Machine Science 150,
https://doi.org/10.1007/978-3-031-47040-0_18

changed as a result of these developments [1]. New rehabilitative practises have improved over the past 20 years as a result of demographic shifts and greater funding for public health [2]. Although rehabilitation is an ancient field of medicine, new telecommunications-based techniques have emerged globally in recent years. These specific rehabilitation techniques are known as telerehabilitation, which is a subsection of telemedicine that consists of a system to manage rehabilitation from a distance [3].

To the greatest extent possible, rehabilitation must be provided to individuals to maximise their functionality and quality of life [4, 5]. While rehabilitation is a thorough, multifaceted and multidisciplinary intervention, the minimum elements needed—which frequently include physical therapy—are determined by the specific health condition and other factors related to the health system or resources available.

When movement and function are in danger, physical therapist interventions are necessary to build, preserve, and restore movement and functional capacity with the understanding that functional movement is essential to health and a high quality of life [6, 7]. Physical therapy and rehabilitation are underutilised despite the benefits being known [8]. If this is combined with the fact that patient or service resources are limited [9], and high demand causes service saturation and the creation of waiting lists [10], the restriction of access becomes a reality.

Alternative rehabilitation models have been developed in this situation, where rehabilitation is required but not adequately implemented, to increase coverage. To increase accessibility and improve continuity of care for vulnerable, geographically remote populations of people with disabilities, telerehabilitation—considered a branch of telehealth—is set up as a system for controlling or monitoring remote rehabilitation using telecommunications technologies. This has the potential to save time and resources in the healthcare industry [11–13].

Controlled studies on rehabilitation have demonstrated that prompt management of an injury or a disease is crucial for achieving positive outcomes in terms of a patient's self-efficacy. Therefore, a rehabilitation programme should begin, as soon as possible, be as intensive as possible, be as lengthy as possible, and continue throughout the recovery phase. The start time of a rehabilitation programme is important, and it should start as quickly as possible. Even if a patient requires precise and intensive treatment, most of the initial phases of rehabilitation following a disease or injury can be performed at home. Consequently, telerehabilitation was developed to accomplish the same outcomes as the standard rehabilitation process in a hospital or in person with a physiotherapist. There have been reports of various types of telerehabilitation interventions and their relative intensities and durations [14].

Health care treatment closer to the needs of the specific pathologies of the patient improves the quality of life and often decreases the duration of treatment.

One instance of a telerehabilitation system is the Rehab@Home framework, which is employed to facilitate rehabilitation in the home environment specifically for individuals who have suffered from strokes [15]. The framework comprises of instrumented insoles that are wirelessly connected to a tablet computer of the third generation, a server, and a graphical web interface designed for medical professionals. The analysis of rehabilitation progress is conducted automatically after the administration of assessment tests on a tablet computer. Both the systems, namely Rehab@Home and the virtual urban environment, were well-received by both patients and doctors due to the positive outcomes they yielded. It is possible that the aforementioned systems may not experience widespread utilisation in the future. However, they are expected to make valuable contributions towards the enhancement of these approaches, particularly in the realm of cardiac telerehabilitation applications.

A telerehabilitation system was used to assess the quality of life in a sample of 99 individuals who had experienced a stroke. The researchers noted a statistically significant alteration in both interventions, namely normal rehabilitation, and robotic rehabilitation. Both modalities demonstrated effectiveness in enhancing the quality of life and alleviating depression outcomes among participants within a timeframe of less than six months following their stroke. The objective of this study was to achieve improved outcomes in the field of robotic rehabilitation. However, the obtained results did not demonstrate statistically significant distinctions between the two experimental groups [16].

The telerehabilitation project known as H-CAD is designed to facilitate upper limb rehabilitation treatment for patients with conditions such as multiple sclerosis, stroke, or traumatic brain injury, within the comfort of their own homes. A help desk was established with the purpose of providing guidance to patients in the development of an appropriate exercise regimen, which involves periodic evaluation of their performance. Patients were afforded the opportunity to engage in interactions with medical professionals at the hospital via a teleconferencing system. The procedure was conducted in two experimental phases. The initial stage involved conducting tests on the system's outcomes by engaging volunteers within a hospital setting. The subsequent stage involved conducting tests on the system in a home environment, using ad hoc patients. The findings were promising, and medical professionals noted a significant enhancement in patients who used this particular system [3].

Musculoskeletal disorders exert a significant influence on the delivery of healthcare, particularly in the context of physiotherapy applied to telerehabilitation. A controlled study was undertaken to evaluate the efficacy of utilising a telerehabilitation methodology as an alternative to the conventional in-person practise. The mentioned study conducted a comprehensive literature review encompassing a total of 898 studies, which were analysed to evaluate the validity and reliability of Internet-based physiotherapy assessment in the context of musculoskeletal disorders. The majority of telerehabilitation approaches demonstrated validity in their application to various physical diseases, except for lumbar spine posture, for which the final score yielded inconclusive results. The study findings indicated that the intervention demonstrated varying levels of effectiveness, ranging from low to moderate scores [17].

A research study utilised Depth sensors to investigate the application of Microsoft Kinect, a motion-sensing input device, in the identification of patients' posture and movement. This technology facilitated the development of personalised exercise routines by caregivers for individual patients. The efficacy of the intervention was verified through testing, which revealed numerous advantages, with a notable emphasis on the development of an individualised physical exercise regimen for the purpose of physical rehabilitation [18].

The aforementioned systems share a common characteristic with REST (REmote Skeleton Telerehabilitation System) in that they enable the execution of rehabilitation activities in the patient's home rather than at a healthcare facility. Specifically, REST also facilitates the quantitative analysis of a patient's movements. This feature enables a novel technical advancement by enabling healthcare professionals to continuously and immediately correct the patient's movements, rather than exclusively focusing on prescribing rehabilitation exercises to be carried out at the patient's residence. However, it should be noted that the REST system does not require the use of any supplementary sensors, such as the Kinect or similar devices [19, 20]. Instead, it uses the webcam of a personal computer to capture images. These images are then transmitted to a remote server, where they are subjected to analysis by a specialised artificial intelligence algorithm.

This paper refers to a topic of biomechanical engineer that Prof. López-Cajún was interested on, as attribute to his expertise.

18.2 Proposed Telerehabilitation System

REST system enables patients to engage in physical-motor rehabilitation sessions from the comfort of their own homes, using their personal devices such as computers, mobile phones, or tablets. REST eliminates the need for patients to travel to a health facility, as the sessions can be conducted remotely through the utilisation of a video camera and optional sensors. The REST methodology integrates the convenience of conducting remote sessions with a quantitative evaluation of the executed movements, facilitated by an artificial intelligence system. Moreover, the significance of the utilised device's power is irrelevant, as the image stream and any supplementary sensor parameters are transmitted to a database employed by the artificial intelligence system to derive the requisite parameters. The conceptual scheme of the proposed system structure is illustrated in Fig. 18.1.

According to Fig. 18.1, REST system (1) captures the movements of individual patients (2) and facilitates the replication of rehabilitation movements prescribed by their respective doctors (9). These movements are typically performed at a rehabilitation facility, but in this scenario, patients (2) can replicate them from the comfort of their own homes. The patient interface facilitates the collection of patient movements, thereby leveraging the system's capabilities. This is achieved by using a video camera integrated into the patient's device. In addition, the installation of the patient interface (4) on the patient's computer (2) enables the connection of a sensory module (5) consisting of a variety of sensors, which can be utilised to detect additional parameters. Multiple patients have the capability to concurrently use the patient interface (4) provided they have scheduled an appointment with the corresponding healthcare professional (9) for the purpose of engaging in telerehabilitation. The patient interface facilitates the transmission of data from individual patients to the database through the

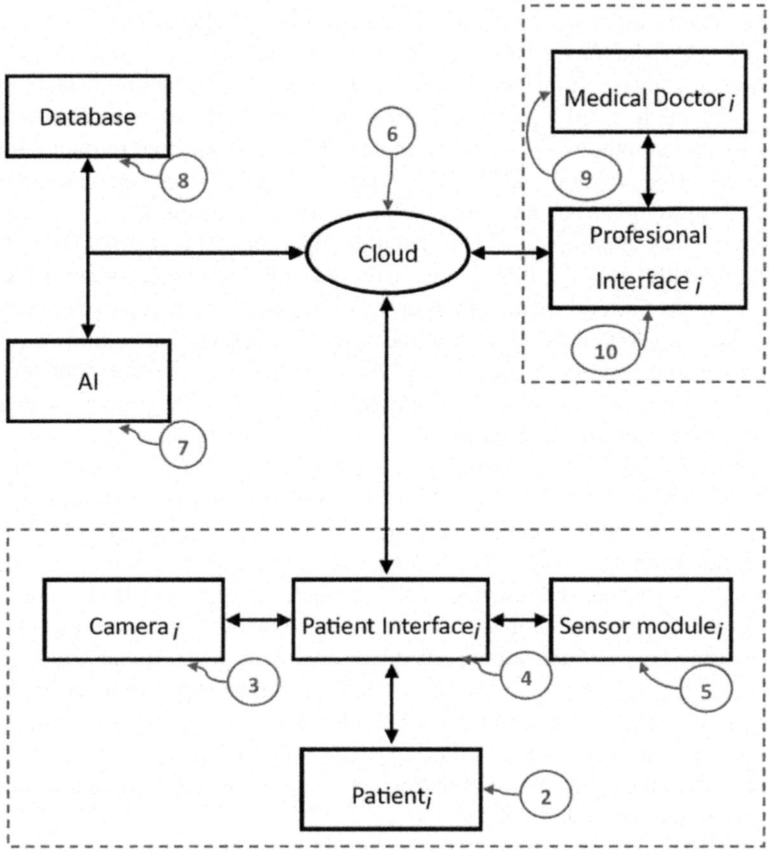

Fig. 18.1 REST conceptual scheme

utilisation of cloud technology. The database has the capability to receive and store data generated by Artificial Intelligence (AI) systems. The AI conducts a quantitative biomechanical analysis of the movements exhibited by individual patients, thereby identifying any instances of incorrectness or errors. This analysis provides valuable information to the respective healthcare professionals, enabling them to promptly diagnose the patient. Moreover, the analysis offers the advantage of incorporating quantifiable numerical parameters that substantiate the diagnostic process. In addition, the artificial intelligence has the capability to indicate the modality by which the correct exercise is completed. Multiple healthcare professional (9) can simultaneously access to the system using an interface named medical doctor interface (10) that can be installed on their devices (such as laptop, desktop, phone, tablet). This interface establishes communication with an artificial intelligence (AI) system (7) through the cloud. During the rehabilitation process, the movements performed by each patient are captured and transmitted in real time to the healthcare professional's device. Simultaneously, an artificial intelligence system analyses the movements and generates a virtual skeleton based on the video feed. This virtual skeleton is then displayed to the doctor in real time. The virtual skeleton can exist in either a two-dimensional or three-dimensional form. In the two-dimensional representation, the patient uses solely a video camera. Conversely, in the three-dimensional representation, the patient employs a sensor module (5) that is connected to their computer. In addition, effective communication between medical practitioners and patients is facilitated through various mediums such as video, text, and voice.

As previously mentioned, REST has the capability to convert a rehabilitation session typically conducted at a health or rehabilitation facility into a session conducted at the patient's residence, thereby minimising the necessity for the patient to travel between locations. Moreover, the utilisation of quantitative movement analysis proves to be appropriate in identifying and characterising errors and deviations in human movements, as well as providing guidance on the proper execution of exercises. The telerehabilitation session can be assured to have the same level of precision as the conventional laboratory methods of analysis, thanks to the quantitative analysis conducted by the AI. Furthermore, the proposed system facilitates telerehabilitation without necessitating intricate and costly apparatus. Moreover, the medical practitioner can intervene promptly by rectifying the patient's activity and concurrently assessing the efficacy of the corrective measures. This entails evaluating whether the patient's movements, as replayed through a virtual skeleton, align with the desired corrections implemented by the medical professional. Consequently, the medical practitioner or rehabilitator will be able to expedite and simplify the process of selecting an appropriate rehabilitation therapy. The implementation of the suggested system ultimately facilitates a decrease in the patient's susceptibility to viral and bacterial pathogens within hospital settings, as they are able to undergo therapy remotely from their own residence.

18.3 Implementation and Testing

A REST prototype has been implemented using Python and MediaPipe, a pretrained AI system offered by Google for Pose estimation, to demonstrate the viability of the proposed approach [21].

The key to REST is establishing communication between the patient, who is the client portion, and the healthcare professional, who is on the server part. The AI elaborates the outputs once the latter sends the video stream to the recipients. It is essential to employ "Sockets" to accomplish this; specifically, "Socket Server", "Socket Client", and "Socket". Figure 18.2 shows a scheme explaining how the procedure works. The server is the component that listens for client requests and fulfils or processes them as necessary. A client, on the other hand, requests this service. A client programme asks the server for some resources, and the server fulfils the request. The socket is where a server and client's two-way communication link ends. Sockets may communicate between processes on the same machine, between processes on separate machines, or within a single process. Any interaction with the remote programme must employ a connection over a socket port.

As previously explained, a socket client requests the socket server for some resources, and the server fulfils the request. As a result, both the server and client models have been created to allow for communication between them. The steps can be thought of as follows:

1. Python socket server program executes at first and wait for any request;
2. Python socket client program will initiate the conversation at first;
3. The server algorithm will response receiving the video stream from the client extracting the virtual skeleton for the analysis;

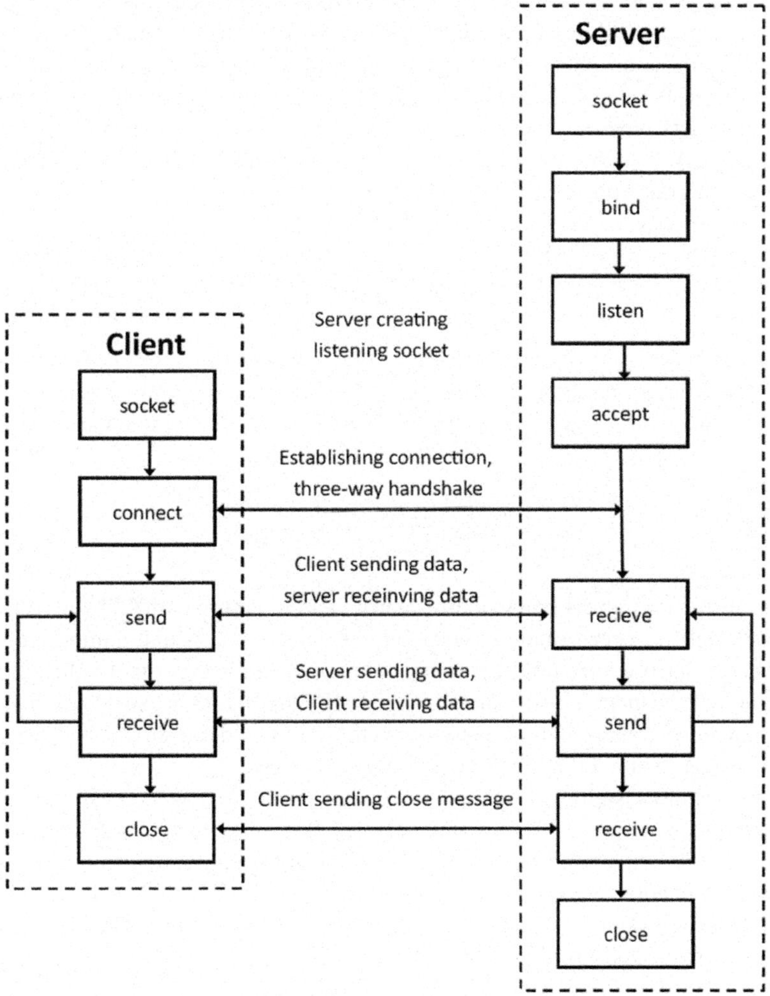

Fig. 18.2 Server-client communication scheme

4. Client program will terminate if user press “esc” button;
5. The server algorithm will output the extracted motion parameter to a spreadsheet for further evaluations.

When the video stream is received by the server algorithm in the third phase, the AI begins to analyse the video stream in order to determine the so-called Pose Estimation. Determining the high degree-of-freedom configuration of a human body with all its extensions, complex self-occlusion, self-similar parts, and large variations due to clothing, body-type, lighting, and many other factors is one of the most difficult tasks in computer vision. Computer vision techniques that predict the location of various human keypoints (joints and landmarks) such as elbows, knees, neck, shoulder, hips, torso, etc. constitute the problem of human pose estimation. MediaPipe and OpenCV libraries, have been used in this article as deep learning-based human pose estimation and video capture technique. MediaPipe Pose (MPP), which is fundamental for pose estimation and is an open-source, cross-platform framework provided by Google, was used to estimate 2D human joint coordinates in each image frame. MediaPipe Pose constructs pipelines and processes video-based cognitive data using machine learning (ML). As depicted in Fig. 18.3, MPP employs a BlazePose [22] that derives 33 2D landmarks from the human body where:

0 - Nose	9 - Mouth (left)	18 - Right pinky	27 - Left ankle
1 - Left eye (inner)	10 - Mouth (right)	19 - Left index	28 - Right ankle
2 - Left eye	11 - Left shoulder	20 - Right index	29 - Reft heel
3 - Left eye (outer)	12 - Right shoulder	21 - Left thumb	30 - Right heel
4 - Right eye (inner)	13 - Left elbow	22 - Right thumb	31 - Left foot index
5 - Right eye	14 - Right elbow	23 - Left hip	32 - Right foot index
6 - Right eye (outer)	15 - Left wrist	24 - Right hip	
7 - Left ear	16 - Right wrist	25 - Left knee	
8 - Right ear	17 - Left pinky	26 - Right knee	

BlazePose is a lightweight machine learning architecture with CPU inference that achieves real-time performance on mobile devices and personal computers. When using normalised coordinates for pose estimation, the y-axis pixel values must be multiplied by the inverse ratio. The estimated MPP landmarks are stored each moment and when the telerehabilitation section ends they are sabed in a spreadsheet for further analysis to understand the effectiveness of the therapy.

To assess the system two laptops have been set-up, one as a server and one as a client and been connected to the same network. The sever has been started and soon after the client connection to the server have been set up configuring the server IP. The communication server-client communication began, and the client video-stream arrived at server and the AI started to execute Pose estimation. The person on the client side was asked to:

1. Run the software;
2. Put in front of the camera of the laptop;

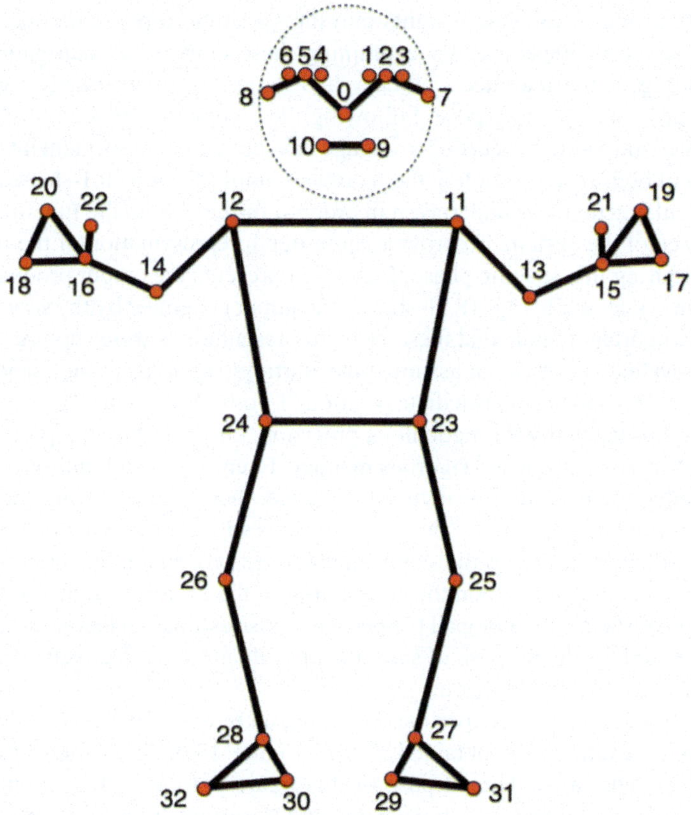

Fig. 18.3 MediaPipe Pose 33 keypoints topology [22]

3. Start the exercise from a standing position;
4. Perform abduction–adduction of both arms reaching a T pose;
5. Stop the exercise in a standing position;
6. Stop the software.

The movement that was executed during the experiment was observable in real-time by both the evaluator as well as the evaluated individual. Furthermore, it is worth noting that the evaluator had the capability to observe the pose estimation skeleton, thereby enhancing the accuracy and precision of the motion analysis. This feature allowed for a more comprehensive and thorough assessment of the executed movements. In this study, the results obtained are presented from the client and server parts of the interface, as depicted in Fig. 18.4a, b, respectively. These figures show the outputs generated during the experimental session, wherein the evaluated individual assumed a standing pose. Figure 18.4a displays the client part output, offering a comprehensive view of the interface's behaviour and functionality as observed by the user. On the other hand, the server part output, as illustrated in

Fig. 18.4b, provides valuable insights into the system's response and performance from the server's perspective. By examining these outputs, we can gain a deeper understanding of the interface's efficacy and its ability to accurately capture and interpret the user's standing pose. Following the completion of the initial exercise, the evaluated individual proceeded to engage in a subsequent movement involving the abduction of both arms, extending them outward until assuming a T-shaped position. This particular action was undertaken in order to further assess the functionality and performance of the system. To provide an in-depth representation of the interface's operation during this specific phase, Fig. 18.5a presents the output produced by the client component, while Fig. 18.5b shows the output generated by the server component. It is important to note that these visual representations were captured precisely when the evaluated individual assumed the starting/ending position (standing position), Fig. 18.5a, and the intermediate position (T pose), Fig. 18.5b. The experimental protocol was executed with meticulous precision, ensuring that the procedure was carried out in a consistent and rigorous manner. To ensure the reliability and validity of the results, the procedure was repeated thrice, thereby minimising the potential for any confounding factors or chance occurrences to influence the outcome. Each repetition adhered strictly to the same standardised procedure, leaving no room for deviation or interruption. Upon the completion of the exercises sections, a comprehensive spreadsheet containing all the pertinent pose estimation landmarks was automatically saved for the purpose of subsequent evaluation to effectively examine and evaluate the movement behaviour.

Upon completion of the experimental phase, an in-depth analysis was conducted to thoroughly examine the obtained results. Figure 18.6 shows the trajectories of the shoulders and arms. The purpose of this figure is to provide a comprehensive representation of the observed movements and positions of the shoulders and arms throughout the experimental procedure. By presenting this visual information, a more detailed understanding of the dynamics and kinematics of the shoulders and arms can be achieved. In the given context, the points A and B are representative of specific

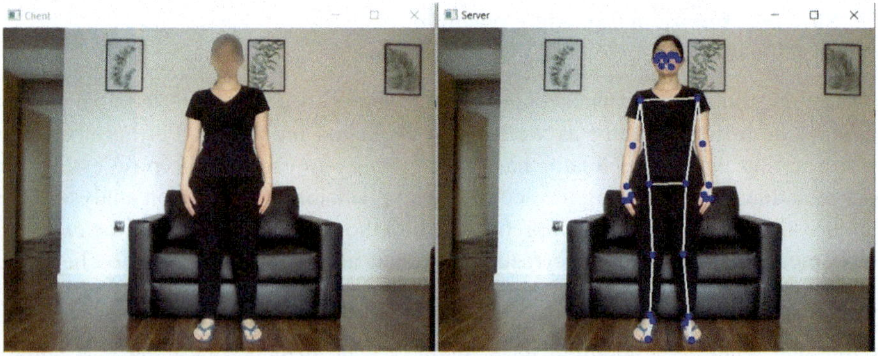

Fig. 18.4 Interface while in standing pose: **a** Patient interface view, **b** Healthcare professional interface view

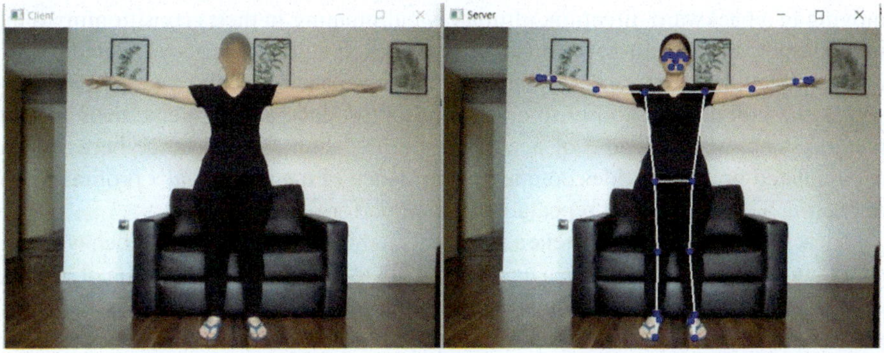

Fig. 18.5 Interface while in T pose: **a** Patient interface view, **b** Healthcare professional interface view

anatomical locations, namely the right and left shoulder regions. The points of curve BC, which is graphically represented, serves to depict the motion of abduction–adduction performed by the right arm. Similarly, the curve EF, serves the purpose of illustrating the abduction–adduction performed by the right arm. Upon analysing the trajectory, it becomes easy to see that the movements exhibit a remarkable degree of symmetry, as they trace an approximate arc. Notably, these movements originate from two distinct points, namely B and E, which are situated directly below A and D, respectively. The current visual representation shows a precise and accurate representation of a standing pose, wherein the positioning of the hands and shoulders is aligned in a manner that ensures optimal symmetry and alignment.

The abduction of both arms brings them to points C and F, wherein a notable observation can be made that both arms attain a position slightly above the shoulder level. The current position shown by the individual in question can be considered as an acceptable representation of the T pose. Following the initial T pose, it is observed that the arms go through a subsequent movement known as adduction, wherein they return to their original standing positions. This adduction motion involves the arms

Fig. 18.6 Shoulders and arms trajectories

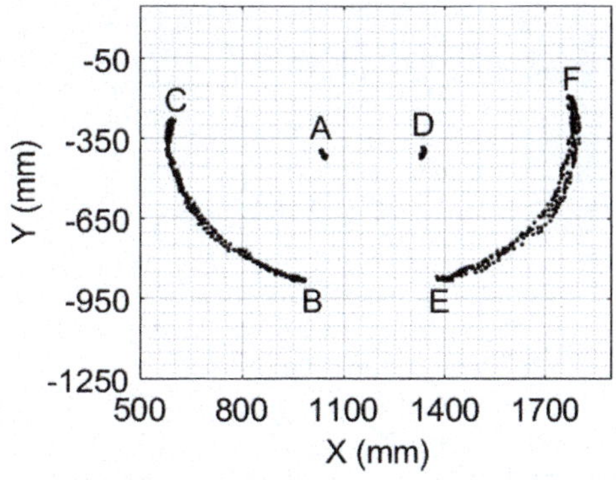

moving in a backward direction, ultimately culminating in their repositioning to the initial stance. Finally, after careful analysis and evaluation of the accumulated points, it becomes evident that the individual's motion is repeatable. This observation underscores the individual's ability to consistently reproduce and replicate their motion patterns with a high degree of accuracy and precision. The distance between the accumulated points provides compelling evidence of the individual's proficiency in delivering a consistent and repeatable motion performance.

To achieve a more comprehensive understanding of the motion behaviour, a supplementary analysis has been undertaken, focusing on the examination of angular displacement related to both the left and right arms. This analysis aims to achieve a greater awareness of the individual arm motion behaviour. The results of this analysis are visually represented in Fig. 18.7, which serves as a graphical representation of the obtained angular displacement data. To assess the angular displacement of the left and right arms, the vertical line that crosses the point of respective shoulder were taken into consideration to evaluate the angular displacement of the left and right arms. It is important to note that when the motion experiment starts the right arm starts with a clockwise movement, while the left arm starts with a counterclockwise movement, both in relation to their respective vertical line passing through the respective shoulder point.

In Fig. 18.7a, the left arm angular displacement is shown, highlighting three distinct repetitions. Throughout the course of the performance, the individual reached a maximum range of motion of 112.63° at second 14.34. Furthermore, it is important to note that the range of motion exhibits a notable characteristic of persistence and repeatability.

Figure 18.7b effectively shows the angular displacement of the right arm. Notably, the graph shows three distinct repetitions as well, which can be easily discerned. It is worth mentioning that throughout the performance, the individual reached a maximum range of motion of 105.00° at second 17.18. It is of great significance to acknowledge that the range of motion, pertaining to the right arm, exhibits a remarkable degree of persistence and repeatability as well.

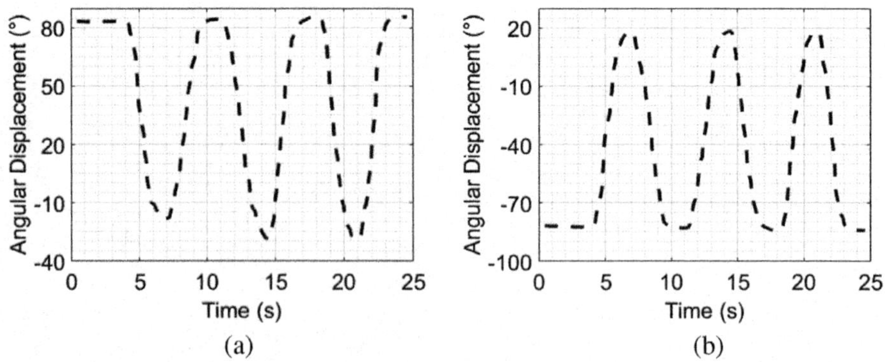

Fig. 18.7 Angular displacement: **a** Left arm, **b** Right arm

18.4 Conclusions

Based on the findings, the REST methodology demonstrates its applicability in the analysis of body movement during telerehabilitation sessions. This approach enables medical professionals and patients to effectively identify and rectify movement patterns that could potentially hinder rehabilitation progress or lead to injuries. Through the provision of immediate feedback, healthcare professionals have the ability to make necessary adjustments to their therapeutic interventions, resulting in enhanced performance outcomes and a lower probability of sustaining injuries. Moreover, this methodology can be employed to conduct a comparative analysis between movements and pre-existing recorded ones, facilitating patients in identifying areas that require improvement and subsequently refining their movements. Hence, the utilisation of REST has emerged as a promising approach to serve as a supplementary tool in facilitating telerehabilitation.

References

1. Ackerman, M.J., Filart, R., Burgess, L.P., Lee, I., Poropatich, R.K.: Developing next-generation telehealth tools and technologies: patients, systems, and data perspectives. Telemed. J. E Health **16**(1), 93–95 (2010). https://doi.org/10.1089/tmj.2009.0153
2. Rogante, M., Grigioni, M., Cordella, D., Giacomozzi, C.: Ten years of telerehabilitation: a literature overview of technologies and clinical applications. Neurorehabilitation **27**(4), 287–304 (2010)
3. Zampolini, M., Todeschini, E., Bernabeu, G.M., Hermens, H., Ilsbroukx, S., Macellari, V., Magni, R., Rogante, M., Scattareggia, M.S., Vollenbroek, M., Giacomozzi, C.: Telerehabilitation: present and future. Ann. Ist. Super. Sanita **44**(2), 125–134 (2008)
4. World Health Organization. World Report on Ageing and Health (2015). https://apps.who.int/iris/handle/10665/186463. Accessed July 2023
5. World Health Organization. Rehabilitation in Health Systems (2023). https://www.who.int/health-topics/rehabilitation#tab=tab_1. Accessed June 2023
6. Peretti, A., Amenta, F., Tayebati, S.K., Nittari, G., Mahdi, S.S.: Telerehabilitation: review of the state-of-the-art and areas of application. JMIR Rehabil. Assist. Technol. **21** 4(2), e7 (2017)
7. World Confederation for Physical Therapy. Policy statement: description of physical therapy. https://www.wcpt.org/policy/ps-descriptionPT (2017). Accessed June 2023
8. Falvey, J.R., Murphy, T.E., Gill, T.M., Stevens-Lapsley, J.E., Ferrante, L.E.: Home health rehabilitation utilization among medicare beneficiaries following critical illness. J. Am. Geriatr. Soc. **68**, 1512–1519 (2020)
9. Zziwa, S., Babikako, H., Kwesiga, D., et al.: Prevalence and factors associated with utilization of rehabilitation services among people with physical disabilities in Kampala, Uganda. A descriptive cross sectional study. BMC Public Health **19**, 1–11 (2019)
10. Deslauriers, S., Déry, J., Proulx, K., et al.: Effects of waiting for outpatient physiotherapy services in persons with musculoskeletal disorders: a systematic review. Disabil. Rehabil. **43**, 611–620 (2019)
11. Rogante, M., Grigioni, M., Cordella, D., Giacomozzi, C.: Ten years of telerehabilitation: a literature overview of technologies and clinical applications. Neuro Rehabil. **27**, 287–304 (2010)

12. Kairy, D., Lehoux, P., Vincent, C., Visintin, M.: A systematic review of clinical outcomes, clinical process, healthcare utilization and costs associated with telerehabilitation. Disabil. Rehabil. **31**, 427–447 (2009)
13. Seron, P., Oliveros, M.J., Gutierrez-Arias, R., Fuentes-Aspe, R., Torres-Castro, R.C., Merino-Osorio, C., Nahuelhual, P., Inostroza, J., Jalil, Y., Solano, R., Marzuca-Nassr, G.N., Aguilera-Eguía, R., Lavados-Romo, P., Soto-Rodríguez, F.J., Sabelle, C., Villarroel-Silva, G., Gomolán, P., Huaiquilaf, S., Sanchez, P.: Effectiveness of telerehabilitation in physical therapy: a rapid overview. Phys Ther. **101**(6), pzab053 (2021)
14. Parmanto, B., Saptono, A.: Telerehabilitation: state-of-the-art from an informatics perspective. Int. J. Telerehabil. **1**(1), 73–84 (2009)
15. Jagos, H., David, V., Haller, M., Kotzian, S., Hofmann, M., Schlossarek, S., Eichholzer, K., Winkler, M., Frohner, M., Reichel, M., Mayr, W., Rafolt, D.: A framework for (tele-) monitoring of the rehabilitation progress in stroke patients: eHealth 2015 special issue. Appl. Clin. Inform. **6**(4), 757–768 (2015)
16. Linder, S.M., Rosenfeldt, A.B., Bay, R.C., Sahu, K., Wolf, S.L., Alberts, J.L.: Improving quality of life and depression after stroke through telerehabilitation. Am. J. Occup. Ther. **69**(2), 6902290020p1–10 (2015)
17. Mani, S., Sharma, S., Omar, B., Paungmali, A., Joseph, L.: Validity and reliability of Internet-based physiotherapy assessment for musculoskeletal disorders: a systematic review. J. Telemed. Telecare **23**(3), 379–391 (2017)
18. Gal, N., Andrei, D., Nemeş, D.I., Nădăşan, E., Stoicu-Tivadar, V.: A Kinect based intelligent e-rehabilitation system in physical therapy. Stud. Health Technol. Inform. **210**, 489–493 (2015)
19. Chaparro-Rico, B.D.M., Cafolla, D.: Test-retest, inter-rater and intra-rater reliability for spatiotemporal gait parameters using SANE (an eaSy gAit aNalysis systEm) as measuring instrument. Appl. Sci. **10**(17), 5781 (2020)
20. Sipari, D., Chaparro-Rico, B.D.M., Cafolla, D.: SANE (Easy Gait Analysis System): towards an AI-Assisted automatic gait-analysis. Int. J. Environ. Res. Public Health **19**(16), 10032 (2022)
21. Lugaresi, C., Tang, J., Nash, H., McClanahan, C., Uboweja, E., Hays, M., Zhang, F., Chang, C.L., Yong, M., Lee, J., Chang, W.T. Mediapipe. A framework for perceiving and processing reality. In: Third Workshop on Computer Vision for AR/VR at IEEE Computer Vision and Pattern Recognition (CVPR) 2019 Jun, vol. 2019
22. Bazarevsky, V., Grishchenko, I.: On-device, real-time body pose tracking with mediaPipe BlazePose, Google Research. https://ai.googleblog.com/2020/08/on-device-real-time-body-pose-tracking.html. Accessed 10 July 2023

Chapter 19
Design of Single/Mixed Chemistry eVTOL Battery Packs

Shuonan Xu and Madhu Raghavan

Abstract In this article, we discuss the design and optimization of battery packs for eVTOL (electrical vertical takeoff and landing) applications. First, we present an overview of eVTOL and the challenges associated with eVTOL battery designs. It is followed by sample designs of a few well-known eVTOL designs. We designed battery packs for each of the applications with a variety of battery cell candidates. This section is followed by a description of tools used in this study. We then discuss the concept of using a mixed chemistry pack for eVTOL battery design, which proves to be a promising method that improves many shortcomings associated with single chemistry designs. At the end, using the simulation tools, we perform battery pack design for multiple eVTOL architectures, using both single chemistry and mixed chemistry battery configurations. The results are analyzed, and a summary of key takeaways are given. The purpose of this study is to establish a process for eVTOL battery pack design, for both single chemistry and mixed chemistry configurations. A tool for eVTOL battery simulation to facilitate battery design optimization is also developed, as part of this study. We also use this as a chance to explore other advanced battery configurations to meet the challenging requirement of eVTOL batteries, in this case the mixed chemistry concept.

Keywords eVTOL · Battery · Mixed chemistry

19.1 Introduction

Electrical vertical takeoff and landing (eVTOL) is a booming sector for Urban Air Mobility (UAM). Such vehicles are capable of taking off and landing vertically, thus a runway is not needed, saving precious land resources. The recent explosive growth of eVTOL development can be attributed to multiple recent technology enablers. Lithium-ion (Li-ion) batteries being one and probably the most important of them.

S. Xu (✉) · M. Raghavan
General Motors, Warren, MI 48092, USA
e-mail: shuonan.1.xu@gm.com

M. Ceccarelli and J. C. Jauregui-Correa (eds.), *State-of-the-Art and Innovations in Mechanism and Machine Science*, Mechanisms and Machine Science 150,
https://doi.org/10.1007/978-3-031-47040-0_19

Both the technology and cost of Li-ion batteries have seen significant improvements in the recent years, thanks to the evolution of transportation electrification [1, 2]. In addition to batteries, recent development of better and more powerful electric motors, SiC inverters, light-weight materials, all contributed as technology enablers. Compared to a VTOL which is powered by internal combustion engines, eVTOLs have less parts, lower energy cost and lower maintenance cost. Compared to helicopters, UberAir estimated eVTOL to have 35% lower operating cost [3].

Major eVTOL types include multirotor, lift and cruise, tiltrotor and tiltwing [4], as shown in Fig. 19.1. The multirotor ones are similar to the consumer drones, where a plurality of rotors is used to provide both vertical and horizontal thrust. This design does not incorporate wings which means all the vertical lift throughout the entire flight will be provided by the rotor only. This design is more common for light applications (1–2 passengers) with short cruise distance. Its application on heavier applications (3+ passengers) and long cruise ranges are limited, due to its high cruise power consumption.

To promote cruise efficiency, many eVTOL designs use wings along with rotors to provide lift during flight. The lift and cruise, tiltrotor and tiltwing types all fall into this category. While all the three types (lift and cruise, tiltrotor and tiltwing) utilize wings, the rotor configuration is quite different. The lift and cruise type uses fixed wings, with certain portion of the motors dedicated for providing vertical lift, while the rest of the rotors are used to generate horizontal thrust. The tiltrotor design also utilizes fixed wing, but some or all the rotors can tilt during the flight to change the direction of the thrust. Last but not least, the tiltwing concept tilts wing and rotor together, to direct the thrust.

One of the many main challenges of eVTOL designs, is its battery system. To design an eVTOL battery system, we need to first consider a few high-level questions: how much energy need to be packed on the vehicle, what is the peak power demand

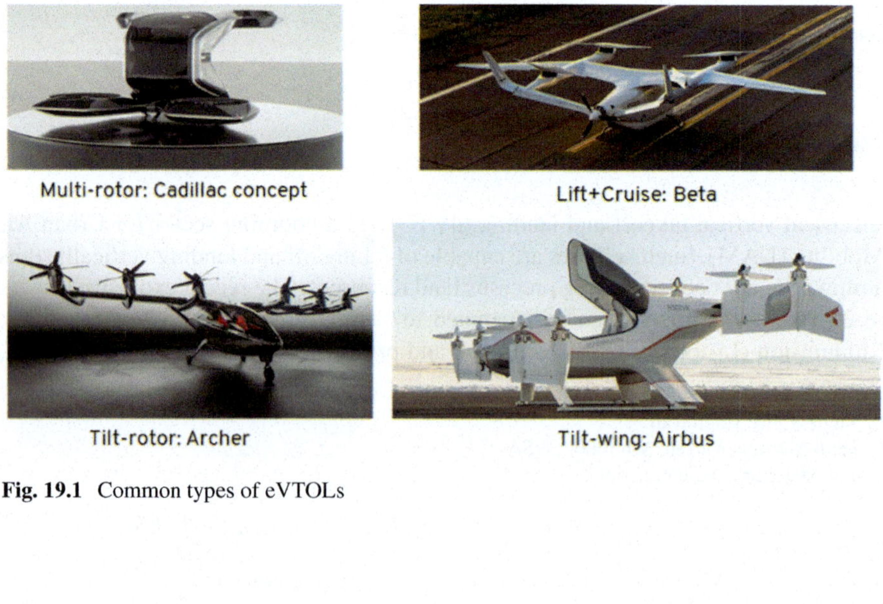

Fig. 19.1 Common types of eVTOLs

and what is the duration of the peak power demand? eVTOLs require high power for takeoff and landing. During the hover period the rotor is responsible to provide all the lift to overcome the vehicle's gravitational force. The battery system on an eVTOL should be designed such that it is capable of providing the high power required, continuously. Due to the high power requirement, an 800 V system is preferred. For a typical 5-seater eVTOL, the hover power is at the level of 1MW. In a 400 V system the total amperage will be 2500 A while in an 800 V system, the peak current will be reduced to 1250 A. The current numbers may feel daunting by first looking at it, however in eVTOLs this total current is divided into multiple rotors and packs. In a system that has 6 rotors and 6 independent packs, the total current of each sub-system is 208 A, the cabling and rest of the electrical system can be designed accordingly to match that requirement.

Another consideration in eVTOL battery design is the selection of power or energy cells. The design and manufacturing of those two types of battery cells are different in many ways. The power cells often have thinner active material coat and thicker current collectors, and they are relatively low in energy density and specific energy. Energy cells on the other hand, often have thicker active material coat and thinner current collectors. The energy density and specific energy of energy cells are higher than that of the power cells.

For eVTOL applications, the battery C rate requirements are typically, around 1C during cruise and up to 5C during takeoff and landing (hover phase). The 5C requirement during takeoff and hover favors towards power cells. However, to obtain a decent cruise range to make the eVTOL commercially feasible, high specific energy is also expected. This poses a unique challenge for eVTOLs since high power and high energy are two conflicting requirements for batteries.

Since an eVTOL is considered as an aircraft, it automatically comes with much higher safety standards than ground vehicles. Those safety standards require redundancy in the propulsion system. The battery system should be designed as a combination of multiple standalone battery packs. In this case, if even one pack fails, the rest of the battery system is still able to safely bring the vehicle down to the ground. This redundancy requirement poses another unique challenge to eVTOLs that are not present on battery electric vehicles (BEV). Essentially it makes it difficult to use large format battery cells on eVTOL. This will be discussed in detail in later sections.

In this work which we respectfully dedicate to the memory of Prof. Carlos Lopez-Cajun, we present the design flow of an eVTOL battery pack and optimize it for maximum flight range given the battery mass constraints, while meeting all the performance requirements. First, we performed an industry prevailing eVTOL study to identify three eVTOL designs that are representative to the major eVTOL types today, then we use those three identified vehicle designs as our base for battery design. In the battery design phase, we identified another three cell candidates for the three eVTOL designs identified in the first step. To arrive at an accurate and realistic battery design, we developed a simulation tool based on equivalent circuit model. In this study we also investigated using mixed chemistry concept to help with eVTOL battery design. We did find out that mixed chemistry batteries to be a promising solution to many eVTOL battery challenges. The designs are summarized and compared with each other. The results and conclusions are given in the last section.

19.2 eVTOL Industry Survey

To cover a wide spectrum of the eVTOL designs, an industry study was conducted to select the design candidates. The first two parameters to investigate are the number of seats and vehicle weight. Those two parameters are connected tightly, thus they are grouped together. Figure 19.2 shows a distribution of vehicle MTGW (maximum takeoff gross weight) versus number of seats, where the pilot is also included. The data is gathered from 11 existing eVTOL designs with publicly published data [3]. It is observed that the designs can be clearly categorized into two groups. One is lighter vehicle with 1 to 2 passengers and approximately 1000 kg of MTGW, while another group is 4–5 seater and around 2500 kg in MTGW. To cover the design of both groups, we selected three designs based on this data. Design 1 is a 2-seater, 1000 kg vehicle, Design 2 is a 4-seater, 2500 kg vehicle and Design 3 is a 5-seater, 3000 kg vehicle.

The next step is to figure out the power requirement. To do this, we need to first take a look into the mission profile of an eVTOL. An example mission profile of typical eVTOL flight is shown in Fig. 19.3.

The profile consists of 5 phases: takeoff hover, climb, cruise, descent, and landing hover. The hover is the phase where the highest power is expected. As is shown in Fig. 19.4, the hover power increases exponentially with the weight of the aircraft. Another important factor that impacts the hover power is the disk loading of the

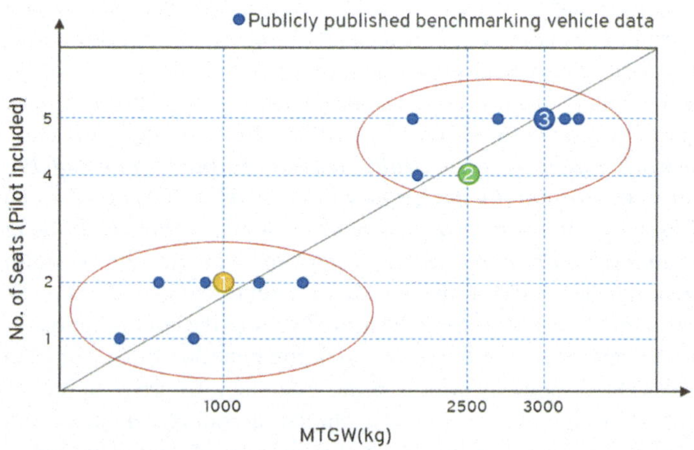

Fig. 19.2 No. of seats versus MTGW

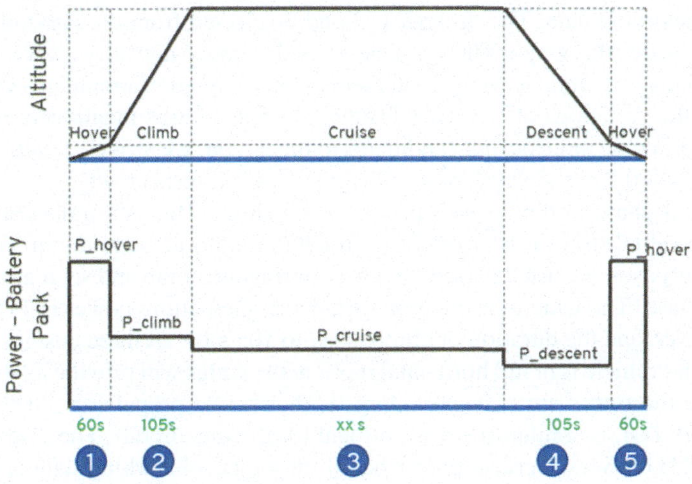

Fig. 19.3 A generic eVTOL mission profile

rotor. The larger the rotor, the less power is expected to provide the lift for hover [5]. However, in the meantime, larger rotor means a smaller number of rotors will be used, which has negative impact on system redundancy. Larger rotors are also less efficient in cruise, where cruise power is a critical parameter for the flight range. Last but not least, larger rotors are a lot noisier [6]. In the end, the selection of the rotor size is a complicated trade off. This is the reason why we are seeing so many drastically different designs today since there is not a true right or wrong answer to this.

The power required for hover is quite consistent during the hover period. In this study, for simplification purposes, we use constant power assumption for hover. Figure 19.4 shows a distribution of vehicle MTGW versus hover power, based on

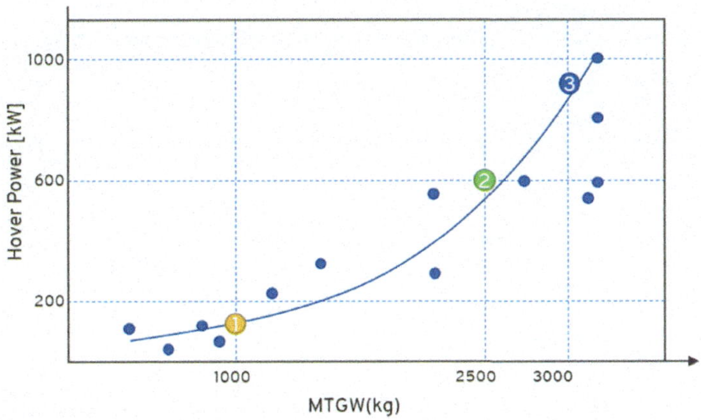

Fig. 19.4 Hover power versus MTGW

the benchmarking data. The designs 1, 2 and 3 selected from the previous step are superimposed in this graph. There is a clear trend showing that hover power increases exponentially with vehicle weight, but there is some deviation around this curve. This is due to the variations of the rotor design selection, as was mentioned previously. As we overlay the three design candidates to this graph, we arrived at a hover power of 150 kW, 600 kW and 900 kW for the three designs, respectively.

The next phase in the mission profile is the climb. This is a transitional period where the vehicle transitions from full vertical to full horizontal motion. For simplification purposes, we use the average power during the climb and set it as a constant for this phase. The generic estimation of the averaged climb power is 115% of the cruise power, and the duration of climb is set to 105 s for all three designs.

Once the vehicle is in full horizontal motion, the wings will provide all the vertical lift, while the rotors are only providing the horizontal propulsion. At this condition, the eVTOL is similar to a conventional fixed wing aircraft. The distribution of vehicle MTGW versus cruise power based on the benchmarking data is shown in Fig. 19.5. It is observed that cruise power increases linearly with vehicle weight, and the percentage-wise increase is a lot smaller compared to that of the hover power. Based on this data, we selected 100 kW, 125 kW and 150 kW as cruise power for the three designs, respectively. The increase of cruise power of design 3 compared to design 1 is 50%, whereas the increase of hover power of design 3 compared to design 1 is 500%, 10 times higher than that of cruise power. Again, the points are quite scattered in this graph, due to various rotor designs. In this study, we want to emphasize cruise efficiency to maximize range, thus the trendline is towards the lower end of the graph.

Note that in this study, cruise range also includes the reserve. Thus, the cruise range is the maximum possible range the vehicle can cover in one charge. In actual use cases, the vehicle will be designed with a portion of that range being reserved, where the claimed useable flight range will be a lot less than the maximum possible range.

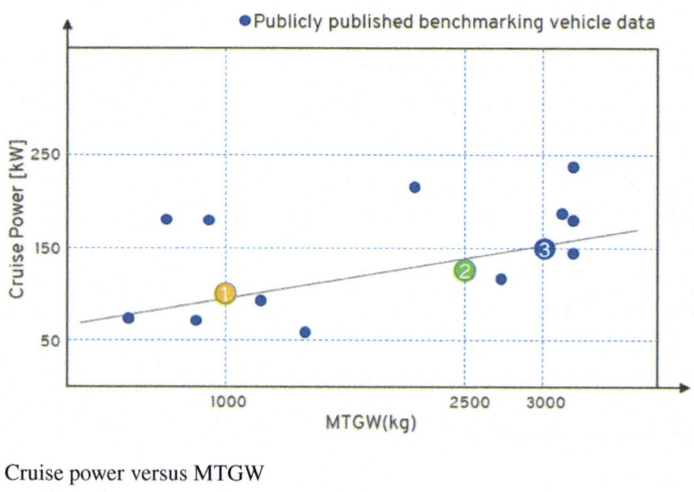

Fig. 19.5 Cruise power versus MTGW

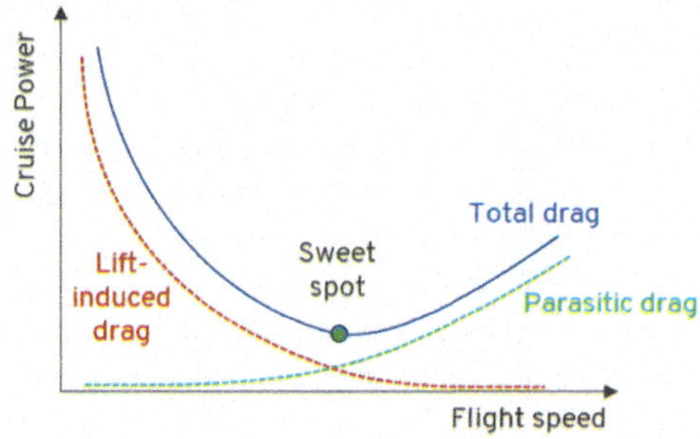

Fig. 19.6 Vehicle drag force versus flight speed

In Fig. 19.3, the duration of the cruise is not shown, since that is a design variable that changes with the ESS (Energy storage system) design. The cruise speed used in this study is selected as 200 km/h for all three designs. The value is determined based on the average value of the benchmarking vehicles. This speed is also considered as the most power efficient cruise speed. If the cruise speed is too low, the wing will not be able to provide all the lift needed, and the rotor will have to come in and assist the vertical lift. On the other hand, if the flight speed gets too high, the parasitic drag will become dominant, and it requires more power to overcome the drag. The tradeoff of flight speed and the power to overcome the drag is illustrated in Fig. 19.6.

After the cruise period, the vehicle will enter the descent phase, which is similar to the climb phase. The descent phase will require less of an averaged power since part of the gravitational energy can be harnessed in that phase, but it can never be zero or net positive since it would still need to transition to hover. In this study, we set descent power as a constant 65% of cruise power, and the descent duration is 105 s.

At the end of descent, the vehicle transitions to full vertical motion and enters another hover phase. The power and duration of this hover phase is the same as the takeoff hover. However, this phase is significantly more challenging than the takeoff hover for the battery, as the battery SOC is at its lowest state in the flight. The battery pack should still be able to deliver the full power, continuously, to ensure a safe landing.

This maneuver is the most challenging part of an eVTOL battery pack design. It is well understood that the battery will experience capacity loss when discharging at high C rate [7], as shown in Fig. 19.7. This is another reason that in an eVTOL with a homogenous pack, we would rarely use the full capacity of the battery since we need to over-design the pack to reserve enough 'juice' for the battery to support the high power discharge at the end. The mixed chemistry pack concept on the other hand, could show improvements to this dilemma. A discussion of this will be covered in the later section.

The next step in the pack design is to determine the total allowable weight of battery cells. Unlike Battery Electric Vehicles (BEVs), eVTOL applications emphasize specific energy (Wh/kg) rather than energy density (Wh/L). The primary reason for this is the hover power being an exponential function of vehicle weight. A 3× weight increase could result in a 5× increase in peak power requirement.

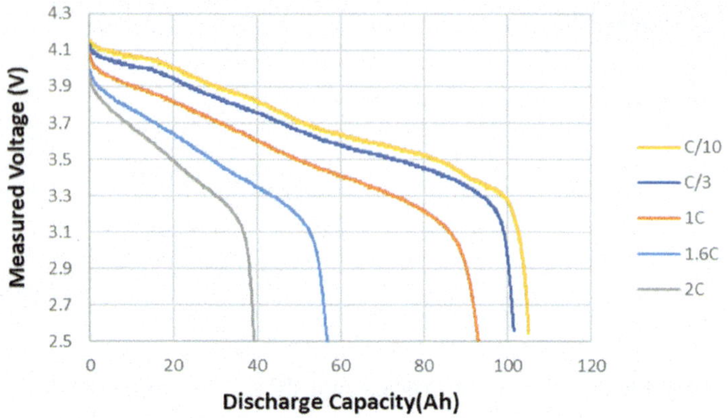

Fig. 19.7 Battery voltage versus capacity at various discharge rate—an illustration

To determine allowable cell mass, we first need to understand the allocation of weight on an eVTOL and what percentage of the total weight can be allocated to batteries. Based on the industry prevailing study, a generic estimation of eVTOL's weight distribution is shown in Fig. 19.8 [1], where the battery pack accounts for 30% of the total weight. The 21% of 'others' include motors, inverters, rotors, interior, avionics and so on.

Based on this data, we set the battery pack weight as 30% of MTGW for all three designs. To derive allowable battery cell mass, an estimate of 25% cell to pack mass overhead is used. The mass includes modules, busbars, BDUs, casing and so on with the assumption that lightweight material is used. Note that this is not a very accurate estimate of cell to pack mass overhead. Doing so will require a much more detailed analysis and it would be beyond the scope of this study.

In summary, three different designs of eVTOLs were selected as design cases, and the specs of those three designs are given in Table 19.1.

Fig. 19.8 An estimation of eVTOL mass allocation

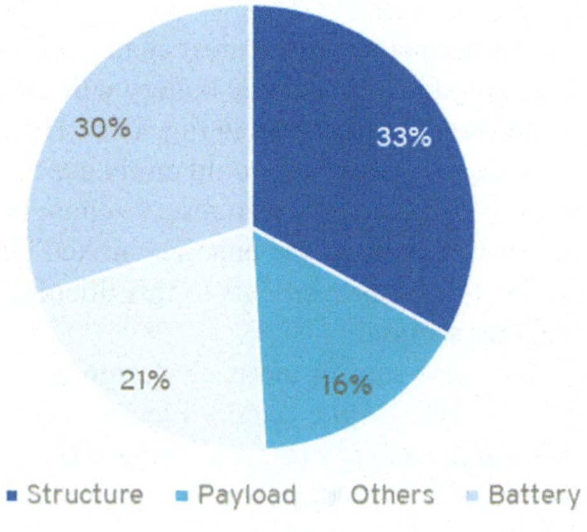

Table 19.1 Summary of the three eVTOL designs

	Design #1	Design #2	Design #3
No. of seats	2	4	5
MTGW (kg)	1000	2500	3000
Hover power (kW)	150	600	900
Cruise power (kW)	100	125	150
Battery pack weight (kg)	300	750	900
Est. total cell mass	225	562	675
Cruise speed (km/h)	200	200	200

19.3 Mixed Chemistry Pack Design

In previous sections we mentioned that a mixed chemistry pack configuration could be a solution to a number of challenges faced by single chemistry designs. In this section we will dive into the mixed chemistry pack design and evaluate its feasibility and benefits. A mixed chemistry pack is basically a battery pack that consists of more than one cell types. The different cell types can be connected in series or in parallel. In this study, we ae referring to the parallel connected type, with a basic circuit topology shown in Fig. 19.9.

A DC-DC converter is necessary in this case since the two cell groups have different terminal voltages, and the SOC of the two cell types are not identical throughout the flight. The DC-DC converter serves as the 'bridge' of the two cell types.

The main benefits of using a mixed chemistry battery pack in an eVTOL are as follows:

1. Flexible power/energy ratio (P/E) for different applications: The benefit of this is two-fold. One is we can match the mission P/E precisely using a mixed chemistry configuration. This is very difficult for single chemistry configurations unless the mission happens to have exactly the same P/E as the cell coincidently. This is a key enabler for mixed chemistry pack to minimize the vehicle weight, otherwise the pack is either overdesigned in energy or overdesigned in power. Another side of the benefit is the pack can be configured to match various missions. Longer mission typically requires lower P/E ratio while short missions favor high P/E. For single chemistry packs, different cells will need to be sourced to match each

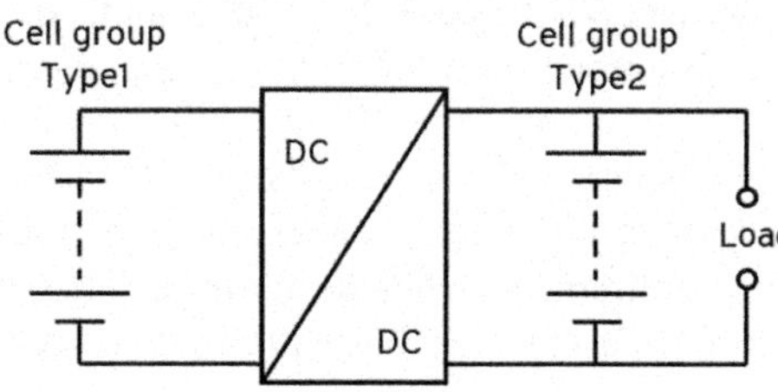

Fig. 19.9 Illustration of a parallel connected mixed chemistry circuit topology

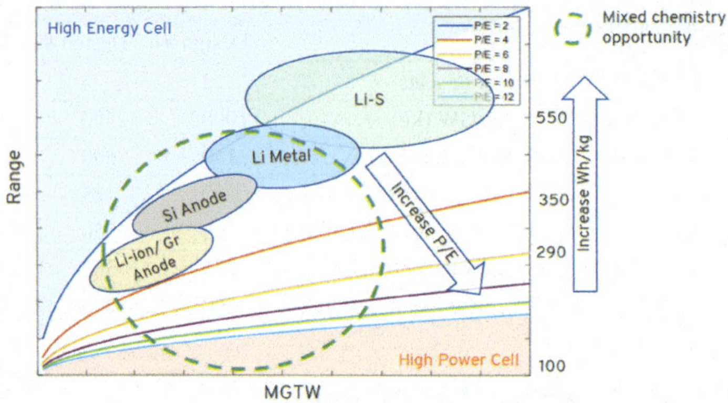

Fig. 19.10 Mixed chemistry pack opportunities for eVTOL battery packs

application, incurring additional cost for cell development or manufacturing, as is illustrated in Fig. 19.10. A mixed chemistry pack on the other hand, can achieve a wide range of pack P/E by only using two types of cells, and for each application, the P/E can be precisely matched to maximize the range or minimize pack weight.

2. Mid-mission energy transfer: The previous sections have emphasized that the most challenging part of the mission is the landing hover. The situation is so challenging because the cells are expected to deliver full power while the SOC is low. This is inherently challenging for Li-ion batteries because the peak discharge C rate reduces with lower SOC. This is one of the primary reasons that a reserve of at least 20% is needed for single chemistry pack to ensure the pack can meet the power demand at landing. The mixed chemistry pack offers a solution to this scenario, by charging up the power cells during the mission. For eVTOL applications, the high power demands only account for a small portion of time in the mission. During takeoff and landing hover, both energy and power cells will provide power to the rotors at the same time. The cruise part accounts for most of the flight time, and the power required for cruise is much lower than that of hover. This provides an opportunity to use the high-energy cells to charge the high-power cells during cruise. Thus, by the time the vehicle is ready for landing, the power cells will be charged to a higher SOC, capable of delivering close to full power again.
3. Flexibility in choice of cell format: The mixed chemistry architecture allows usage of large format cells and advanced high energy density cells. When designing an eVTOL using a single chemistry pack, a lot of constraints need to be considered. First is the power density. The cells need to have enough discharge C rating to be able to handle the power for hover. This inherently reduces the cell's energy density since energy and power are two conflicting requirements when designing a cell. Another constraint is the cell format. Larger format cells are basically out of question for single chemistry eVTOL packs,

due to the conflicting requirement of system voltage and redundancy. In a mixed chemistry pack, since the DC-DC converter is part of the configuration, it offers flexibility of the pack's terminal voltage, thus allowing larger format cells to be used, improving the overall energy density. A mixed chemistry pack also enables the usage of advanced cell chemistry i.e., Lithium metal batteries (LMB) since the drawbacks of those cell types can be compensated for by another chemistry.

19.4 Simulation Tool Development

In this study, we developed a 6p equivalent circuit battery model capable of simulating dual-chemistry packs. The equivalent circuit model is formulated with an Ohmic resistance followed in series by a parallel-connected RC pair (charge-transfer and double-layer) and a variable parallel RC pair (diffusion) [8–10], as shown in Fig. 19.11.

$$V_{terminal} = V_{OCV} + IR_{ohm} + V_h + V_{dl} + V_{diff} \tag{19.1}$$

where $V_{terminal}$ is terminal voltage, V_{OCV} is the cell's open circuit voltage, I is the cell current, R_{ohm} is cell's ohmic resistance, V_h is the hysteresis voltage, V_{dl} is the dual-layer voltage and V_{diff} is the diffusion voltage.

The hysteresis voltage V_h is updated in a discrete form as:

$$V_{h_k} = V_{h_{k-1}} + \beta I_k \Delta t[V_h^{max} - sgn(I_k)V_{h_{k-1}}] \tag{19.2}$$

where β is the hysteresis constant derived from an empirical fit of charge and discharge data, V_h^{max} is the upper limit of hysteresis voltage derived from the difference between the charge-increasing and charge-decreasing OCV curves.

The double layer circuit can be described as:

$$V_{dl_k} = \left(1 - \frac{\Delta t}{R_{ct}C_{dl}}\right)V_{dl_{k-1}} + \frac{\Delta t}{C_{dl}}I_{k-1} \tag{19.3}$$

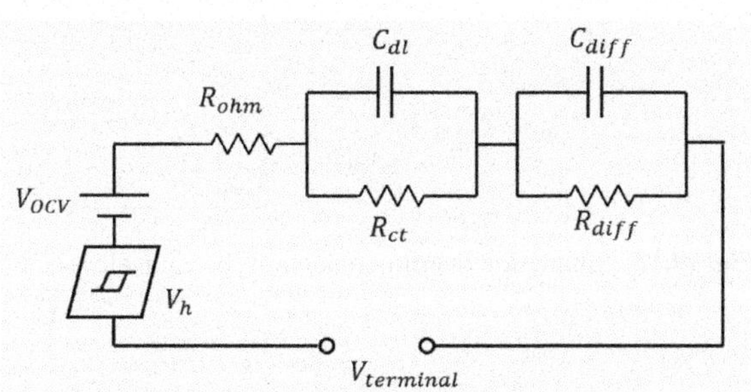

Fig. 19.11 Equivalent circuit battery model

where R_{ct} is the SOC and temperature dependent charge transfer resistance and C_{dl} is the SOC and temperature dependent double layer capacitance.

The diffusion circuit can be described as:

$$V_{diff_k} = \left(1 - \frac{\Delta t}{R_{diff} C_{diff}}\right) V_{diff_{k-1}} + \frac{\Delta t}{C_{diff}} I_{k-1} \tag{19.4}$$

where R_{diff} is the SOC and temperature dependent diffusion resistance and C_{diff} is the SOC and temperature dependent diffusion capacitance.

The cell SOC is derived as:

$$SOC_k = SOC_{k-1} + \eta \frac{I_k}{Q} \Delta t \tag{19.5}$$

where η is the coulombic efficiency and Q is capacity of the cell.

The equivalence circuit is then instrumented in Simulink. The model is instrumented in a way that it can simulate a parallel connected mixed chemistry pack, as shown in Fig. 19.12.

To better monitor the battery status on the fly, a monitoring GUI is instrumented as part of the tool, as shown in Fig. 19.13. By monitoring this window, decisions can be quickly made while the model is running, to help users of the tool to make model changes more efficiently.

The simulation tool also integrates a full setup to configure the eVTOL mission profile. The profile is simplified down to a power profile as a function of time. After the tool is executed, the user will be able to review the SOC, terminal voltage and current draw of each cell type during the mission profile, as shown in Fig. 19.14, and a lot more.

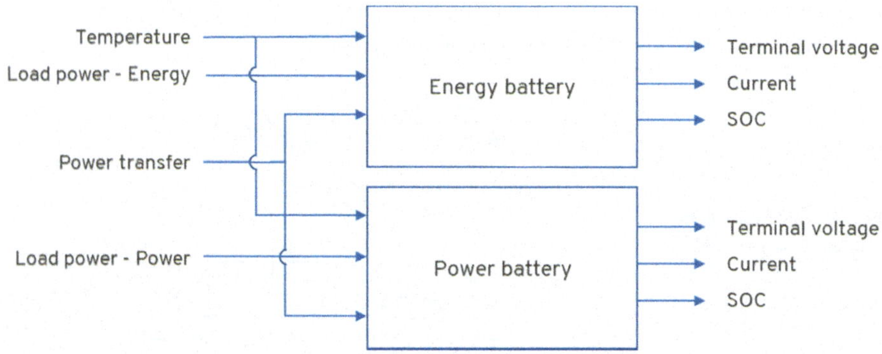

Fig. 19.12 Simulation instrumentation of the equivalent circuit model

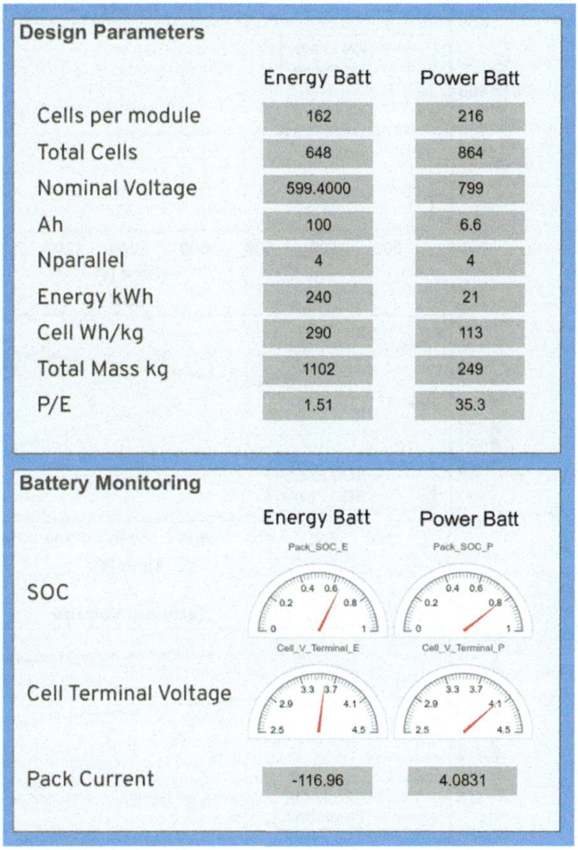

Fig. 19.13 Battery status monitoring panel

19.5 Application to eVTOL Design #2

A mixed chemistry pack design was explored, as we realized that a mixed chemistry pack has the potential to improve or overcome multiple challenges associated with a single chemistry pack. To evaluate the mixed chemistry pack concept, we use the eVTOL Design #2 described in the previous section as the base vehicle, shown in Table 19.2, and designed a mixed chemistry pack around it.

The cell candidates used in this study consist of a selection of energy cell and a selection of power cell. The energy cell selected is a 101 Ah pouch cell, while the power cell is a 6.8 Ah pouch cell. The cell specs are given in Table 19.3.

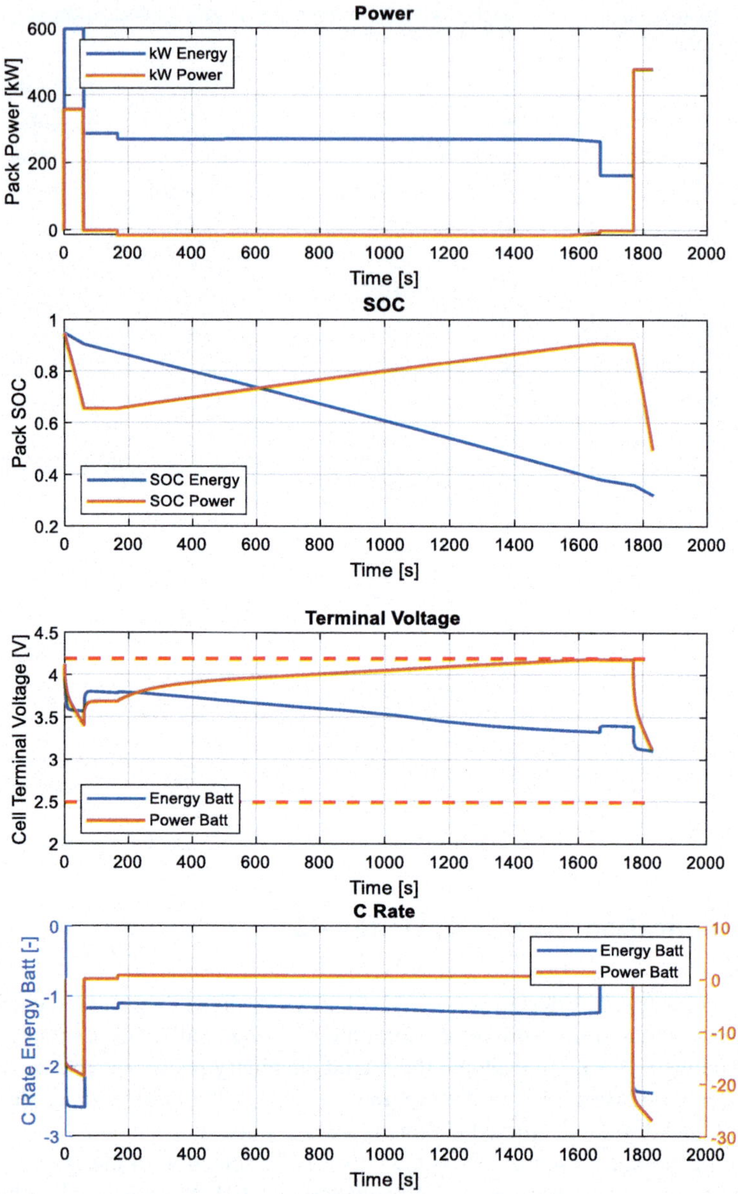

Fig. 19.14 Plotting key performance data of simulation results

Table 19.2 Summary of eVTOL design used for mixed chemistry pack assessment

	Design #2
No. of seats	4
MTGW (kg)	2500
Hover power (kW)	600
Cruise power (kW)	125
Battery pack weight (kg)	**750**
Est. total cell mass	**562**
Cruise speed (km/h)	200

Table 19.3 Summary of specs of mixed chemistry pack cells

	Energy cell	Power cell
Type	Pouch	Pouch
Anode	Gr	Gr
Cathode	NCM	NCM
Capacity	101Ah	6.8Ah
Nominal voltage	3.7 V	3.68 V
Specific energy	290 Wh/kg	113 Wh/kg
Max discharge C rate	3.5C	28C

19.5.1 Energy Cell-Only Design

First, we evaluate the feasibility of designing a homogeneous pack using either of the cell candidates only. The first candidate evaluated is the energy cell. The same design process used in the prior section was applied, and the results are given in Table 19.4 and Fig. 19.15.

It is observed that the energy-only pack does not meet the mission requirements. At takeoff hover, the cells violate the 3.5C discharge limit. This is an indication that the P/E of energy cell is too low for this vehicle. Using only energy cells for this application is not a feasible solution.

Table 19.4 Summary of performance of using an energy cell only pack

Performance summary	
No. of pack	2
Pack config	1p220s (800 V system)
Total cell mass	562 kg
Total ESS energy	163 kWh
Max cruise duration	N/A
Cruise range	N/A

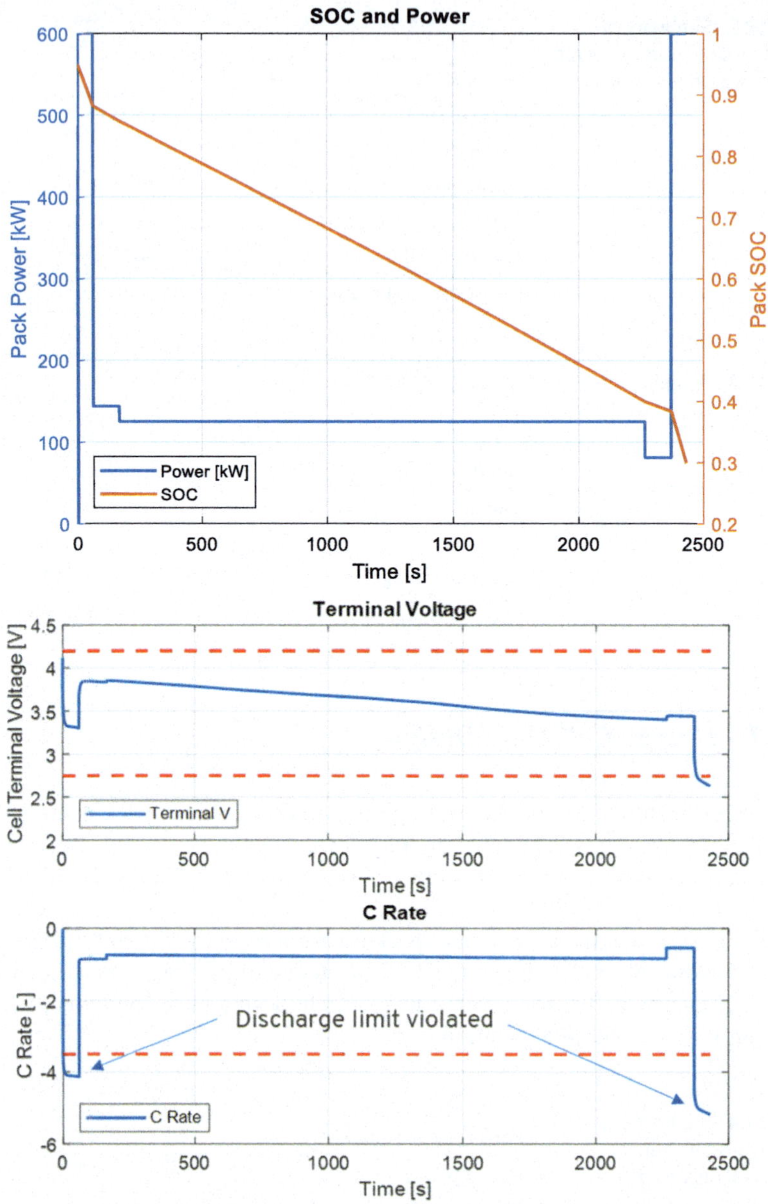

Fig. 19.15 Results of energy cells with 2500 kg 4-seater eVTOL

Table 19.5 Summary of performance of using a power cell only pack

Performance summary	
No. of pack	6
Pack config	2p216s (800 V system)
Total cell mass	560 kg
Total ESS energy	63 kWh
Max cruise duration	9 min
Cruise range	30 km

19.5.2 Power Cell-Only Design

Now on to the high power cells. The results of using a power cell only pack are given in Table 19.5 and Fig. 19.16.

It is shown that a power cell pack is able to complete the mission with no problem. The max discharge C rate during the mission is 11C which is well within the cell's limit. However, the cruise range is very short (30 km). This is an indication that the P/E of this cell is too high of P/E for the application, and the pack is overdesigned in power.

19.5.3 Mixed Chemistry Design

To design a mixed chemistry battery pack, the following design constraints need to be respected:

$$m_{cell_E} + m_{cell_P} = m_{cell_{total}} \tag{19.6}$$

$$\eta_{cell}(E_E + E_P) > E_{mission} \tag{19.7}$$

$$C_{E_{max}} < C_{E_{lim}}, C_{P_{max}} < C_{P_{lim}} \tag{19.8}$$

$$E_P > 2\int P_{hover\_P}dt \tag{19.9}$$

where m_{cell_E} is the total cell mass of the energy cells, $m_{cell_{total}}$ is total cell mass of the power cells, m_{cell} is total cell mass, η_{cell} is percentage of useable cell energy, E_E is total cell energy in energy cells, E_P is total cell energy in power cells, $E_{mission}$ is the energy required to complete the mission, $C_{E_{max}}$ is the max discharge C rate of the energy cell during the mission, $C_{P_{max}}$ is the max discharge C rate of the power cell during the mission, $C_{E_{lim}}$ is the energy cell discharge C rate limit, $C_{P_{lim}}$ is the power cell discharge C rate limit and P_{hover_P} is the hover power expected from the power cells.

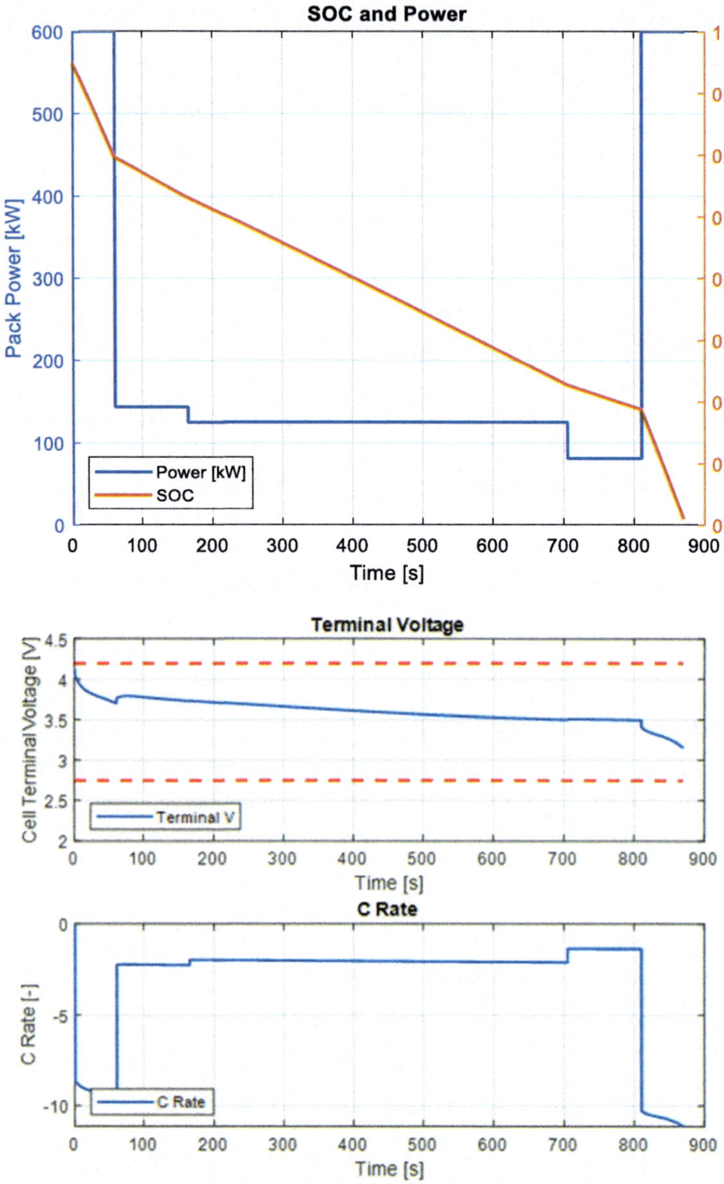

Fig. 19.16 Results of power cells with 2500 kg 4-seater eVTOL

Equations (19.6) to (19.8) are self-explanatory. The purpose of Eq. (19.9) is to ensure that the power cells have the enough energy to support two consecutive hover maneuvers. This is to protect for a condition where an emergency landing is requested right after the takeoff. The power cells will not have a chance to get any charge from the energy cells, and it should still be able to deliver the power to support a safe landing.

Total cell mass can be derived as:

$$m_{cell} = \frac{E_{cell}}{e_{cell}} \tag{19.10}$$

where E_{cell} is the cell energy for either cell type and e_{cell} is the specific energy of the cell for either cell types.

Substitute Eq. (19.10) to Eq. (19.6):

$$\frac{E_E}{e_E} + \frac{E_P}{e_P} = m_{cell_{total}} \tag{19.11}$$

where e_E and e_P are the specific energy of the energy cell and power cell, respectively.

Rearrange Eq. (19.11):

$$E_E = \left(m_{cell_{total}} - \frac{E_P}{e_P}\right)e_E \tag{19.12}$$

Substitute Eq. (19.12) to Eq. (19.7):

$$\left(m_{cell_{total}} - \frac{E_P}{e_P}\right)e_E + E_P > \frac{E_{mission}}{\eta_{cell}} \tag{19.13}$$

Rearrange Eq. (19.13):

$$E_P < \frac{\frac{E_{mission}}{\eta_{cell}} - m_{cell_{total}} \cdot e_E}{1 - \frac{e_E}{e_P}} \tag{19.14}$$

Combining (19.14) and (19.9):

$$2\int P_{hover\_P} dt < E_P < \frac{\frac{E_{mission}}{\eta_{cell}} - m_{cell_{total}} \cdot e_E}{1 - \frac{e_E}{e_P}} \tag{19.15}$$

Equation (19.15) can be used as a starting point for determining the sizing of the power pack. The total number of power cells to be used can be derived from the energy:

$$n_P = \frac{E_P}{Ah_P \cdot V_{nom_P}} \tag{19.16}$$

The final numbers for n_P is determined based on the result of Eq. (19.16), system voltage requirement, packaging constraints and redundancy requirements. Table 19.6 gives a summary of the mixed chemistry pack designed for the 2500 kg 4-seater eVTOL. Figure 19.17 gives power and SOC of energy and power packs during the mission and Fig. 19.18 gives terminal voltage and C rate of energy and power cells during the mission.

Figure 19.19 gives a comparison of cruise range of the three designs: energy cell only, power cell only and mixed chemistry. The energy cell only design fails the mission with 0 cruise range, power cell only pack is able to support the mission but

Table 19.6 Summary of a mixed chemistry battery pack design

Performance summary			
	Power cell	Energy cell	Total
No. of pack	2	4	6
Pack config	1p226s (800 V system)	1p90s (800 V using DCDC)	
Total cell mass	98 kg	464 kg	562 kg
Total ESS energy	11 kWh	135 kWh	146 kWh
Max cruise duration	33 min		
Cruise range	110 km		

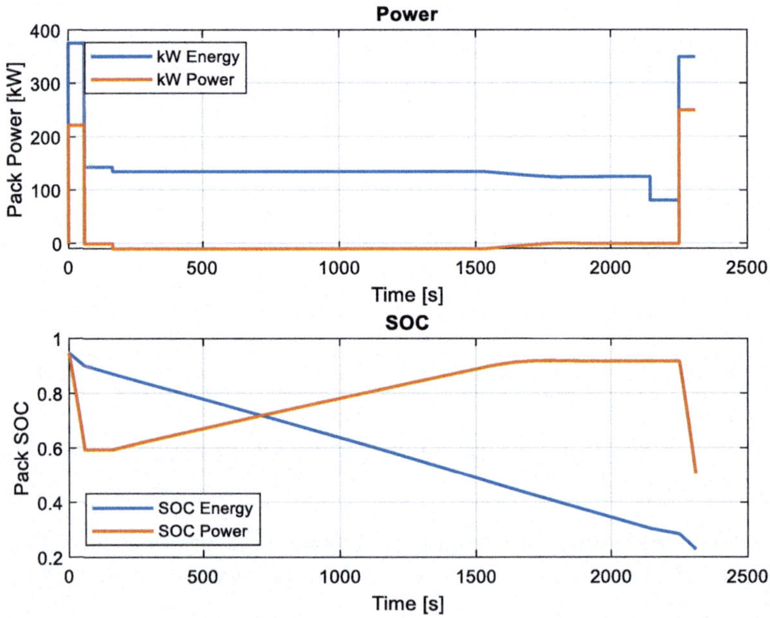

Fig. 19.17 Power and SOC of energy and power packs during the mission

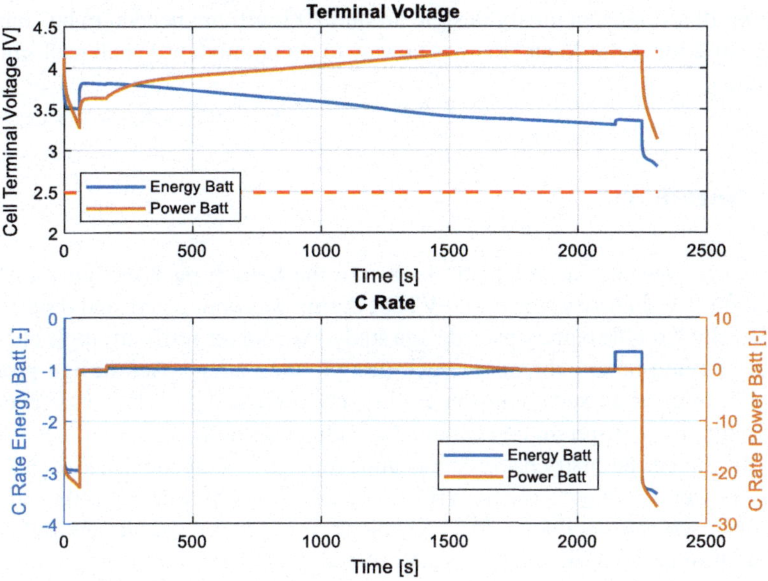

Fig. 19.18 Cell terminal voltage and C rate during the mission

only resulted in a 30 km range. The mixed chemistry pack won by a large margin with a cruise range of 110 km.

Based on the results, a few interesting findings are observed for the mixed chemistry architecture. The max C rate of the energy cell during the mission is 3.3C while the max C rate of the power cell is 28C, both within the cells' limits. The end SOC of the energy cell at the end of the mission is close to 20%, which is an indication that most of the cell's capacity is used, and the pack's P/E matches well with the application. The mixed chemistry pack also enables the energy cell to be used in this application even though it is a large format cell. In this case we have energy cells in 1p90s configuration with a total of 4 such sub-packs, thanks to the DC-DC

Fig. 19.19 Cruise range comparison of energy cell only, power cell only and mixed chem designs

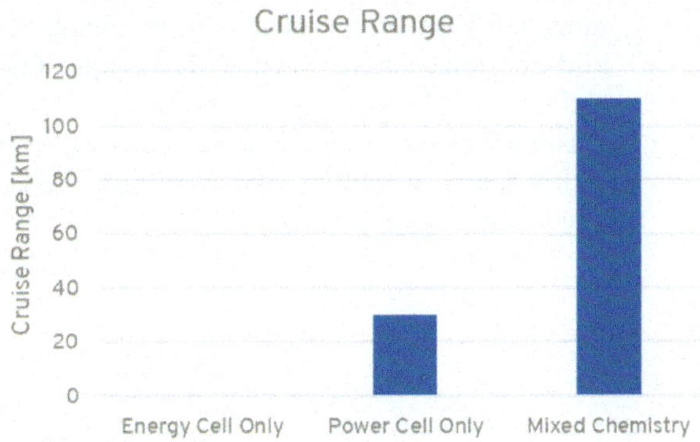

converter. In a single chemistry design, around 200 cells are needed to be connected in series to achieve 800V, and there is no way to accommodate 4 such packs into this application.

19.6 Summary

In this work, we took a systematic look into the battery pack design for eVTOL applications. An industry survey of eVTOL design was conducted, and three designs were identified for the battery pack design study. We looked into both single chemistry pack design and mixed chemistry pack design, and battery pack design processes for both single and mixed chemistry packs were proposed. For the single chemistry pack design, three different cell candidates were evaluated.

For the mixed chemistry pack, energy and power cells were identified as the energy cell and power cell, a design comparison of energy only, power cell only and mixed chemistry pack was conducted. An equivalent circuit-based simulation tool capable of simulating both single chemistry and mixed chemistry pack configurations was also developed. The findings and key takeaways from this study are summarized as follows:

1. An eVTOL pack should have both high energy density and high power capability.
2. Hover power requirements for eVTOLs grow exponentially with vehicle mass.
3. The best way to minimize the battery pack weight is to have a battery pack that matches exactly to the mission P/E ratio
4. Smaller format cells (in terms of Ah) are preferred for redundancy reasons, allowing numerous parallel sub-packs to be incorporated in the architecture. Alternatively, the judicious use of DC-DC converters, allows the mixing of large format energy cells and smaller format power cells to achieve redundancy.
5. The most challenging aspect of eVTOL pack design is the landing, where full power is expected while battery SOC is at its lowest.
6. The mixed chemistry pack is a promising solution for eVTOL applications with three main benefits:

    a. tunable P/E and consequent mass minimization,
    b. mid-mission energy transfer and consequent hover capability during landing, and
    c. feasible inclusion of large format automotive BEV cells in the eVTOL application space, as well as advanced high energy density cells with inherently lower C rates.

References

1. Etacheri, V., Marom, R., Elazari, R., Salitra, G., Aurbach, D.: Challenges in the development of advanced Li-ion batteries: a review. Energy Environ. Sci. **4**(9), 3243 (2011). https://doi.org/10.1039/c1ee01598b
2. Goodenough, J.B., Park, K.-S.: The Li-Ion rechargeable battery: a perspective. J. Am. Chem. Soc. **135**(4), 1167–1176 (2013). https://doi.org/10.1021/ja3091438
3. Wang, J.: 2022 eVTOL design short course
4. Bacchini, A., Cestino, E.: Electric VTOL configurations comparison. Aerospace **6**(3), 26 (2019). https://doi.org/10.3390/aerospace6030026
5. Yang, X.-G., Liu, T., Ge, S., Rountree, E., Wang, C.-Y.: Challenges and key requirements of batteries for electric vertical takeoff and landing aircraft. Joule **5**(7), 1644–1659 (2021). https://doi.org/10.1016/j.joule.2021.05.001
6. Weitsman, D., Greenwood, E.: Parametric study of eVTOL rotor acoustic design trades. In: AIAA Scitech 2021 Forum, VIRTUAL EVENT (2021). https://doi.org/10.2514/6.2021-1987
7. Gao, L., Liu, S., Dougal, R.A.: Dynamic lithium-ion battery model for system simulation. IEEE Trans. Compon. Packag. Technol. **25**(3), 495–505 (2002). https://doi.org/10.1109/TCAPT.2002.803653
8. He, H., Xiong, R., Fan, J.: Evaluation of lithium-ion battery equivalent circuit models for state of charge estimation by an experimental approach. Energies **4**(4), 582–598 (2011). https://doi.org/10.3390/en4040582
9. Grazioli, D., Magri, M., Salvadori, A.: Computational modeling of Li-ion batteries. Comput. Mech. **58**(6), 889–909 (2016). https://doi.org/10.1007/s00466-016-1325-8
10. Wang, Y., et al.: A comprehensive review of battery modeling and state estimation approaches for advanced battery management systems. Renew. Sustain. Energy Rev. **131**, 110015 (2020). https://doi.org/10.1016/j.rser.2020.110015

Index

M. Ceccarelli and J. C. Jauregui-Correa (eds.), *State-of-the-Art and Innovations in Mechanism and Machine Science*, Mechanisms and Machine Science 150,
https://doi.org/10.1007/978-3-031-47040-0

SPRINGER NATURE

GPSR Compliance

The European Union's (EU) General Product Safety Regulation (GPSR) is a set of rules that requires consumer products to be safe and our obligations to ensure this.

If you have any concerns about our products, you can contact us on ProductSafety@springernature.com

In case Publisher is established outside the EU, the EU authorized representative is:

Springer Nature Customer Service Center GmbH
Europaplatz 3
69115 Heidelberg, Germany

The manufacturer's authorised representative in the EU is Springer Nature Customer Service Centre GmbH, Europaplatz 3, 69115 Heidelberg, Germany. If you have any concerns regarding our products, please contact ProductSafety@springernature.com

Printed and bound by CPI Group (UK) Ltd, Croydon, CR0 4YY
01/10/2026
02225907-0009